Universitext

Universitext

Universitext is a series of textbooks that presents material from a wide variety of mathematical disciplines at master's level and beyond. The books, often well class-tested by their author, may have an informal, personal, even experimental approach to their subject matter. Some of the most successful and established books in the series have evolved through several editions, always following the evolution of teaching curricula, into very polished texts.

Thus as research topics trickle down into graduate-level teaching, first textbooks written for new, cutting-edge courses may make their way into *Universitext*.

More information about this series at http://www.springer.com/series/223

Roger Godement

Introduction to the Theory of Lie Groups

Translated by Urmie Ray

 Springer

Roger Godement (Deceased)
Paris
France

Translation from the French language edition: *Introduction à la théorie des groupes de Lie* by Roger Godement, © Springer-Verlag GmbH Berlin Heidelberg 2004. All Rights Reserved.

ISSN 0172-5939   ISSN 2191-6675 (electronic)
Universitext
ISBN 978-3-319-54373-4   ISBN 978-3-319-54375-8 (eBook)
DOI 10.1007/978-3-319-54375-8

Library of Congress Control Number: 2017933553

Mathematics Subject Classification (2010): 22E15, 22E40, 58A40

Printed on acid-free paper

This Springer imprint is published by Springer Nature
The registered company is Springer International Publishing AG
The registered company address is: Gewerbestrasse 11, 6330 Cham, Switzerland

Preface

Except for the occasional terminology, this book is a word for word reproduction of the notes drafted in 1973–1974 for a course given at the *Écoles Normales Supérieures* of Paris and Sèvres (subsequently merged into one). They were published in 1982 by the Department of Mathematics of the University of Paris VII for a necessarily limited distribution. Twenty years later, it seemed appropriate to entrust Springer-Verlag with the task of turning them into a book, its means of distribution being of an altogether different order. Strictly speaking, it only requires some basic knowledge of general topology and linear algebra, but it is mainly aimed at master's students in mathematics. On the whole, the subject matter falls under the general theory of Lie groups or, in the first two chapters, of topological groups. The reader may find some parts of chapters IX, X, XI and XII of my Analysis III and IV (Springer, 2015) helpful—referred to as MA IX, MA X, etc. To probe further, in other words to delve into the theory of semisimple Lie groups and algebras, would have required hundreds of additional pages. There are several books on the subject written by excellent specialists, notably those of Varadarajan and Knapp mentioned at the beginning of Chap. 4, as well as the relevant chapters in N. Bourbaki; attempting to improve on these would be somewhat foolish.

It is now about fifty years since I discovered Claude Chevalley's *Theory of Lie Groups* (Princeton UP, 1947), a book that continues to be available and relevant, and in which he once and for all established this theory and more generally that of manifolds from a decidedly global viewpoint. This viewpoint is still adopted in a virtually unchanged form after more than half a century of often rather nebulous developments. Some forerunners should nonetheless be mentioned: Von Neumann (see Chap. 3), Hermann Weyl and his theory of compact Lie groups, and to some extent, Elie Cartan, and Lev Pontryagin's book on topological groups. Apart from being a great mathematician, Chevalley embodied for me all that is anti-pedantic, and above all was a great friend. This book, the outcome of a great revelation, is dedicated to his memory.

Paris, France   Roger Godement
March 2003

Contents

Chapter 1
Topological Groups

1.1 Topological Groups: Examples

A topological group is a group G together with a *separable* topology such that the map $(x, y) \mapsto xy^{-1}$ from the product space $G \times G$ to G is continuous. Equivalently, the maps $x \mapsto x^{-1}$ and $(x, y) \mapsto xy$ can be required to be continuous. The translations $x \mapsto ax$ and $x \mapsto xa$ are then homeomorphisms of G. Immediate examples: the additive group $\mathbb{R}$, more generally the additive group of a finite-dimensional real vector space with the obvious topology, the multiplicative group $\mathbb{R}^*$ or $\mathbb{C}^*$ with the topology induced from that of $\mathbb{R}$ or $\mathbb{C}$, more generally the group $GL_n(\mathbb{R})$ or $GL_n(\mathbb{C})$, the latter being an open subset of the vector space $M_n(\mathbb{R})$ or $M_n(\mathbb{C})$ endowed with the induced topology.

Other examples are found by observing that if G is a topological group, then every subgroup H of G, endowed with the induced topology, is in turn a topological group. For instance, let $G = GL_n(\mathbb{R})$, and take H to be the group $SL_n(\mathbb{R})$ defined by $\det(g) = 1$, or the orthogonal group of a quadratic form, i.e. the set of $g \in G$ such that $gsg' = s$, where s is a non-degenerate symmetric matrix, or, if n is even, the symplectic group, i.e. the set of $g \in G$ such that $gsg' = s$, where $s = -s'$ is a skew-symmetric non-degenerate matrix, or the group of upper triangular matrices, or that of strictly upper triangular matrices (i.e. whose diagonal entries are all equal to 1), etc.

Note that the induced topology on a *non*-closed subgroup H can be strange. Let $\mathbb{T} = \mathbb{R}/\mathbb{Z}$. $\mathbb{T}$ can be identified with the multiplicative group of complex numbers having absolute value 1 via the map $x \mapsto e^{2\pi i x}$. Endowed with the topology of $\mathbb{C}$, the latter is a compact topological group, and thus so is $\mathbb{T}^n = \mathbb{T} \times \cdots \times \mathbb{T}$ endowed with the algebraic and topological product structure. Now, $\mathbb{T}^n = \mathbb{R}^n/\mathbb{Z}^n$, and if p denotes the canonical map from $\mathbb{R}^n$ onto $\mathbb{T}^n$, it is somewhat obvious that a set $U \subset \mathbb{T}^n$ is open if and only if $p^{-1}(U)$ is open. Hence, if D is a 1-dimensional vector subspace of $\mathbb{R}^n$, the image $p(D) = H$ is a subgroup of $\mathbb{T}^n$ (a "1-parameter subgroup" of $\mathbb{T}^n$), and every vector $X \neq 0$ in D obviously gives rise to a continuous homomorphism $\gamma_X : t \mapsto p(tX)$ from the additive group $\mathbb{R}$ to $\mathbb{T}^n$. There are two possibilities. If

© Springer International Publishing AG 2017
R. Godement, *Introduction to the Theory of Lie Groups*,
Universitext, DOI 10.1007/978-3-319-54375-8_1

D contains non-zero elements of $\mathbb{Z}^n$, then X may be assumed to be in $\mathbb{Z}^n$. In this case, $tX \in \mathbb{Z}^n$ for all $t \in \mathbb{Z}$, and so clearly $\gamma_X(t + q) = \gamma_X(t)$ for all $q \in \mathbb{Z}$, so that $\gamma_X(\mathbb{R}) = \gamma_X(I)$, where I is the compact interval $[0, 1]$. The map γ_X being continuous, it follows that in this case the subgroup $H = \gamma_X(\mathbb{R}) = p(D)$ is compact, and hence necessarily closed. Assume X is a primitive element of the lattice $\mathbb{Z}^n$ (i.e. that $tX \in \mathbb{Z}^n \Longleftrightarrow t \in \mathbb{Z}$). This we can always assume, if need be by dividing X by an appropriately chosen integer.[1] Then, for given t and q, $\gamma_X(t + q) = \gamma_X(t)$ requires $q \in \mathbb{Z}$, and hence γ_X is the composition of the canonical map from $\mathbb{R}$ onto $\mathbb{R}/\mathbb{Z} = \mathbb{T}$ and of an injective and continuous homomorphism from $\mathbb{T}$ to $\mathbb{T}^n$. Since $\mathbb{T}$ is compact, *it is a homeomorphism from $\mathbb{T}$ onto the subgroup H*, whose topology, in this case, is therefore also as "normal" as desired in a clinical sense. If, on the contrary, $D \cap \mathbb{Z}^n = \{0\}$, the situation becomes "pathological". Taking $n = 2$ for simplicity's sake, the subgroup H is then found to be *everywhere dense in $\mathbb{T}^2$*. In other words, the subgroup $D + \mathbb{Z}^2$ is everywhere dense in $\mathbb{R}^2$, the map γ_X from $\mathbb{R}$ onto H is bijective (which makes it possible to identify the set H with the set $\mathbb{R}$, hence to transfer to $\mathbb{R}$ the topology of H induced by that of $\mathbb{T}^2$). Once this has been done, for any open subset U of $\mathbb{T}^2$, the open subset $U \cap H$, transferred to $\mathbb{R}$, becomes a countably *infinite* union of mutually disjoint intervals. This is far from being the usual topology on $\mathbb{R}$ (MA IX, Sect. 13). We will return to this example later in connection with the structure of closed subgroups of $\mathbb{R}^n$ (pp. 11–13).

Every group G can be trivially turned into a topological group by endowing it with the discrete topology in which every set is open—an easy device to obtain non-trivial results in some fields (non-commutative harmonic analysis).

A topological group is said to be *locally compact* (resp. *connected*) if this is the case for its topology. As the translations $x \mapsto ax$ are homeomorphisms, they clearly map open sets (resp. neighbourhoods, closed sets, compact sets) onto open sets (resp. ...). To check that a group G is locally compact it, therefore, suffices to show that the identity element e has a compact neighbourhood. This is the case in all the above examples, including obviously discrete groups, the only exception being the group $\mathbb{R}$ endowed with the strange topology obtained by identifying it with a non-closed subgroup of $\mathbb{T}^2$. Besides, not only is it clear that a closed subgroup H of a locally compact group G is itself locally compact, but the converse also holds: *a locally compact subgroup H of a topological group G is necessarily closed*. Replacing G by $\bar{H}$ (which is also a subgroup for trivial reasons), H may be assumed to be everywhere dense in G. By assumption, there is a neighbourhood U of e in G such that $H \cap U$ is compact. As H is dense in G, every interior point of U is the limit of points of $U \cap H$, which is *closed* since it is compact. So H contains the interior of U, i.e. contains an open neighbourhood V of e in G. Hence, H is the union of the open sets $hV, h \in H$, i.e. is open in G; but an *open subgroup H is necessarily closed* (thus $H = G$ in the case at hand) because $G - H$, being the union of the cosets gH, which are open, is open.

The groups $\mathbb{R}^n$ or $\mathbb{T}^n$ are obviously connected, but $GL_n(\mathbb{R})$ is not since its image under the continuous map $g \mapsto \det(g)$ is $\mathbb{R}^*$, and this is not connected. The subgroup

---

[1]Exercise: if D is a line, $D \cap \mathbb{Z}^n$ is the set of all integral multiples of a unique vector up to sign.

$GL_n^+(\mathbb{R})$ of matrices for which $\det(g) > 0$ will later be shown to be connected; it is obviously a normal subgroup of index 2 in $GL_n(\mathbb{R})$, and is open. *If the group G is connected, then for any neighbourhood U of e in G the subgroup generated by U* (in a purely algebraic sense) *is the whole of G*—this subgroup contains U, and so contains an open subset; thus it is open, and so also closed, and as G is connected, it must be the whole of G. If in particular G is locally compact, G can be deduced to be *countable at infinity*, i.e. the countable union of a family of compact sets: take a compact neighbourhood U of e, replace it by $K = U \cap U^{-1}$ so as to obtain a compact neighbourhood K such that $K = K^{-1}$, and then the subgroup it generates, i.e. G, is the union of the sets $K^n = K \cdots K$; but K^n is the image of $K \times \cdots \times K$ under the continuous map $(x_1, \cdots, x_n) \mapsto x_1 \cdots x_n$ from $G \times \cdots \times G$ to G, and so is compact, qed.

It is often the case that G is not connected but has neighbourhoods of the identity (hence, by translation, of any of its points) that are; G is then said to be *locally connected* (this will be the case for Lie groups—in this case there are neighbourhoods of e homeomorphic to Euclidean balls). If G° denotes the identity component in G, i.e. the union of all the connected parts of G containing e, it is then a *connected open subgroup* of G. First, G° is a subgroup because the continuous map $(x, y) \mapsto xy^{-1}$ transforms the connected set $G^\circ \times G^\circ \subset G \times G$ into a connected subset of G which, since it contains e, is necessarily contained in G°. On the other hand, G being locally connected, G° contains a neighbourhood of e, and so is open.

There are groups in nature (that of mathematicians) whose behaviour, from this point of view, is totally at odds with that suggested by the examples which have already been mentioned. It may well happen that $G^\circ = \{e\}$; G is then said to be *totally disconnected*. For instance, let us take a preferably infinite, arbitrary family of finite groups, say $(G_i)_{i \in I}$, and let us consider the group $G = \prod G_i$ endowed with the product topology. If F is an arbitrary finite subset of I, and if E_i denotes the trivial subgroup of G_i, then (by definition of a product topology) the products

$$U_F = \prod_{i \in F} E_i \times \prod_{i \notin F} G_i$$

form a fundamental system of neighbourhoods of the identity element in G. The neighbourhoods of an arbitrary point follow by translation from the sets U_F. Next, let C be a connected subset of G containing e, and let pr_i be the projection $G \mapsto G_i$. It is continuous, and so maps C to a connected subset of G_i containing e; as G_i is finite, it follows that $pr_i(C) = \{e\}$ for all i, and thus $C = \{e\}$, qed.

Another example, fundamental in arithmetic, is that of the additive group of p-adic integers. Choose a prime number p and for all n, set $G_n = \mathbb{Z}/p^n\mathbb{Z}$; this is a finite additive group. The cosets mod p^{n+1} are canonically mapped to the cosets mod p^n, thereby giving a sequence of canonical homomorphisms

$$G_o = \{0\} \leftarrow G_1 \leftarrow G_2 \leftarrow G_3 \leftarrow \cdots,$$

that are in fact surjective. Let π_n denote the homomorphism from G_n onto G_{n-1} and consider the set of families $x = (x_n)$ with

$$x_n \in G_n \text{ and } x_{n-1} = \pi_n(x_n) \text{ for all } n.$$

The set G of these families is non-empty; in fact, *the map $x \mapsto x_n$ from G to G_n is surjective*: $x_n, \cdots, x_0$ is determined by x_n; besides, once x_n is fixed, it suffices to choose some $x_{n+1} \in G_{n+1}$ mapped onto x_n, then some $x_{n+2} \in G_{n+2}$ mapped onto x_{n+1} and so on. The set G can be endowed with the structure of a group by setting

$$(x_n) + (y_n) = (x_n + y_n),$$

which turns G into a subgroup of the product $\prod G_n$, and this subgroup is *closed* (hence compact, like $\prod G_n$) in the product topology: indeed, it consists of the elements $x \in \prod G_n$ for which

$$pr_{n-1}(x) = \pi_n \circ pr_n(x)$$

for all n. Now, the maps pr_{n-1} and $\pi_n \circ pr_n$ are continuous, so that the previous condition is "preserved by passing to the limit", and hence defines a closed subset. This compact group G is the group known as the *ring (why?) of p-adic integers.* For each n, denote by U_n the set of $x \in \prod G_k$ such that $x_k = 0$ for all $k \leq n$. These U_n clearly form a fundamental system of neighbourhoods of 0 in $\prod G_k$, and hence the subsets $U_n \cap G$ form a fundamental system of neighbourhoods of 0 in G, and each of them is an open subgroup of finite index in G. It is now obvious that since G is the product of the subgroups G_n, it is totally disconnected. Further information about *p*-adic numbers can notably be found in Z.I. Borevich and I.R. Shafarevich, *Number Theory* (Academic Press, 1966), pp. 18–32.

The totally disconnected groups G that have been constructed satisfy the property that *every neighbourhood of e in G contains a non-trivial subgroup*, and in fact even an open subgroup of G. The group $GL_n(\mathbb{R})$, and perforce every subgroup G of $GL_n(\mathbb{R})$, has the opposite property: *every sufficiently small subgroup of G is trivial.* It suffices to prove this for $GL_n(\mathbb{R})$, and hence for $GL_n(\mathbb{C})$. Let $\|g\|$ denote the usual norm of an $n \times n$ complex matrix, and let H be a subgroup of $G = GL_n(\mathbb{C})$ contained in the ball $\|g - 1\| < 1$. As the norms of the eigenvalues of g are bounded above by $\|g\|$, the eigenvalues of a matrix g for which $\|g - 1\| < 1$ also satisfy the inequality $|\lambda - 1| < 1$, and hence are non-trivial. Consequently, the ball in question is fully in G (another proof: write $g = 1 - x$ with $\|x\| < 1$ and observe that the series $\sum x^n$ converges to the inverse matrix of g). Besides, for $n \in \mathbb{Z}$ and g invertible, the eigenvalues of g^n are known to be obtainable by taking the nth power of those of g. Since $g \in H$ implies that $g^n \in H$ and thus that $\|g^n - 1\| < 1$, we get $|\lambda^n - 1| < 1$ for every $n \in \mathbb{Z}$ and every eigenvalue λ of any matrix $g \in H$. As the powers of λ remain bounded, $|\lambda| = 1$ follows. Applying a trivial geometric argument, this in turn gives $\lambda = 1$. Hence 1 is the only eigenvalue of a matrix $g \in H$. But then there is a basis $e_1, \cdots e_n$ of $\mathbb{C}^n$ satisfying

$$g(e_1) = e_1,$$
$$g(e_2) = e_2 + g_{12}e_1,$$
$$g(e_3) = e_3 + g_{23}e_2 + g_{13}e_1,$$

etc. As r varies in $\mathbb{Z}$, g^r remains in a fixed compact set—namely in the ball $\|x - 1\| \leq 1$—so that the vectors $g^r(e_2) = e_2 + rg_{12}e_1$ remain in a fixed compact subset of $\mathbb{C}^n$; this obviously requires $g_{12} = 0$. We then get $g^r(e_3) = e_3 + rg_{23}e_2 + rg_{13}e_1$, and as these vectors need to stay in a fixed compact set, it follows that $g_{23} = g_{13} = 0$, and so on.

The subgroups of the linear groups $GL_n(\mathbb{R})$ are not the only ones to admit "arbitrarily small" subgroups. Indeed, let G be a locally compact group and suppose there is a neighbourhood U of e in G and an *injective continuous* map ϕ from U to a linear group $H = GL_n(\mathbb{R})$ such that

$$\phi(xy^{-1}) = \phi(x)\phi(y)^{-1} \text{ whenever } x, y \text{ and } xy^{-1} \text{ are in } U. \qquad (*)$$

There is a neighbourhood V of the identity in H which does not contain any non-trivial subgroups, and as ϕ is continuous, we may assume that $\phi(U) \subset V$. Then ϕ clearly maps every subgroup of G contained in U to a subgroup of H contained in V, and hence trivial. Since ϕ is injective, U does not contain any non-trivial subgroup of G.

The previous trivial property admits a rather difficult but rarely used converse: *if a locally compact group G does not have arbitrarily small subgroups, then the identity component G° is open in G and there is a local embedding of G into a linear group*. In other words, it is possible to find a neighbourhood U of e in G and an injective continuous map ϕ from U to a group $GL_n(\mathbb{R})$ satisfying the property used above. This result is more or less equivalent to the resolution of "Hilbert's fifth problem". The latter consists in showing that if in a topological group G, there is a neighbourhood of the identity homeomorphic to a ball in a Euclidean space, then G is a Lie group (in other words, it is possible to introduce local coordinates in the neighbourhood of e with respect to which multiplication is expressed by *analytic* formulas). See D. Montgomery and L. Zippin, *Topological Transformation groups* (Interscience, 1955), or exercises 5–9 of Chap. XIX, §8, in J. Dieudonné's *Eléments d'Analyse*.

Condition $(*)$ enables us to define the more general concept of a *local homomorphism* from a topological group G to another. For example, taking $G = \mathbb{T}$, $H = \mathbb{R}$ and for U the set of elements $z = \exp(i\theta)$ of G such that $|\theta| < \pi/4$, we get a local homomorphism ϕ defined on U, by associating to each $z \in U$ its argument between $-\pi/4$ and $+\pi/4$. Note that in this example ϕ cannot be extended to a "global" homomorphism, i.e. defined everywhere, from G to H (the image of $\mathbb{T}$ under a continuous homomorphism to $\mathbb{R}$ would be a compact subgroup of $\mathbb{R}$, and hence reduced to $\{0\}$). Nonetheless, there do exist (connected) groups G for which any local homomorphism from G to another group can always be extended to G; these are the *simply connected* groups of Chap. 2.

1.2   Homogeneous Spaces

Let G be a topological group which, as an abstract group (i.e. without topology), acts on the left on a topological space X; G is said to *act continuously* on X if the map $(g, x) \mapsto gx$ from $G \times X$ to X is continuous. In particular, it follows that if X is separable,[2] the stabilizer G_x in G of any point x of X is a *closed* subgroup of G.

For example, consider the topological group G and a *closed* subgroup H of G. Let p be the canonical map from G onto $X = G/H$. A set $U \subset G/H$ will be said to be open if $p^{-1}(U)$ is open in G. This obviously defines a topology on X, and X, endowed with this topology, is *separable* (remember that G is assumed to be separable and H closed). Indeed, let $a = p(x)$ and $b = p(y)$ be two distinct points of X; as $xH \neq yH$, $y^{-1}x \notin H$. Hence there is a neighbourhood V of e in G such that $Vy^{-1}x$ does not meet H since H is closed. So $Vy^{-1}x = y^{-1}V'x$, where $V' = yVy^{-1}$ is again a neighbourhood of e since, for all $a \in G$, the automorphism $g \mapsto aga^{-1}$ is obviously a homeomorphism from G onto itself. Now, if a neighbourhood W of e in G is chosen so that $W^{-1}W \subset V'$, then $y^{-1}W^{-1}Wx \cap H$ is empty. Hence so is $WxH \cap WyH$, and the images under p of Wx and Wy are then disjoint neighbourhoods of a and b in X, and the Hausdorff axiom follows.

The map p from G onto X is continuous—by construction of the topology on X—and it is also *open*; indeed, if U is an open subset in G, then $p^{-1}(p(U)) = UH$. Now, UH, being the union of open subsets Uh, is open. Hence $p(U)$ is open in X. In particular, the image under p of a neighbourhood of $g \in G$ is a neighbourhood of $p(g)$, and if G is locally compact, so is X.

Let us finally show that G acts continuously on $X = G/H$. We need to show that, for any open subset U of X, the set of pairs (g, x) such that $gx \in U$ is open in $G \times X$. Now, these are the pairs $(g, p(g'))$ for which $g.p(g') \in U$, i.e. $p(gg') \in U$, i.e. $gg' \in p^{-1}(U)$; but as $p^{-1}(U)$ is open in G, pairs (g, g') such that $gg' \in p^{-1}(U)$ form an open subset of $G \times G$. Therefore, the map $(g, g') \mapsto (g, p(g'))$ from $G \times G$ to $G \times X$ remains to be shown to be open. Now, an open subset of $G \times G$ is a union of open sets of the form $A \times B$, where A and B are open in G; as $p(B)$ is open in X, the image of $A \times B$ under $(g, g') \mapsto (g, p(g'))$ is open in $G \times X$. This obviously gives the desired result since a union of open sets is open.

Conversely, let X be a topological space acted on by G. Fix an element $a \in X$ and let $H = G_a$ be the stabilizer of a in G. The map $m : g \mapsto ga$ is clearly constant on the H-cosets, and hence can be factorized into the canonical map $p : G \to G/H$ and a map $f : G/H \to X$ which is in fact a bijection from G/H onto the orbit Ga of the point a of X. The map f is continuous; more generally, it is clear that *a map from G/H to a topological space X is continuous if and only if so is the composite map from G to X* ("universal" property of quotients).

It would be really nice if it were possible to deduce that f is in fact a homeomorphism from G/H onto the orbit of a, but this is far from being the case without

---

[2]This assumption will not always be stated in the rest of the book; namely, unless explicitly stated, *all the spaces considered will be assumed to be separable.*

additional assumptions. This hope is nonetheless justified in a context which notably includes the theory of Lie groups:

Theorem 1 *Let G be a locally compact group, countable at infinity, and let X be a locally compact space on which G acts continuously and transitively. Let H be the stabilizer of a point a of X. Then the natural bijection $f : G/H \to X$ is bicontinuous.*

As we know that f is continuous and bijective, it suffices to show that f is open, in other words that $f(U)$ is open for any open subset U of G. For this, it suffices to check that $f(U)$ is a neighbourhood of $f(g) = ga$ for all $g \in U$, and as $x \mapsto gx$ is a homeomorphism from X onto itself, it suffices to check that $f(g^{-1}U)$ is a neighbourhood of a. In other words, it amounts to showing that f maps every neighbourhood U of e in G to a neighbourhood of a in X.

Let V be a neighbourhood of e in G such that $V^{-1}V \subset U$; set $U(a) = f(U)$ and $V(a) = f(V)$ so that $U(a) \supset f(V^{-1}V) = \bigcup v^{-1} \cdot V(a)$. To show that a is in the interior of $U(a)$, it therefore suffices to show that it is in the interior of $v^{-1} \cdot V(a)$ for at least some $v \in V$ for which $v(a)$ is an interior point of $V(a)$. It is namely a matter of showing that $V(a)$ has at least one interior point.

Now, by assumption G is the union of a sequence of compact groups K_n; as g varies in K_n, the sets gV cover K_n. It follows that each K_n is covered by a finite number of left translates of V. Hence there is a sequence (g_n) of points of G such that $G = \bigcup g_n V$. Consequently, $X = \bigcup g_n \cdot V(a) = \bigcup A_n$, where each $A_n = g_n \cdot V(a)$ is the image of V under a continuous map, namely $v \mapsto g_n va$, and so is compact and thus necessarily closed if we suppose that V itself is compact, which we can. Besides, if $g_n \cdot V(a)$ has an interior point, clearly so does $V(a)$. Thus, the proof of the theorem reduces to that of the following result:

Lemma *– Let A_n be a sequence of closed subsets in a locally compact space X. If X is the union of the sets A_n, at least one of them has an interior point.*

Indeed, let us suppose that the lemma does not hold. Then $A_1 \neq X$, and so there is a relatively compact non-trivial set U_1 in X such that $\bar{U}_1 \cap A_1 = \emptyset$ (for U_1 take a sufficiently small neighbourhood of some point in $X - A_1$). As A_2 does not have an interior point, the intersection $\bar{U}_1 \cap A_2$ does not contain the whole of U_1. Choosing a sufficiently small neighbourhood of some point in U_1 not contained in the closed set $\bar{U}_1 \cap A_2$ gives us a second relatively compact non-trivial set $U_2 \subset U_1$ such that $\bar{U}_2 \cap A_k = \emptyset$ for $k \leq 2$. Continuing in this way, we obtain relatively compact non-trivial open sets $U_1 \supset U_2 \supset \cdots$ such that, for all n, $\bar{U}_n \cap A_k = \emptyset$ for all $k \leq n$. By compactness, the intersection of the $\bar{U}_n$ is not empty, and so meets some A_k, leading to a contradiction, qed.

Corollary 1 *Let G and G' be two locally compact groups and j a bijective continuous homomorphism from G onto G'. If G is countable at infinity, j is bicontinuous.*

Indeed, G can be made to act on the space G' via the map $(x, y) \mapsto j(x)y$, and the stabilizer of $e \in G'$ in G being trivial, the result follows from the theorem.

Corollary 2 *Let G be a group and T and T' two topologies on G with respect to which G is locally compact and countable at infinity. Then* $T = T'$.

Endow the product $G \times G$ with the topology $T \times T'$ and the diagonal D of $G \times G$ with the induced topology. The projections $D \to G$ are bijective and continuous regardless of whether G is endowed with T or T'. Since D is countable at infinity as a closed subset of $G \times G$, Corollary 1 shows that the two projections $D \to G$ are bicontinuous. Considering the "difference", the identity map from G endowed with T onto G endowed with T' is seen to be bicontinuous, qed.

Corollary 2 shows that if an "abstract" group G can be endowed with a topology compatible with its algebraic structure and with respect to which it is locally compact and countable at infinity (for example connected), then this topology is *unique*. This result no longer holds if the countability condition is dropped: the discrete topology enables us to endow every group G with a locally compact topology compatible with the structure of G; but naturally, it is countable at infinity only if G is countable.

As an application of Theorem 1, we show that *the group* $G = GL_n^+(\mathbb{R})$ *is* connected by making it act on the space $X = \mathbb{R}^n - \{0\}$. For a, let us take the first basis vector e_1, so that the stabilizer H is the subgroup of matrices $h = \begin{pmatrix} 1 & u \\ 0 & x \end{pmatrix}$, where u denotes a row matrix of order $n - 1$ and x is a matrix in $GL_{n-1}^+(\mathbb{R})$ (why?). The assumptions of the theorem are obviously satisfied, so that $\mathbb{R}^n - \{0\}$ can be topologically identified with the quotient G/H.

To infer that the group $GL_n^+(\mathbb{R})$ is connected, we use induction on n, the case $n = 1$ being clear. Since

$$\begin{pmatrix} 1 & u \\ 0 & 1 \end{pmatrix} \begin{pmatrix} 1 & 0 \\ 0 & x \end{pmatrix} = \begin{pmatrix} 1 & ux \\ 0 & x \end{pmatrix},$$

$H = UM$, where U is the subgroup of matrices $\begin{pmatrix} 1 & u \\ 0 & 1 \end{pmatrix}$ and M that of matrices $\begin{pmatrix} 1 & 0 \\ 0 & x \end{pmatrix}$; the former is homeomorphic to $\mathbb{R}^{n-1}$, and hence connected, and the latter to $GL_{n-1}^+(\mathbb{R})$, and so is connected by the induction hypothesis. As H is the image of the connected space $U \times M$ under the continuous map $(u, x) \mapsto ux$, H must be connected. On the other hand, $G/H = \mathbb{R}^n - \{0\}$ is also connected.

We still need to show that *if a topological group G has a closed subgroup H such that H and G/H are connected, then G is connected.* Let U be an open and closed subset of G. For all $g \in G$, gH is connected like H, and $U \cap gH$ is open and closed in gH. Hence either $gH \subset U$ or $U \cap gH = \emptyset$, so that $UH = U$, the same result holding for $V = G - U$, which is open and closed like U. But then the images of U and V in G/H form a partition of G/H into two open sets; as G/H is connected, the conclusion readily follows.

As an aside, note that the group $G = GL_n^+(\mathbb{R})$ is in fact path-connected. Being open in the vector space $M_n(\mathbb{R})$, G clearly contains a path-connected neighbourhood U of e. It generates G since G is connected. All $g \in G$ can therefore be written as $g = g_1 \cdots g_n$, where each g_i is either in U or U^{-1}, and so can be connected to the origin e by a path drawn in G. The identity e can then be connected to g by connecting e to g_1, then g_1 to $g_1 g_2$ (connect e to g_2 and then take its left translate by g_1), then $g_1 g_2$ to $g_1 g_2 g_3$ (connect e to g_3 and take its left translate by $g_1 g_2$), etc.. We assume that the reader will have no difficulty in extending these arguments to more general groups satisfying easily discovered assumptions.

The case $G = GL_n(\mathbb{C})$ may be dealt with by the same methods. This time the group is connected because $\mathbb{C}^* = GL_1(\mathbb{C})$, in contrast to $\mathbb{R}^*$, is connected, and induction shows that, as above, $GL_n(\mathbb{C})$ admits a connected subgroup for which the homogeneous space is homeomorphic to $\mathbb{C}^n - \{0\}$, which is connected.

Similar arguments show that the subgroups $SL_n(\mathbb{R})$ and $SL_n(\mathbb{C})$, obtained by restricting ourselves to matrices with determinant 1, are connected.

To conclude these general remarks on homogeneous spaces, we state some lemmas whose proofs will be given later.

Lemma 1 *Let G be a locally compact group and H a closed subgroup of G. Then every compact subset of G/H is the image of a compact subset of G.*

Let p be the canonical map from G onto G/H and K a compact subset of G/H. As p is open, images under p of relatively compact open subsets of G form an open cover of G/H, and so perforce of K. Hence there are finitely many relatively compact open subsets $U_1, \cdots U_n$ in G such that $K \subset \bigcup p(U_i)$. The union of all $\bar{U}_i$ is a compact set $A \subset G$ such that $p(A) \supset K$, and then the compact set $B = A \cap p^{-1}(K)$ satisfies $p(B) = K$. The lemma follows (which naturally *does not* mean that $p^{-1}(K)$ is compact for every compact subset K of G/H).

Lemma 2 *Let G be a topological group acting continuously on a topological space X. Then, for any compact subset $A \subset X$ and any closed subset $B \subset X$, the set $C \subset G$ of $g \in G$ such that[3] $gA\#B$ is closed in G.*

Setting p to be the map $(g, x) \mapsto gx$ from $G \times A$ to X, and pr_G the map $(g, x) \mapsto g$ from $G \times A$ onto G,

$$C = pr_G[p^{-1}(B)] = pr_G(M)$$

clearly holds, with M a closed subset of $G \times A$. Hence what requires showing is that *if X is a topological space and Y a compact space, then the map pr_X from $X \times Y$ onto X transforms every closed subset M of $X \times Y$ into a closed subset of X.* We give a heuristic proof (which holds if X and Y are metrizable): supposing that the elements $x_n \in pr_X(M)$ converge to a limit x, take elements $y_n \in Y$ such that $(x_n, y_n) \in M$, and, if need be by extracting a subsequence thanks to the compactness of Y, observe that the elements y_n can be assumed to converge to some $y \in Y$; then $(x_n, y_n) \in M$ converge to $(x, y) \in M$ since M is closed, and so $x \in pr_X(M)$, which is therefore closed. The general case is dealt with similarly using ultrafilters instead of sequences; N. Bourbaki, *Gen. Top.: Chaps. 5–10*, §10 may prove helpful (?).

---

[3] The notation $A\#B$ is an effective substitute for $A \cap B \neq \emptyset$. Read "A meets B".

Lemma 3 *Let A and B be subsets of a topological group G. If A is compact and B is closed, then AB is closed.*

Indeed, $g \in AB$ means that $A^{-1}g$ meets B. It then suffices to apply Lemma 2 by making G act on itself on the right.

Lemma 4 *Let H and M be two closed subgroups of a locally compact group G. Suppose that $H/H \cap M$ is compact. Then the image of M in G/H is closed and it is moreover discrete if so is M in G.*

Let p be the canonical map from G to G/H. Then $p^{-1}(p(M)) = MH$, so that the first thing to check is that MH is closed in G. By Lemma 1, there is a compact subset $K \subset H$ which is mapped onto $H/H \cap M$, so that $H = K \cdot (H \cap M)$ and hence $H = (H \cap M)K'$, where $K' = K^{-1}$ is compact. This gives $MH = MK'$, and Lemma 3 shows that MH is closed.

Suppose that M is discrete. To show that $p(M)$ is discrete, it suffices to show that $A \cap p(M)$ is finite for any compact subset $A \subset G/H$. As $A = p(B)$ for some compact subset $B \subset G$, we obviously get $A \cap p(M) = p(BH \cap MH) = p(B \cap MH) = p(B \cap MK')$, where $K' = K^{-1}$ is the compact subset of H used above. The equality $p(B \cap MK') = p(BK \cap M)$ trivially holds. As M is discrete and BK is compact, $BK \cap M$ is finite, giving the desired result.

1.3   Elementary Abelian Groups

The simplest locally compact groups are those that are abelian; examples are readily found: $\mathbb{R}^n$, $\mathbb{T}^n$, $\mathbb{Z}^n$ and, more generally, any discrete linear group. A group is said to be *elementary* if it is a topological group of the form $G = \mathbb{R}^p \times \mathbb{T}^q \times F$, where F is a discrete abelian group of *finite type*, i.e. a quotient of some group $\mathbb{Z}^n$ or equivalently, as is well known (see S. Lang, Algebra, Chap. I, §10, or the exercises of §31 in the author's *Cours d'Algèbre*), the product of some group $\mathbb{Z}^d$ by some finite abelian group. In this section, our intention is to prove the following result:

Theorem 2 *Let G be an elementary group and H a closed subgroup of G. Then H and G/H are elementary groups.*

The statement about G/H assumes that G/H is endowed with the structure of a topological group. For this it suffices to observe that if a closed subgroup H of a topological group G is normal in G, then the quotient topology is compatible with the group structure of G/H, which can be seen by applying the above arguments.

Let us first show that, *if G is an elementary group, then there exist a space $\mathbb{R}^n$ and closed subgroups L and M of $\mathbb{R}^n$, with $M \subset L$ such that $G = L/M$.* For this, write $G = \mathbb{R}^p \times \mathbb{T}^q \times (\mathbb{Z}^r/D)$, where D is a subgroup of $\mathbb{Z}^r$, and observe that $\mathbb{T}^q = \mathbb{R}^q/\mathbb{Z}^q$ not only algebraically, but also topologically. Then $G = (\mathbb{R}^{p+q} \times \mathbb{Z}^r)/(\mathbb{Z}^q \times D)$, both algebraically and topologically (for example, apply Theorem 1).

However, $\mathbb{R}^{p+q} \times \mathbb{Z}^r$ is clearly a closed subgroup of $\mathbb{R}^{p+q+r}$, and our assertion follows. The proof of Theorem 2 now essentially consists in showing that, conversely, every quotient L/M is elementary, and this requires precise information about closed subgroups of a vector space.

Theorem 2a – *Every closed subgroup L of $\mathbb{R}^n$ is isomorphic to $\mathbb{R}^q \times \mathbb{Z}^p$ with $p + q < n$.*

To show this let us consider the union V of all lines (1-dimensional vector subspaces) contained in L (if there are none, set $V = \{0\}$). V is a vector subspace: indeed, if D' and D'' are contained in the subgroup L, so is the set $D' + D''$ of elements $x' + x''$, with $x' \in D'$ and $x'' \in D''$. Let W be a complement of V in $\mathbb{R}^n$ and let $M = L \cap W$ be a closed subgroup of W. Then $L = V + M$. For if $x = v + w$ is the decomposition of some $x \in L$, then $v \in V \subset L$ and therefore $w \in L \cap W = M$. The map $(v, m) \mapsto v + m$ from $V \times M$ to L is obviously a bijective continuous homomorphism, and thus is bicontinuous by Theorem 1, so that $L = V \times M$ both algebraically and topologically (Theorem 1 can be bypassed by observing that the projection operators corresponding to the decomposition of $\mathbb{R}^n$ into the direct sum of V and W are continuous...).

It remains to be shown that M is isomorphic to a group $\mathbb{Z}^q$. By construction, M is clearly a closed subgroup of $\mathbb{R}^n$ not containing any lines. From this we first deduce that it is discrete by showing that *every non-discrete subgroup M of $\mathbb{R}^n$ contains at least one line.*

M not being discrete, it indeed contains a sequence of *non-trivial* points x_n converging to 0, and if necessary by extracting a subsequence, the unit vectors $u_n = x_n/\|x_n\|$ may be assumed to approach a limit u. We show that the line D through u is contained in M (this type of argument will also be used later to show that every closed subgroup of a Lie group is a Lie group—using similar arguments, we will start by proving that the subgroup at hand contains "sufficiently" many "1-parameter subgroups"—in the case of $\mathbb{R}^n$, these happen to be the lines through the origin, a strange coincidence). Indeed, let t be a real number so that $t u_n = t_n x_n$ with $t_n = t/\|x_n\|$. We can write $t_n = q_n + r_n$ with $q_n \in \mathbb{Z}$ and $0 \le r_n < 1$. Then,

$$tu = \lim t u_n = \lim(q_n x_n + r_n x_n) = \lim q_n x_n$$

since $\|r_n x_n\| < \|x_n\|$ converges to 0. As $q_n x_n \in M$ and as M is closed, the conclusion that the line through u lies in M follows as claimed.

To finish the proof that *every closed subgroup of $\mathbb{R}^n$ is of the form $\mathbb{R}^p \times \mathbb{Z}^p$*, it remains to show that every discrete subgroup M of $\mathbb{R}^n = E$ is of the form $\mathbb{Z}^q$ and is in fact (algebraically) generated by q linearly independent vectors. Replacing E by the vector subspace generated by M, M can be assumed to generate E as a vector space. If $n = 1$, then M is a non-trivial discrete subgroup of $\mathbb{R}$. As M is discrete, the elements of $M \cap \mathbb{R}_+^*$ have a lower bound $u > 0$. It belongs to M since M is closed,[4]

---

[4]Let H be a subgroup of a topological group G; suppose that the topology of G induces the discrete topology on M; then H is closed. Indeed, there is a neighbourhood U of e in G such that $H \cap U$

and $M = \mathbb{Z}u$ (whence the result in this case) which follows from Euclidean division (any $x \in M$ is the sum of an integral multiple of u and of an element of M contained between 0 and u, and so is trivial by definition of u).

In the general case, choose a line D in E such that $D \cap M$ is non-trivial (take a line connecting 0 to a non-trivial element of M). As $D \cap M$ is a non-trivial discrete subgroup of $D \simeq \mathbb{R}$, there is a vector u_1 such that $D \cap M$ is the set of integral multiples of u_1, and $D/D \cap M \simeq \mathbb{R}/\mathbb{Z}$ is clearly compact. Lemma 4 of the preceding section then shows that *the image M' of M in the quotient $E' = E/D$ is a discrete subgroup of $E' \simeq \mathbb{R}^{n-1}$*.

Since M moreover generates the vector space $\mathbb{R}^n$, it is clear that M' generates the quotient vector space E'. By induction on n, M' can then be supposed to be a subgroup generated by a basis of E'. Let $u_2, \cdots, u_n$ be the elements of M mapped onto the basis elements of E'. For all $m \in M$, there clearly exists a linear combination m' of $u_2, \cdots, u_n$ with integral coefficients having the same image as m in E', and thus such that $m - m' \in D \cap M$, so that $m = m' + r_1 u_1$ for some integer r_1. The subgroup M can therefore be generated by the vectors $u_1, \cdots, u_n$ which evidently form a basis for $\mathbb{R}^n$. Hence it can be concluded that, if M is a discrete subgroup of a finite-dimensional vector space E, then, as a subgroup, M is generated by a basis of the vector subspace of E generated by M. Therefore, $M \simeq \mathbb{Z}^q$ with $q \leq \dim(E)$, and this completes the proof of Theorem 2a.

Theorem 2 itself is still in need of a proof. Let us first show that *if there are two closed subgroups L and M in $\mathbb{R}^n$ with $M \subset L$, then the quotient group $G = L/M$ is elementary*.

However, the proof of Theorem 2a shows that $L = V \times D$, where V is the union of lines contained in L, and the discrete set D is the intersection of L and a complement of V in $\mathbb{R}^n$. Then consider the union V (resp. W) of lines contained in L (resp. M), so that $W \subset V$, and fix a complement V' of W in V as well as a complement U of V in $\mathbb{R}^n$, so that $V' \oplus U$ is a complement of W in $\mathbb{R}^n$. From the above, it follows that there are isomorphisms

$$M \approx W \times (M \cap (V' + U)), \quad L = V + (L \cap U) \approx W \times V' \times (L \cap U). \quad (*)$$

As these identifications of L and M with products are compatible with the topological group structures of all the objects considered, there is likewise an identification

$$L/M \approx (V' \times (L \cap U))/(M \cap (V' + U)).$$

(Footnote 4 continued)
is trivial. Next take an adherent point x of H. Then x is also an adherent point of $xV \cap H$ for any neighbourhood V of e in G. But if xV contains $h', h'' \in H$, then clearly

$$h'^{-1}h'' \in V^{-1}V \cap H.$$

The latter is contained in $U \cap H$ if V is sufficiently small. So $h' = h''$, in other words $xV \cap H$ contains only one point if V is sufficiently small, whence $x \in H$.

As will be shown later, V' and U may be chosen in such a way that

$$M \cap (V' + U) = (M \cap V') + (M \cap U) \approx (M \cap V') \times (M \cap U). \qquad (**)$$

Admitting this point for now, we then get an isomorphism[5]

$$L/M \approx [V' \times (L \cap U)]/[(M \cap V') \times (M \cap U)]$$
$$\approx [V'/(M \cap V')] \times [(L \cap U)/(M \cap U)].$$

As $L \cap U$ is a group $\mathbb{Z}^q$, as a topological group, the quotient $(L \cap U)/(M \cap U)$ is discrete of finite type. Besides, $M \cap V'$ is a discrete subgroup of V', and so generated by linearly independent vectors. Completing these to a basis for V', it becomes clear that $V'/(M \cap V')$ is isomorphic to $\mathbb{R}^p \times (\mathbb{R}^q/\mathbb{Z}^q) = \mathbb{R}^p \times \mathbb{T}^q$ for properly chosen p and q. Hence, pending the proof of $(**)$, the quotient L/M is indeed seen to appear as the product of some $\mathbb{R}^p$, some $\mathbb{T}^q$ and a quotient of some $\mathbb{Z}^r$, i.e. it is, as claimed, an *elementary group* for all L and M.

To deduce the assertion of Theorem 2 that if H is a closed subgroup of an elementary group G, then so are H and G/H, start by observing that G can be written as L/M with closed subgroups of a space $\mathbb{R}^n$. Denoting by L' the inverse image of H under the canonical map from L onto G, we then obviously get bijective homomorphisms from L'/M onto H and from L/L' onto G/H, and the proof that H and G/H are elementary then reduces to checking that these bijections are bicontinuous.

Let us start with the first one. The open subsets of H are the intersections of H and open subsets of G. Their inverse images in L' are therefore the intersections of L' and the inverse images in L of the open subsets of G, i.e. M-invariant open subsets of L. This gives exactly all the M-invariant open subsets of L', and since these correspond precisely to open subsets of L'/M, as claimed, there is a homeomorphism between H and L'/M.

In the second case, the open subsets of G/H correspond exactly to the H-invariant open subsets of G, which are obviously the images of open L'-invariant subsets of L. But these correspond precisely to the open subsets of L/L', whence the homeomorphism between G/H and L/L'. The fact that L is a closed subgroup of some $\mathbb{R}^n$ is clearly not needed.

Hence everything has been shown *except* that V' and U can be chosen in such a way that $(**)$ holds. Choosing a complement E of W in $\mathbb{R}^n$ gives $M = W + (E \cap M)$ and $L = W + (E \cap L)$ as W is contained in M and in L. Hence we can work in E, which is going to replace the subspace $V' \oplus U$ appearing, for example, in $(*)$. The proof then consists in showing that *if there are two closed subsets L and M in E, with M discrete and contained in L, then there is a complement U in E of the union V of lines of L (which now replaces V' of $(*)$) such that $M = (M \cap V) + (M \cap U)$.* Let S be an arbitrary complement of U in E, and p the canonical map from E onto

---

S corresponding to the decomposition $E = S \oplus V$. As $V \subset L$, $p(L) = L \cap S$, a discrete subgroup of S containing $p(M)$ since $M \subset L$. So there is a basis $s_1, \cdots, s_r$ of $p(L)$ and integers $n_1, \cdots, n_r$ such that, $p(M)$ is generated by the elements $n_i s_i$, according to the theorem of elementary divisors (structure of subgroups of $\mathbb{Z}^r$: *Cours d'Algèbre* by the author, §31, exercise 8, or else S. Lang, *Algebra*, Chap. XV, §2, theorem 5), and it may be assumed that $n_i \neq 0$ for $i \leq k$, and that $n_i = 0$ for $k < i$. For each $i \leq k$, let us choose $m_i \in M$ such that $p(m_i) = n_i s_i$. Every $m \in M$ is then the sum of an integral linear combination of these m_i and of a zero of p, i.e. in V, and even necessarily in $M \cap V$. Hence $M = (M \cap V) \times (M \cap U)$ whenever U is chosen to be a complement of V containing the m_i. This is obviously possible since, by construction, the m_i are linearly independent mod V, completing the proof.

To conclude, we make a few remarks concerning *the determination of closed subgroups of* $\mathbb{R}^n$ *using systems of equations.* Let L be such a subgroup. Then $\mathbb{R}^n = U \oplus V \oplus W$ can be decomposed into a direct sum of three vector spaces such that, on the one hand, $L = (L \cap U) \times (L \cap V) \times (L \cap W)$, and, on the other, $L \cap U = U$ (i.e. $U \subset L$), $L \cap V$ is the lattice generated by a basis of V, and finally $L \cap W = \{0\}$. This follows readily from the structure of L, which was determined above. Hence there is a basis $(a_1, \cdots, a_n)$ of $\mathbb{R}^n$ and integers p and q, with $p + q \leq n$, such that L is the set of vectors $\sum x_i a_i$ for which

$$x_i \in \mathbb{R} \text{ if } 1 \leq i \leq p, \quad x_i \in \mathbb{Z} \text{ if } p + 1 \leq i \leq p + q,$$
$$x_i = 0 \text{ if } p + q + 1 \leq i.$$

Then let $L^\sim$ be the set of linear forms f in the dual of the vector space $\mathbb{R}^n$ with integral values on L. These forms are obvious characterized by

$$f(e_i) = 0 \quad \text{for} \quad i \leq p, \quad f(e_i) \in \mathbb{Z} \quad \text{for} \quad p < i \leq p + q.$$

In other words, with respect to the dual basis $(u_1, \cdots, u_n)$ of the given basis, these are the forms $f = \sum y_i u_i$ whose coordinates satisfy

$$y_i \in \mathbb{R} \quad \text{if} \quad i > p + q, \quad y_i \in \mathbb{Z} \quad \text{if} \quad p < i \leq p + q, \quad y_i = 0 \text{ if } i \leq p.$$

It is clear that these two situations are mutually symmetrical, and that consequently the vectors $x \in L$ are, for their part, characterized by

$$f(x) \in \mathbb{Z} \text{ for all } f \in L^\sim,$$

which can be expressed more strikingly as $(L^\sim)^\sim = L$.

This result enables us to complete what we said in Chap. 1 about homomorphisms from $\mathbb{R}$ to $\mathbb{T}^n = \mathbb{R}^n/\mathbb{Z}^n$ or, if preferred, about the image in $\mathbb{T}^n$ of a line $D \subset \mathbb{R}^n$ through the origin, i.e. consisting of the scalar multiples tX of a fixed vector X. Let p be the canonical map from $\mathbb{R}^n$ onto $\mathbb{T}^n$ and $H = \overline{p(D)}$ the closure of the 1-parameter subgroup considered. Clearly, $p^{-1}(H)$ is the smallest *closed* subgroup

L of $\mathbb{R}^n$ containing D and $\mathbb{Z}^n$, i.e. L is the closure of $D + \mathbb{Z}^n$. As integral-valued linear forms on a subgroup are also integral-valued on its closure (a limit of integers is an integer!), the subgroup $L^{\sim}$ of the dual space is the set of forms $f(x) = \sum \xi_i x_i$ on $\mathbb{R}^n$ whose values on $\mathbb{Z}^n$ and D are integral, i.e. whose coefficients ξ_i are integers, and vanish at X. Setting $X = (a_1, \cdots, a_n)$, it is then a matter of finding the integral solutions of the linear equation

$$a_1 \xi_1 + \cdots + a_n \xi_n = 0.$$

For the subgroup $p(D)$ to be dense $\mathbb{T}^n$, it is necessary and sufficient for $\xi_1 = \cdots = \xi_n = 0$ to be the only integral solution, or what comes to the same, *rational* solution, of the previous equation, in other words for the leading coefficients of the line D to be *linearly independent over* $\mathbb{Q}$. As mentioned in Chap. 1, for $n = 2$, this means that the slope of D is irrational.

Finally, note that given a vector $x \in \mathbb{R}^n$ and a linear form f on $\mathbb{R}^n$, the relation $f(x) \in \mathbb{Z}$ can also be written as

$$e^{2\pi i f(x)} = 1.$$

However, for given f, the map $x \mapsto e^{2\pi i f(x)}$ is clearly a continuous homomorphism from the additive group $G = \mathbb{R}^n$ to the multiplicative group $\mathbb{T}$ of complex numbers with absolute value 1. Such a homomorphism is called a *character* (or a *unitary character*) of the locally compact abelian group G. These are the only ones except for those we have just written down. It obviously suffices to prove this for $n = 1$, i.e. for $\mathbb{R}$, but then a continuous homomorphism from $\mathbb{R}$ to $\mathbb{C}^*$ is necessarily differentiable ("adjustment" by a C^∞ function with compact support), hence satisfies a differential equation of the form $\chi'(x) = \lambda.\chi(x)$, and so is an exponential function $e^{\lambda x}$, with λ purely imaginary if the character χ considered takes values having absolute value 1. Note that, on $\mathbb{R}$, these characters form the starting point of the Fourier transform. As has been well known since about 1940, the same is true for all locally compact abelian groups (MA XI, §26).

If every linear form f on $\mathbb{R}^n$ is identified with the character $x \mapsto e^{2\pi i f(x)}$ it defines, we get a bijection from the dual of the space $\mathbb{R}^n$ onto the set of unitary characters of $\mathbb{R}^n$. For a closed subgroup L of $\mathbb{R}^n$, the subgroup $L^{\sim}$ of the dual is then transformed into the set of characters χ of $\mathbb{R}^n$ that are equal to 1 on L. Hence

$$x \in L \Longleftrightarrow \chi(x) = 1 \text{ for any character}$$

$$\chi \text{ of } \mathbb{R}^n \text{ such that } \chi(L) = \{1\}.$$

This statement can be shown to be extendable to all locally compact abelian groups.

1.4   Proper Group Actions

Let H be a locally compact group acting continuously on a locally compact space X.[6] The quotient topological space $H\backslash X$ can be defined (even without the local compactness assumption) by setting its open subsets to be those whose inverse images in X are open. This topology is not always separable (if it is, all the orbits in X are closed, which immediately gives counterexamples: make the multiplicative group $\mathbb{R}_+^*$ act on $\mathbb{R}$ by the map $(t, x) \mapsto tx$).

A necessary and sufficient condition for the quotient to be separable is the existence of neighbourhoods U and V for any two points x and y of X on distinct orbits such that $h'U \cap h''V = \emptyset$ for all $h', h'' \in H$, or equivalently, such that $hU \cap V = \emptyset$ for all h. Let us then argue heuristically by supposing (which is not restrictive when X and H are metrizable) that it is possible to argue in terms of converging sequences. If it is not possible to "separate" Hx and Hy, then, for all U and V there is some $u \in U$, $v \in V$ and $h \in H$ such that $v = hu$. Applying this argument to sequences of smaller and smaller neighbourhoods, we find a sequence x_n converging to x, a sequence y_n converging to y, and a sequence h_n defined by $y_n = h_n x_n$ for all n. The elements x_n belong to a compact subset A of X and the y_n to a compact subset B of X. Hence the h_n belong to a set $C \subset H$ of h such that $hA\#B$. Suppose that C is relatively compact: then, if need be by extracting subsequences, suppose that h_n converges to a limit h. But then $h_n x_n = y_n$ converges to hx, and it follows that $Hx = Hy$, contrary to the assumption.

Note that, by Lemma 2, in the previous argument, the set C is necessarily closed. Hence saying that C is relatively compact in fact means that C is compact. This remark prompts us to introduce the next definition.

A group H is said to *act properly on* X if, for all compact subsets A and B of X, the set C of $h \in H$ such that $hA\#B$ is compact. The following result then holds:

Theorem 3 *Let H be a locally compact group acting properly on a locally compact set X. Then the quotient space $H\backslash X$ is locally compact (hence separable).*

Let us consider two distinct orbits Hx and Hy and let U be a compact neighbourhood of x in X. The set of h such that $hy \in U$ is compact since H acts properly. Let A denote this compact set. Being the image of A under the continuous map $h \mapsto hy$, the set Ay of hy with $h \in A$ is compact. Since $x \notin Hy$, $x \notin Ay \subset U$, and as a consequence, there is a compact neighbourhood $V \subset U$ of x in X which does not meet Ay. As $hy \in V$ implies that $hy \in U$ and thus that $hy \in Ay$, replacing U by V, we may assume that A is not empty, i.e. that U does not meet Hy (which by the way proves that each orbit is closed). Let V now be a compact neighbourhood of y in X. The elements h such that $V\#hU$ form a compact subset B in H, so that $V \cap HU = V \cap BU$ is compact and does not contain y since $Hy \cap U = \emptyset$. Replacing V by a smaller neighbourhood, it is therefore possible to suppose that $V \cap HU = \emptyset$, whence $HV \cap HU = \emptyset$, which proves that $H\backslash X$ is separable.

---

[6]See also MA XI, §15.

To show that $H \backslash X$ is locally compact, it then suffices to check that the canonical map $X \mapsto H \backslash X$ is open, i.e. that $HU = \bigcup hU$ is open for all open subsets U of X. This being clearly the case, the proof is now complete.

These considerations notably apply to the case $X = G/K$, where G is locally compact and K is *compact*, with H a *closed* subgroup of G acting on X in the obvious manner. To show that H acts properly, let us take two compact subsets A and B in X. Their inverse images $A' = p^{-1}(A)$ and $B' = p^{-1}(B)$ in G are also compact. Indeed, by Lemma 1, there is a compact set $A'' \subset G$ such that $p(A'') = A$. Then $A' = A''K$, whence the compactness of A'. Now, the set C of $h \in H$ such that $hA \# B$ is obviously the intersection of H and $B'A'^{-1}$, and as the latter set is compact in G, the compactness of C follows readily. *Every closed subgroup of G is thus seen to act properly on G/K whenever K is compact.* In particular, this makes it possible to define a locally compact quotient $H \backslash (G/K)$, i.e. a locally compact topology on the set

$$H \backslash G / K$$

of double cosets HgK in G. (Note that this set can also be obtained by making $H \times K$ act on G via the formula $(h, k)g = hgk^{-1}$.)

Let us for example consider the upper half-plane P, i.e. the set of $z \in \mathbb{C}$ such that $Im(z) > 0$, and let us make the group $G = SL_2(\mathbb{R})$ act on P by the conformal maps $z \mapsto \frac{az+b}{cz+d}$. Obviously G acts continuously and transitively on P. The stabilizer K of the point $z = i$, consisting of the matrices such that $i(ci + d) = ai + b$, i.e. of the real matrices $\begin{pmatrix} a & b \\ -b & a \end{pmatrix}$ with $a^2 + b^2 = 1$, is compact and is in fact isomorphic (as a topological group) to the group $\mathbb{T}$. This can be seen by associating the previous complex number $ai + b$ to the matrix. The map $g \mapsto z = g(i) = \frac{ai+b}{ci+d}$ thus enables us to identify (both setwise *and* topologically: Theorem 1) P with the homogeneous space G/K. In particular, it follows that *every discrete subgroup Γ of $SL_2(\mathbb{R})$ acts properly on P and that, for such a subgroup, the quotient $\Gamma \backslash P$ is locally compact.*

It is then obvious that the, let us say continuous, Γ-invariant functions of P, i.e. satisfying $f(\gamma(z)) = f(z)$ for all $\gamma \in \Gamma$, can be identified with the continuous functions ϕ on G such that

$$\phi(\gamma g k) = \phi(g) \tag{$*$}$$

for all $\gamma \in \Gamma$, $g \in G$, $k \in K$: it suffices to set $\phi(g) = f(\frac{ai+b}{ci+d})$ for $g = \begin{pmatrix} a & b \\ c & d \end{pmatrix}$. More generally, let us take an integer n and consider the continuous functions f of P such that

$$f(\gamma(z)) = (cz + d)^n f(z) \text{ if } \gamma = \begin{pmatrix} a & b \\ c & d \end{pmatrix} \in \Gamma.$$

Example (MA XII, §15): take the modular group $\Gamma = SL_2(\mathbb{Z})$ and for f an "integral modular form" of weight n (there are many other examples if the holomorphy condition is omitted!). Associate to f the function ϕ on G given by

$$\phi \begin{pmatrix} a & b \\ c & d \end{pmatrix} = (ci + d)^{-n} f \begin{pmatrix} ai + b \\ ci + d \end{pmatrix}$$

for $\begin{pmatrix} a & b \\ -b & a \end{pmatrix} \in G$. Relation $(*)$ is then easily verified to be equivalent to the condition

$$\phi(\gamma g k) = \phi(g)\rho_n(k) \quad \text{for all} \quad \gamma \in \Gamma, g \in G, k \in K,$$

where ρ_n is the homomorphism from K to $\mathbb{C}^*$, in fact to $\mathbb{T}$, given by

$$\rho_n \begin{pmatrix} a & b \\ -b & a \end{pmatrix} = (a + ib)^n.$$

These operations naturally raise the problem of transferring the holomorphy condition of its corresponding function f to ϕ, i.e. on the group G. It can be resolved by endowing $G = SL_2(\mathbb{R})$ with the structure of a Lie group and replacing the operator $\partial/\partial\bar{z}$ whose null set includes all holomorphic functions by a properly chosen "differential operator" on the group G (Chap. 5, Sect. 5.12).

We next make some general remarks about discrete groups. Let Γ be a *discrete group acting properly on a locally compact space* X. This means that, for any compact subsets A and B of X, the set of elements $\gamma \in \Gamma$ such that $\gamma A \# B$ is compact in Γ, i.e. is finite. In particular, the stabilizer of every $x \in X$ in Γ is a finite subgroup Γ_x, and there is a Γ_x-invariant fundamental system of neighbourhoods of x: indeed if U is an arbitrary neighbourhood of x, then so is the set

$$U' = \bigcap_{\gamma \in \Gamma_x} \gamma U \tag{**}$$

and U' is clearly Γ_x-invariant, proving our assertion. Besides, note that, *if x and y are given points of X, then it is possible to find neighbourhoods U and V of x and y in X such that*

$$\gamma U \# V \quad \text{if and only if} \quad \gamma x = y. \tag{$**$}$$

To prove this, start with arbitrary compact neighbourhoods U' and V' of x and y, and consider the elements $\gamma \in \Gamma$ such that $\gamma U' \# V'$. There are finitely many of them. Let $\gamma_1, \cdots, \gamma_q$ be those for which $y \neq \gamma x$. For each k, there are neighbourhoods $U'_k \subset U'$ and $V'_k \subset V'$ of x and y such that $V'_k \cap \gamma_k U'_k = \emptyset$ (Hausdorff axiom for X). Let U (resp. V) be the intersection of the sets U'_k (resp. V'_k). Then $\gamma U \# V$ implies $\gamma U' \# V'$, so that either $\gamma x = y$ or $\gamma = \gamma_k$ for some index k. This is precluded since U and $\gamma_k V$ are contained in U'_k and $\gamma_k V'_k$, which do not meet. Condition $(***)$ therefore indeed holds.

In particular, for $x = y$, x is seen to have a neighbourhood U such that $\gamma U \# U$ if and only if $\gamma \in \Gamma_x$, and obviously—use $(**)$ to modify U—U may be assumed to be Γ_x-invariant. Since any two Γ-equivalent points of U are Γ_x-equivalent, the canonical map $p : X \to \Gamma\backslash X$, which takes U onto a neighbourhood U' of the point

$p(x)$, induces a bijective map from the quotient $\Gamma_x \backslash U$ onto U'. We show that it is a *homeomorphism* if U is open.

To do so, let as associate the set $V = p^{-1}(V') \cap U$ to every subset V' of U'. As $p^{-1}(U')$ is the union of the sets γU, $p^{-1}(V')$ is likewise the union of the sets γV. An immediate implication is that $p^{-1}(V')$ is open if and only if so is V. However, identifying U' with $\Gamma_x \backslash U$ clearly identifies the set V with the inverse image of V'. Consequently, V' is open in U' if and only if its inverse image under the map from U onto $U' = \Gamma_x \backslash U$ is open in U, giving the desired result.

This result is of great practical important because it shows that if we want to study the structure of the quotient $\Gamma \backslash X$ in the neighbourhood of the point $p(x)$, we can replace X by U and Γ by Γ_x, i.e. by a *finite* group.

A seemingly simple particular case follows by supposing that Γ acts *freely* on X, i.e. that $\Gamma_x = \{e\}$ for all x. For every $x \in X$, the canonical map p then induces a *homeomorphism* from a sufficiently small neighbourhood of x onto a neighbourhood of $p(x)$, and the following even more precise property holds: *for every sufficiently small subset V of $\Gamma \backslash X$, the inverse image $p^{-1}(V)$ is the disjoint union of open sets homeomorphically mapped onto V under p.* This is exactly the situation one comes across in topology in the theory of *covering spaces*, which will be expounded later (Chap. 2) when we will define simply connected spaces. The easiest example is obtained by making the additive group $\mathbb{Z}$ act on $\mathbb{R}$ by translations, the quotient then being $\mathbb{T}$. Figure 2.1.1 in M. Berger and G. Gostiaux, *Differential Geometry* (Springer, 1988) explains what is going on with sufficient clarity so that there is no need for comments.

The modular group $\Gamma = SL_2(\mathbb{Z})$ does not act freely on the upper half-plane P. We know (MA XII, §17) that $\Gamma_x \neq \{e\}$ if and only if x belongs to the orbit of the point i or that of $\rho = e^{2\pi i/3}$ (since -1 acts trivially on P, strictly speaking, what we have just said is correct only if Γ is replaced by the quotient group $PSL_2(\mathbb{Z})$ of $SL_2(\mathbb{Z})$ by the closed subgroup consisting of the two matrices $+1$ and -1). To obtain a group acting freely, replace Γ by the subgroup $\Gamma(2)$ of matrices $\gamma = \begin{pmatrix} a & b \\ c & d \end{pmatrix}$ such that $a \equiv d \equiv 1$ and $c \equiv d \equiv 0 \pmod 2$. This is a normal subgroup of index $[\Gamma : \Gamma(2)] = 6$ in Γ, and it is easily seen to act freely on the upper half-plane.

It is even possible to construct discrete subgroups Γ of $SL_2(\mathbb{R})$ acting freely on P, and for which the quotient $\Gamma \backslash P$ is *compact*. They arise either in the uniformization theory of compact Riemann surfaces of genus ≥ 2 (for such a group, such a surface is necessarily of the form $\Gamma \backslash P$), or in the arithmetic study of indefinite quaternion fields over $\mathbb{Q}$ (non-commutative fields of dimension 4 over $\mathbb{Q}$ that become isomorphic to $M_2(\mathbb{R})$ when the base field is extended from $\mathbb{Q}$ to $\mathbb{R}$). The former situation is definitely beyond the scope of this presentation, but the latter one can be easily explained and is a beautiful application of the ideas expressed in this chapter.

1.5 Discrete Groups with Compact Quotient: Arithmetic Examples

Let K be a *finite-dimensional algebra over the field* $\mathbb{Q}$ of rational numbers, i.e. a vector space endowed with an associative composition law $(x, y) \mapsto xy$ associative, bilinear with respect to the vector space structure of K over $\mathbb{Q}$, having an identity, and finally such that $(\lambda 1)x = x(\lambda 1) = \lambda x$ for $\lambda \in \mathbb{Q}, x \in K$. The algebra K is said to be a *non-commutative field* (or a *division algebra*) over $\mathbb{Q}$ when every non-trivial element of K is invertible.

Let (a_i) be a basis for K over $\mathbb{Q}$. Then there are multiplication formulas

$$a_i a_j = \sum \gamma_{ijk} a_k \tag{1.5.1}$$

with "structure constants" $\gamma_{ijk} \in \mathbb{Q}$. Multiplying the a_i by a common denominator of the rationals γ_{ijk} whenever necessary, these may be assumed to be *integers*. The subgroup o (small Gothic o) generated by the a_i, i.e. the set of linear combinations of the a_i with coefficients in $\mathbb{Z}$, is then a subring of K. Any subring of K obtained in this manner, or equivalently generated as a subgroup by a basis of K, is called an *order*. For example, if K is a field of algebraic numbers (a finite degree extension of $\mathbb{Q}$, i.e. a commutative division algebra), then the set of algebraic integers of K is easily shown to be an order of K (see the author's *Cours d'Algèbre*, §34, exercise 47, or else S. Lang, *Algebraic Number Theory*, Chap. I), but this result will not be needed. By definition, the *units* of an order o are the invertible elements of the ring o. They form a multiplicative group o*. When K is a *division* algebra, o* will later be shown to be a discrete subgroup with compact quotient of a "continuous" group G that will also be defined.

Apart from algebraic number fields, the easiest examples of division algebras over $\mathbb{Q}$ are the *quaternion fields* (see *Cours d'Algèbre*, §15, exercise 10, or else, the obviously far more complete, André Weil, *Basic Number Theory*, Springer, 1974). To define such a field K, we choose two rational numbers p and q and set K to have a basis $1, i, j, k$ satisfying the following multiplication formulas:

$$
\begin{array}{c|c|c|c}
 & i & j & k \\
\hline
i & p1 & k & pj \\
\hline
j & -k & q1 & -qi \\
\hline
k & -pj & qi & -pq1 \\
\end{array}
\tag{1.5.2}
$$

Associativity needs to be checked—which poses no problem as it is highlighted by the above multiplication table—, then p and q need to be chosen in such a way that K is indeed a *division* algebra. But defining the *conjugate* of a quaternion $u = x + yi + zj + tk$ as $\bar{u} = x - yi - zj - tk$, we get

$$u\bar{u} = \bar{u}u = (x^2 - py^2 - qz^2 + pqt^2)1. \tag{1.5.3}$$

Easy arguments then show that u is invertible if and only if $x^2 - py^2 - qz^2 + pqt^2 \neq$ 0. Consequently, K is a division algebra if and only if

$$x^2 - py^2 - qz^2 + pqt^2 = 0 \Longrightarrow x = y = z = t = 0$$

for $x, \cdots, t \in \mathbb{Q}$. This means that (i) p is not the square of an element of $\mathbb{Q}$ and (ii) the equation $x^2 - py^2 = q$ has no solution with $x, y \in \mathbb{Q}$, in other words that q is not the "norm" of a number of the quadratic extension $\mathbb{Q}(\sqrt{p})$. In practice, p and q are always assumed to be integers (the general case can be reduced to this by an obvious change of basis), so that *integral* linear combinations of $1, i, j, k$ form an order $\mathfrak{o}$ of K (there are many others). As $u \mapsto \bar{u}$ is an automorphism of the ring $\mathfrak{o}$, if u is invertible in $\mathfrak{o}$, then so is $\bar{u}$, and hence so is $u\bar{u}$. An immediate consequence is the invertibility in the ring $\mathbb{Z}$ of the coefficient of 1 in (1.5.3); namely

$$u \in \mathfrak{o}^* \Longleftrightarrow x^2 - py^2 - qz^2 pqt^2 \in \mathbb{Z}^* = \{+1, -1\}$$

for $x, \cdots, t \in \mathbb{Z}$.

Let us return to a finite-dimensional algebra K over $\mathbb{Q}$. Associate to each $a \in K$ the endomorphism $u_a : x \mapsto ax$ of the vector space K; obviously,

$$u_{a+b} = u_a + u_b, \quad u_{ab} = u_a \circ u_b. \tag{1.5.4}$$

Lemma a *An element of K is invertible if and only if so is u_a.*

By the second equality in (1.5.4), the necessity of the condition is clear. Conversely, suppose that u_a is invertible, and hence subjective. There exists a $b \in K$ such that $u_a(b) = 1$, i.e. such that $ab = 1$. Then $bx = 0 \Rightarrow a(bx) = 0 \Rightarrow x = 0$, so that u_b is injective, hence invertible since we are in finite dimensions, and hence also surjective. So there exists a $c \in K$ such that $bc = 1$. But then b is invertible and $b^{-1} = a = c$, qed.

Besides, the proof shows that every zero divisor $a \in K$ (i.e. such that u_a is injective) is invertible. In what follows, we set

$$\mathrm{Tr}(a) = \mathrm{Tr}(u_a),$$

and so (1.5.4) implies that

$$\mathrm{Tr}(a + b) = \mathrm{Tr}(a) + \mathrm{Tr}(b), \mathrm{Tr}(ab) = \mathrm{Tr}(ba).$$

Lemma b *Let K be a division algebra. Then the symmetric bilinear form $\mathrm{Tr}(xy)$ is non-degenerate.*

Indeed, suppose that $\mathrm{Tr}(ay) = 0$ for all $y \in K$. This implies that $\mathrm{Tr}(a^n) = 0$ for all n, so that the linear operator $u = u_a$ satisfies $\mathrm{Tr}(u^n) = 0$ for all n. In characteristic 0, this shows that $u^n = 0$ for large n (*Cours d'Algèbre*, §34, exercise 28), which forces $a^n = 0$, and as K is a field, we get $a = 0$, qed.

(In the general case, the property in the statement clearly means that K has no left ideal all of whose elements are nilpotent. Such an algebra is called semisimple, but this notion will not be needed here.)

Lemma c *For all $a \in K$, let u_a and v_a be the endomorphisms $x \mapsto ax$ and $x \mapsto xa$ of the vector space K. Suppose that K is a division algebra. Then $\det(u_a) = \det(v_a)$ for all a.*

Indeed, since $\mathrm{Tr}(axy) = \mathrm{Tr}(xya)$, identifying K with its dual through the non-degenerate bilinear form $\mathrm{Tr}(xy)$ identifies the operator v_a with the transpose of the operator u_a. It therefore has the same determinant.

Next set

$$N(a) = \det(u_a) = \det(v_a) \quad (\textit{norm of } a) \tag{1.5.5}$$

for all $a \in K$. This is obviously a polynomial function of the coordinates of a with respect to an arbitrary basis of K (exercise: find it using the structure constants in (1.5.1)). For the algebra (1.5.2), $N(u) = (u\bar{u})^2$.

Lemma d *Let K be a division algebra and $\mathfrak{o}$ an order of K. For all non-trivial $a \in \mathfrak{o}$,*

$$\mathrm{Card}(\mathfrak{o}/a\mathfrak{o}) = |N(a)|, \tag{1.5.6}$$

and

$$a \in \mathfrak{o}^* \iff N(a) \in \{+1, -1\}. \tag{1.5.7}$$

Let (a_i) be a basis of $\mathfrak{o}$. This enables us to identify K with $\mathbb{Q}^n$ and $\mathfrak{o}$ with $\mathbb{Z}^n$ as additive groups. The operator u_a of K is then identified with an endomorphism u of $\mathbb{Q}^n$; $u(\mathbb{Z}^n) \subset \mathbb{Z}^n$ since $a \in \mathfrak{o}$ is equivalent to $a\mathfrak{o} \subset \mathfrak{o}$, i.e. to $u_a(\mathfrak{o}) \subset \mathfrak{o}$. Consequently, the matrix of u is integral and invertible in the ring $M_n(\mathbb{Q})$ since a is non-trivial and hence defines (K is a field) an automorphism of the vector space K. So all that remains to be proved is the following:

Lemma d' *Let u be an automorphism of $\mathbb{Q}^n$ such that $u(\mathbb{Z}^n) \subset \mathbb{Z}^n$. Then*

$$\mathrm{Card}(\mathbb{Z}^n/u(\mathbb{Z}^n)) = |\det(u)|. \tag{1.5.8}$$

Identify u with an $n \times n$ matrix whose entries are in the principal ring $\mathbb{Z}$, and let us apply exercise 9 of §31 from my *Cours d'Algèbre* (elementary divisor theorem). We find that there are two automorphisms v and w of $\mathbb{Z}^n$ (hence represented by matrices of determinant $+1$ or -1) such that

$$v \circ u \circ w = \mathrm{diag}(d_1, \ldots, d_n)$$

with d_i integers. It is clear that replacing u by $v \circ u \circ w$ leaves both sides of (1.5.8) unchanged. Hence it is possible to assume that $u = \mathrm{diag}(d_1, \ldots, d_n)$; but then

$$\mathbb{Z}^n / u(\mathbb{Z}^n) = (\mathbb{Z}/d_1\mathbb{Z}) \times \cdots \times (\mathbb{Z}/d_n\mathbb{Z}),$$

proving the lemma in this case.

In (1.5.7), the implication $\Rightarrow$ follows from $N(ab) = N(a)N(b)$. Conversely, if $|N(a)| = 1$, then $a\mathfrak{o} = \mathfrak{o}$, and so there exists a $b \in \mathfrak{o}$ such that $ab = 1$, when $a \in \mathfrak{o}^*$.

Lemma e *Let $\mathfrak{o}_\nu$ be the set of $a \in \mathfrak{o}$ for which $|N(a)|$ takes fixed values $\nu \neq 0$. Then $\mathfrak{o}_\nu$ is the finite union of $\mathfrak{o}^*$-cosets.*

In other words, there exist finitely many $a_i \in \mathfrak{o}_\nu$ such that $\mathfrak{o}_\nu = \bigcup a_i \mathfrak{o}^*$ (by (1.5.7), $\mathfrak{o}^* \mathfrak{o}_\nu = \mathfrak{o}_\nu \mathfrak{o}^* = \mathfrak{o}_\nu$ clearly hold).

Indeed, (1.5.6) shows that for every element a of $\mathfrak{o}_\nu$, the quotient $\mathfrak{o}/a\mathfrak{o}$ is a finite group of order ν and so the order of all its elements is divisible by ν. Consequently, $\nu x \in a\mathfrak{o}$ for all $x \in \mathfrak{o}$, or $\nu\mathfrak{o} \subset a\mathfrak{o} \subset \mathfrak{o}$. But the quotient $\mathfrak{o}/\nu\mathfrak{o}$ is obviously finite. Hence there are only finitely many possibilities for the right ideal $a\mathfrak{o}$ of $\mathfrak{o}$ as a varies in $\mathfrak{o}_\nu$. It then remains to check that, for non-trivial a, b, $a\mathfrak{o} = b\mathfrak{o}$ if and only if $a\mathfrak{o}^* = b\mathfrak{o}^*$; but setting $a = bc$ with $c \in K$, $a\mathfrak{o} = b\mathfrak{o}$ is clearly equivalent to $\mathfrak{o} = c\mathfrak{o}$, which implies that $c \in \mathfrak{o}$, and then that there exists a $d \in \mathfrak{o}$ such that $cd = 1$, whence $c \in \mathfrak{o}^*$, and finally $a\mathfrak{o}^* = bc\mathfrak{o}^* = b\mathfrak{o}^*$, proving the lemma.

After these essential arithmetic preliminaries (which the reader can check by hand for the field of quaternions), we can now return to topological groups. First note that every $\mathbb{Q}$-algebra K canonically defines an $\mathbb{R}$-algebra $K_\mathbb{R}$, obtained either by setting

$$K_\mathbb{R} = \mathbb{R} \otimes_\mathbb{Q} K,$$

or by requiring a $\mathbb{Q}$-basis (a_i) of K to also be a $\mathbb{R}$-basis of $K_\mathbb{R}$ with the same multiplication table. As will be seen, an algebra that is a division algebra over $\mathbb{Q}$ may no longer be so over $\mathbb{R}$—taking $K = \mathbb{Q}(\sqrt{2})$ shows this to be obvious since 2 is a square in $\mathbb{R}$. Whatever be the case, operators u_a and v_a can still be defined for $a \in K_\mathbb{R}$, as well as $Tr(a)$, and if K is a division algebra over $\mathbb{Q}$, $\det(u_a) = \det(v_a)$ continues to hold in $K_\mathbb{R}$. Indeed, if two polynomials with real coefficients have the same values when the indeterminates take rational values, then they are identical. (In fact Lemma b holds in $K_\mathbb{R}$ because if a system of homogeneous linear equations with coefficients in a field k has a non-trivial solution in a field $k' \supset k$, then it has a solution in k...) Hence definition-relation (1.5.5) can be applied in $K_\mathbb{R}$.

Theorem 4 *Let K be a division algebra over $\mathbb{Q}$, G the set of elements $x \in K_\mathbb{R}$ such that $|N(x)| = 1$, $\mathfrak{o}$ an order of K, and $\Gamma = \mathfrak{o}^*$ the set of units in $\mathfrak{o}$. Then G, endowed with the multiplicative composition law and topology of $K_\mathbb{R}$, is a locally compact group and Γ is a discrete subgroup of G with compact quotient.*

If $|N(x)| = 1$, then $\det(u_x) \neq 0$, and so u_x is invertible and x has an inverse in $K_\mathbb{R}$, namely the unique element y such that $u_x(y) = 1$. Since moreover $N(xy) = N(1) = 1 = N(x)N(y)$, we get $y \in G$. Trivial arguments then show that G is a subgroup of the multiplicative group of invertible elements of the ring $K_\mathbb{R}$. Since in addition xy is a polynomial function of the pair x, y, the map $(x, y) \mapsto xy$ is continuous with respect to the topology of $K_{\mathbb{R}}$. The same holds for $x \mapsto x^{-1}$. Indeed, fixing a basis for $K_{\mathbb{R}}$, the coordinates y_j of $y = x^{-1}$ are obtained by applying the Cramer formulas to solve a system of linear equations whose coefficients are themselves linear functions of the coordinates x_j of x. The y_j are, therefore, rational functions, and so perforce continuous, of the x_j. Finally, $|N(x)| = 1$ defined a closed subset in $K_{\mathbb{R}}$. So the group G is indeed locally compact.

By (1.5.7),

$$\Gamma = G \cap \mathfrak{o},$$

and as $\mathfrak{o}$ is discrete in $K_{\mathbb{R}}$ (the situation is similar to that of $\mathbb{Z}^n$ in $\mathbb{R}^n$), Γ is indeed a discrete subgroup of G.

To establish the *compactness* of G/Γ (which is, or at least implies, an existence theorem for $\mathfrak{o}^*$), a much more powerful argument than the above is required:

Minkowski's Lemma – *Let m be the Lebesgue measure on $\mathbb{R}^n$ and U a measurable subset of $\mathbb{R}^n$ such that $m(U) > 1$ Then the set $U - U$ of differences $x - y$, with $x, y \in U$, contains a non-trivial element of $\mathbb{Z}^n$.*

Indeed, suppose that $(U - U) \cap \mathbb{Z}^n = \{0\}$. Then the translations $U + \xi$ of U by the vectors $\xi \in \mathbb{Z}^n$ are mutually disjoint. Let K be the cube formed by the $(x_i) \in \mathbb{R}^n$ such that $0 \leq x_i < 1$ for all i, and for each $\xi \in \mathbb{Z}^n$, let K_ξ be the set of $x \in K$ such that $x + \xi \in U$. Then

$$K_\xi = (U - \xi) \cap K,$$

so that the sets K_ξ are mutually disjoint, which implies that $\sum m(K_\xi) \leq m(K) = 1$. On the other hand, $\mathbb{R}^n$ is the union of the mutually disjoint sets $K + \xi$, so that U is the union of the mutually disjoint sets $U \cap (K + \xi) = K_\xi + \xi$; hence

$$m(U) = \sum m(K_\xi + \xi) = \sum m(K_\xi) \leq 1, \quad \text{qed.}$$

Note that the assumption $m(U) > 1$ continues to hold if U is replaced by $g(U)$, where g is an automorphism of $\mathbb{R}^n$ such that $|\det(g)| = 1$: indeed the Lebesgue measure is preserved by such an automorphism.

To apply Minkowski's Lemma, let us choose a basis for $\mathfrak{o}$ to enable us to identify $\mathfrak{o}$ with $\mathbb{Z}^n$ and $K_{\mathbb{R}}$ with $\mathbb{R}^n$, and let us choose a *compact* subset U of $K_{\mathbb{R}}$ whose measure (after identification of $K_{\mathbb{R}}$ with $\mathbb{R}^n$ and of $\mathfrak{o}$ with $\mathbb{Z}^n$) is > 1. For all $a \in G$, the map $x \mapsto xa^{-1}$ preserves the measure of U, so that Minkowski's Lemma applies to the set Ua^{-1}. Taking into consideration the compact set $V = U - U$, it follows that for all $a \in G$, there is some non-trivial ξ such that

$$\xi \in \mathfrak{o} \cap Va^{-1}. \tag{1.5.9}$$

The norm of ξ is integral since ξ belongs to an order of K, and, on the other hand, its absolute value is equal to that of an element of the compact set V, on which the norm function is obviously bounded. Consequently, when $a \in G$ varies, the norms of the elements $\xi \neq 0$ satisfying (1.5.9) remain within a finite subset of $\mathbb{Z}$, and by the lemma, the elements ξ themselves remain in the union of *finitely* many cosets $\xi_i \mathfrak{o}^*$ with non-zero $\xi_i \in \mathfrak{o}$. As each set Va^{-1} meets at least one of the sets $\xi_i^{-1} \mathfrak{o}^*$, in conclusion, for all $a \in G$, the set $\mathfrak{o}^* a$ meets one of the sets $\xi_i^{-1} V$. As the union of these latter sets is compact in $K_{\mathbb{R}}$, the set $M = G \cap \bigcup \xi_i^{-1} V$ is a compact subset of G meeting all $\mathfrak{o}^*$-cosets, proving Theorem 4.

Example. Begin with a field K of quaternions over $\mathbb{Q}$, given by multiplication table (1.5.2). Formula (1.5.3) continues to hold over $\mathbb{R}$, and so does the consequence drawn from it, namely that for $u \in K_{\mathbb{R}}$, u is invertible if and only if $x^2 - py^2 - qz^2 + pqt^2 \neq 0$. It follows that

$$K_{\mathbb{R}} \text{ is a field} \Leftrightarrow x^2 - py^2 - qz^2 + pqt^2 \text{ is a definite quadratic form}$$
$$\Leftrightarrow p < 0 \text{ and } q < 0.$$

A fairly simple basis change shows that in this case $K_{\mathbb{R}}$ is obviously just the usual field of quaternions over $\mathbb{R}$ (a far from surprising result given that it is the *only* finite-dimensional non-commutative division algebra over $\mathbb{R}$...). In this case the group G is compact—it is the rotation group in 4 variables—and Γ is finite, a seemingly obvious result because of the signs of p and q.

Hence the interesting case occurs when p and q are not both negative (the "indefinite" case). *The algebra $K_{\mathbb{R}}$ is then isomorphic to the matrix algebra $M_2(\mathbb{R})$, and G to the subgroup* of elements g of $GL_2(\mathbb{R})$ satisfying $|\det(g)| = 1$. Note that the second statement follows from the first because in the algebra $M_n(\mathbb{R})$ the norm of an element u is $\det(u)^n$—indeed, the decomposition of a matrix $x \in M_n(\mathbb{R})$ into its *columns* $x_1, \cdots, x_n \in \mathbb{R}^n$ transforms the map $x \mapsto ux$ from $M_n(\mathbb{R})$ to $M_n(\mathbb{R})$ into the map

$$(x_1, \cdots, x_n) \mapsto (u(x_1), \cdots, u(x_n))$$

from $\mathbb{R}^n \times \cdots \times \mathbb{R}^n$ to itself, whence the result.

To show that $K_{\mathbb{R}} \simeq M_2(\mathbb{R})$, we may assume that $p > 0$ (otherwise permute p and q). The matrices

$$E = \begin{pmatrix} 1 & 0 \\ 0 & 1 \end{pmatrix}, \ I = \begin{pmatrix} -\sqrt{p} & 0 \\ 0 & \sqrt{p} \end{pmatrix}, \ J = \begin{pmatrix} 0 & 1 \\ q & 0 \end{pmatrix}, \ K = \begin{pmatrix} 0 & -\sqrt{p} \\ q\sqrt{p} & 0 \end{pmatrix}$$

are then found to satisfy multiplication table (1.5.2). The isomorphism is then given by

$$x + yi + zj + tk \mapsto \begin{pmatrix} x - y\sqrt{p}, & z - t\sqrt{p} \\ q(z + t\sqrt{p}), & x + y\sqrt{p} \end{pmatrix} = xE + yI + zJ + tK.$$

By confining ourselves to the identity component $G^\circ = SL_2(\mathbb{R})$ of G and to the subgroup $\Gamma^\circ = G^\circ \cap \Gamma$ of $\gamma \in \mathfrak{o}^*$ such that $N(\gamma) = 1$, we, therefore, get a *discrete* *subgroup with compact quotient* of $SL_2(\mathbb{R})$. Numerical example: $p = 5, q = 2$ gives the group of matrices

$$\gamma = \begin{pmatrix} x - y\sqrt{5} & z - t\sqrt{5} \\ 2(x + t\sqrt{5}) & x + y\sqrt{5} \end{pmatrix},$$

where $x, \cdots, t$ are all *integers* such that

$$\det(\gamma) = x^2 - 5y^2 - 2z^2 + 10t^2 = 1.$$

1.6   Applications to Algebraic Number Fields

In the statement of Theorem 4, the division algebra K may well be commutative. It is then an algebraic number field (finite extension of $\mathbb{Q}$), and, in this case, Theorem 4 leads to a well-known theorem due to Dirichlet on *units* (in the arithmetic sense) of K. To be precise, if we consider the ring $\mathfrak{o}$ of integers of K, the *multiplicative group* $\Gamma = \mathfrak{o}^*$ (whose elements are by definition the units of the field K) will be the *direct product of a finite group and a group* $\mathbb{Z}^d$, where the exponent d can be calculated. Besides, this will hold for any order $\mathfrak{o}$ of K.

Let d be a *positive* integer which is not a square and let us first take K to be the corresponding quadratic extension. It has a basis $(1, \omega)$ with $\omega^2 = d$, and trivial calculations show that in this instance $N(x + y\omega) = x^2 - dy^2$. For $\mathfrak{o}$ choose the set[7] of elements $x + y\omega$ with $x, y \in \mathbb{Z}$. Then Γ is the set of solutions of the equation $x^2 - dy^2 = +1$ or -1 (with $x, y \in \mathbb{Z}$) named after Pell. They form a group whose operation is given by the multiplication in K:

$$(x, y).(u, v) = (xu + dyv, xv + yu). \tag{1.6.1}$$

The algebra $K_{\mathbb{R}}$ is similarly defined. It obviously is no longer a field, and in fact a canonical isomorphism j can be readily defined from $K_{\mathbb{R}}$ onto the algebra $\mathbb{R} \times \mathbb{R}$, endowed with the ordinary multiplication

$$(x, y)(u, v) = (xu, yv), \tag{1.6.2}$$

by setting

$$j(x + y\omega) = (x + y\sqrt{d}, x - y\sqrt{d}), \tag{1.6.3}$$

where $\sqrt{d}$ is a root of d chosen once and for all in $\mathbb{R}$ (not to be confused with the letter ω...). With multiplication given by (1.6.2), G is clearly the set of $(x, y) \in \mathbb{R}^2$ such that

---

[7]The ring $\mathfrak{o}$ is obviously an order of K, but not necessarily the ring of all integers of K. If d is assumed not to be divisible by the square of a prime—which we are allowed to do—, $\mathfrak{o}$ can be easily shown to be the set of integers of K only if $d \equiv 2$ or $3 \mod 4$. If $d \equiv 1 \mod 4$, as a $\mathbb{Z}$-module, the ring of integers is generated by the numbers 1 and $\frac{1}{2}(1 + \sqrt{d})$.

$|xy| = 1$. To somewhat simplify the situation, it is better to replace G by its identity component $G°$, i.e. the set of (x, y) in G such that $x > 0$, $y > 0$, $xy = 1$. $G°$ is clearly isomorphic to $\mathbb{R}$—under $(x, y) \mapsto \log x$—of index 4 in G. "Abstract nonsense" style arguments show that $\Gamma° = G° \cap \Gamma$ is a discrete subgroup with compact quotient of G. Now, G is isomorphic to the additive group $\mathbb{R}$. It follows that $\Gamma°$ *is isomorphic to* $\mathbb{Z}$.

However, $\Gamma°$, with multiplication (1.6.1), can clearly be identified with the set of integral solutions of Pell's equation $x^2 - dy^2 = 1$ such that $x + y\sqrt{d} > 0$ (they are known as its "positive" solutions). Hence this gives us a method for solving Pell's equation in a finite number of steps: search for a *minimal* solution x_0, y_0 such that $x_0 + y_0\sqrt{d} > 1$. The other solutions are then obtained by choosing a rational integer $n \in \mathbb{Z}$ and by writing

$$(x_0 + y_0\omega)^n = x_n + y_n\omega.$$

For K, let us now take an *algebraic number field* and for $\mathfrak{o}$ its *ring of integers*. Using arguments generalizing those of the previous example, we then get Dirichlet's unit theorem, which sets out the structure of the group $\mathfrak{o}^*$. Proceed as follows.[8]

By definition, K is a commutative extension of $\mathbb{Q}$ of finite degree n. The ring $\mathfrak{o}$ consists of the elements $x \in K$ satisfying an equation of type $x^r + a_{r-1}x^{r-1} + \cdots + a_0 = 0$ with coefficients $a_{r-1}, \cdots, a_0 \in \mathbb{Z}$. We know (*Cours d'Algèbre*, exercises 47 and 48 of §34, or else S. Lang, *Algebraic Number Theory*, Chap. I), that $\mathfrak{o}$ is indeed a subring of K isomorphic to $\mathbb{Z}^n$ as a $\mathbb{Z}$-module. So it is an order of K considered as a $\mathbb{Q}$-algebra (in fact, any other order of K is readily seen to be contained in $\mathfrak{o}$). To apply Theorem 4 to this situation, we need to describe the extended algebra $K_\mathbb{R} = \mathbb{R} \otimes_\mathbb{Q} K$ and the group G of $x \in K_\mathbb{R}$ such that $|N(x)| = 1$.

To do so, first note that K being a separable extension of $\mathbb{Q}$ (*Cours d'Algèbre*, §26, exercise 4) since we are in characteristic zero, there exist n distinct homomorphisms $j_1, \cdots, j_n$ from K to $\mathbb{C}$ (op. cit., §33, exercises 26 and 27). Such a homomorphism (perforce injective—there are no ideals in a field...), being a $\mathbb{Q}$-linear map from K to $\mathbb{C}$, can be canonically extended to an $\mathbb{R}$-linear map from $K_\mathbb{R}$ to $\mathbb{C}$. It will be denoted by the same letter as the homomorphism considered. If (a_i) is a $\mathbb{Q}$-basis of K, the fact that a $\mathbb{Q}$-linear map j from K to $\mathbb{C}$ is a ring homomorphism is reflected by the formulas $j(a_p a_p) = j(a_p)j(a_p)$. From this it immediately follows that the extension of j to $K_\mathbb{R}$ is also a homomorphism from the $\mathbb{R}$-algebra $K_\mathbb{R}$ to $\mathbb{C}$, considered as a 2-dimensional $\mathbb{R}$-algebra. Since $j(K) \supset \mathbb{Q}$, $j(K_\mathbb{R})$, a ($\mathbb{R}$-) vector subspace of $\mathbb{C}$, is clearly just $\mathbb{R}$ or $\mathbb{C}$. The former occurs if $j(K) \subset \mathbb{R}$ (j is then said to be a *real embedding* of K, and the subfield $j(K)$ a *real conjugate* of K), and the latter if $j(K) \not\subset \mathbb{R}$ (j is then said to be an *imaginary embedding*, and the subfield $j(K)$ an *imaginary conjugate* of K). If j is imaginary, the map $x \mapsto \overline{j(x)}$ is a homomorphism from K to $\mathbb{C}$ distinct from j, which we denote by $\bar{j}$. Hence, there is an even number $2r_2$ of imaginary embeddings of K, and if r_1 denotes the number of real embeddings

---

[8]The following points are exercises for the interested reader, and do not in any way impact on the theory of Lie groups.

of K, then $r_1 + 2r_2 = n$, using the traditional notation of the (very visible) Invisible College.

For example, the field $K = \mathbb{Q}(\sqrt{2})$ only has real embeddings, namely

$$x + \omega y \mapsto x + y\sqrt{2} \quad \text{and} \quad x + \omega y \mapsto x - y\sqrt{2}.$$

(K is represented as the set of $x + y\omega$ with $x, y \in \mathbb{Q}$ and $\omega^2 = 2$. The symbol $\sqrt{2}$ denotes the usual real number, whereas ω is the element of an "abstract" extension of $\mathbb{Q}$, which is not necessarily defined as a subfield of $\mathbb{R}$.) On the contrary, $\mathbb{Q}(\sqrt{-1})$ only has imaginary embeddings, namely

$$x + \omega y \mapsto x + iy \quad \text{and} \quad x + \omega y \mapsto x - iy,$$

with the obvious notation.

Returning to the general case, the structure of $K_\mathbb{R}$ is then given by the following result. Let us denote the real embeddings of K as $j_1, \cdots, j_{r_1}$, and let us number the others in such a way that $j_{r_1+1}, \cdots, j_{r_1+r_2}$ and their conjugates are all the imaginary embeddings of K. Then there is a linear map

$$j : x \mapsto (j_1(x), \cdots, j_{r_1+r_2}(x)) \tag{1.6.4}$$

from K to $\mathbb{R}^{r_1} \times \mathbb{C}^{r_2}$, and it is in fact a homomorphism of $\mathbb{R}$-algebras if the product in $\mathbb{R}^{r_1} \times \mathbb{C}^{r_2}$ is defined by the general formula enabling the definition of the Cartesian product of algebras, namely

$$(x_1, \cdots, x_r)(y_1, \cdots, y_r) = (x_1 y_1, \cdots, x_r y_r), \tag{1.6.5}$$

where $r = r_1 + r_2$. Accordingly, the homomorphism (1.6.4) is *bijective*, in other words the structure of $K_\mathbb{R}$ is given by

$$K_\mathbb{R} \simeq \mathbb{R}^{r_1} \times \mathbb{C}^{r_2}. \tag{1.6.6}$$

To prove this result, first observe that both sides of (1.6.6) have the same dimension over $\mathbb{R}$. So it suffices to show that the map is *injective*. The equality $\bar{j}(x) = \overline{j(x)}$ defining the conjugate of an imaginary embedding for $x \in K$ can obviously be extended to $K_\mathbb{R}$. So if $x \in K_\mathbb{R}$ is in the kernel of homomorphism (1.6.4), then clearly

$$j_p(x) = 0 \quad \text{for} \quad 1 \leq p \leq n. \tag{1.6.7}$$

If, for $a \in K$, $Tr(a)$ denotes the trace of the endomorphism $x \mapsto ax$ of K regarded as a $\mathbb{Q}$-vector space (so that, in this particular case, the Tr function becomes the one introduced above for all $\mathbb{Q}$-algebras following the proof of Lemma a), then it is known (*Cours d'Algèbre*, §33, exercise 29) that

$$Tr(a) = \sum j_p(a) \quad \text{for all} \quad a \in K. \tag{1.6.8}$$

Defining $Tr(a)$ in a similar manner for all $a \in K_{\mathbb{R}}$, this relation continues to hold by linearity, whence it follows that (1.6.7) implies

$$Tr(xy) = \sum j_p(xy) = \sum j_p(x)j_p(y) = 0 \text{ for all } y \in K_{\mathbb{R}}. \tag{1.6.9}$$

But the bilinear form $(x, y) \mapsto Tr(xy)$ is not degenerate when it is considered on K. Hence neither is it degenerate when the base field is extended to $\mathbb{R}$. Consequently, (1.6.7) implies that $x = 0$ as claimed.

Thus, as claimed, map (1.6.4) is an isomorphism of the real algebra $K_{\mathbb{R}}$ over the real algebra $\mathbb{R}^{r_1} \times \mathbb{C}^{r_2}$. It follows that the multiplicative group of invertible elements of $K_{\mathbb{R}}$ can be identified (with its topology) with the group $(\mathbb{R}^*)^{r_1} \times (\mathbb{C}^*)^{r_2}$. To determine its subgroup G defined by $|N(x)| = 1$, the norm in the algebra (1.6.5) needs to be computed. If $x = (x_1, \cdots, x_r)$, where $r = r_1 + r_2$, then $N(x)$ will clearly be the product of the determinants of the operators $y \mapsto x_p y$ $(1 \le p \le r)$ on $\mathbb{R}$ (if $p \le r_1$) or on $\mathbb{C}$ considered as a 2-dimensional $\mathbb{R}$-vector space (if $p > r_1$). But if $\mathbb{C}$ is considered as a real vector space, then, for all $x \in \mathbb{C}$, the determinant of the linear map $y \mapsto xy$ from $\mathbb{C}$ to $\mathbb{C}$ is just $|x|^2$. As a consequence,

$$N(x) = x_1 \cdots x_{r_1} |x_{r_1+1}|^2 \cdots |x_{r_1+r_2}|^2 \text{ if } x = (x_1, \cdots, x_{r_1+r_2}), \tag{1.6.10}$$

which, in $K_{\mathbb{R}}$, can also be written as

$$N(x) = j_1(x) \cdots j_n(x) \tag{1.6.11}$$

since if j_k is a complex embedding of K, then $|j_k(x)|^2 = j_k(x)\bar{j}_k(x)$ for all $x \in K$, and so also for all $x \in K_{\mathbb{R}}$.

Whatever the case, it is now clear that the subgroup G of the multiplicative group of $K_{\mathbb{R}}$ defined by $|N(x)| = 1$, together with its topology, is here identified with the subgroup of $(\mathbb{R}^*)^{r_1} \times (\mathbb{C}^*)^{r_2}$ defined by the equality

$$|x_1| \cdots |x_{r_1}| \cdot |x_{r_1+1}|^2 \cdots |x_{r_1+r_2}|^2 = 1. \tag{1.6.12}$$

This enables us to immediately determine its structure:

$$G = F \times \mathbb{T}^{r_2} \times \mathbb{R}^{r_1+r_2-1}, \tag{1.6.13}$$

where F is a finite group (namely the product of r_1 cyclic factors of order 2). To obtain this result, we have obviously used the fact that the log function provides an isomorphism between multiplicative $\mathbb{R}_+^*$ and additive $\mathbb{R}$.

Theorem 4 then tells us that, in the case of the ring $\mathfrak{o}$ of integers of the number field K, the group $\mathfrak{o}^*$ can be identified with a discrete subgroup with *compact quotient* of the above group G. As G is the product of a *compact* group by the factor $\mathbb{R}^{r-1}$, the projection of $\mathfrak{o}^*$ on G is also a discrete subgroup with compact quotient of $\mathbb{R}^{r-1}$, i.e. isomorphic to $\mathbb{Z}^{r-1}$. The kernel of the homomorphism $\mathfrak{o}^* \to \mathbb{Z}^{r-1}$ thereby defined is the intersection of $\mathfrak{o}^*$ and the factor $F \times \mathbb{T}^{r_2}$ of the decomposition. It is clearly the set of $u \in \mathfrak{o}^*$ such that $|j_p(u)| = 1$ for $1 \le p \le n$. Since it is a finite group (as the intersection of a discrete group and a compact group), all its elements have finite order, and so are roots of unity in K. Conversely, if some $u \in K$ satisfies $u^N = 1$ for some integer N, then u is clearly an algebraic integer, and also $u^{-1} = u^{N-1}$, so that u is an element of finite order of $\mathfrak{o}^*$, and hence is a zero of the above homomorphism from $\mathfrak{o}^*$ to $\mathbb{R}^{r-1}$. As a result, *if U denotes the finite group of roots of unity contained in K, then there is an exact sequence* (1) $\to U \to \mathfrak{o}^* \to \mathbb{Z}^{r-1} \to$ (0) *and hence a decomposition* $\mathfrak{o}^* = U \times \mathbb{Z}^{r-1}$ *into a direct product of U and a subgroup isomorphic to $\mathbb{Z}^{r-1}$* (take the subgroup of $\mathfrak{o}^*$ generated by the $r - 1$ elements whose images in $\mathbb{Z}^{r-1}$ form a basis of $\mathbb{Z}^{r-1}$). This is Dirichlet's unit Theorem. Remember that we have set

$$r = r_1 + r_2 \tag{1.6.14}$$

(so that, for example, $r = 2$ if K is a real quadratic field and $r = 1$ if K is an imaginary quadratic field). In the case of an arbitrary order $\mathfrak{o}$, the same result holds as long as U is taken to be the set of roots of unity contained in $\mathfrak{o}$.

Before ending this chapter, we give a completely different proof, possibly simpler, of isomorphism (1.6.6).

Let A be an (associative) algebra (with unit element) over a commutative field k. We assume that A is commutative and finite-dimensional over k. The reader certainly knows what an *ideal* of A is, and that any ideal other than A is contained in at least one *maximal ideal* (here ideals are vector spaces, and as A is finite-dimensional, it is not necessary to use Zorn's lemma or transfinite induction to prove this point). Let us first show that, given the stated assumptions, *the intersection of all maximal ideals of A consists of the nilpotent elements of A.*

First of all, the ring A/I is clearly a field if the ideal I is maximal, since it does not contain any non-trivial ideal. As a field does not have any zero divisors, never mind nilpotent elements, I contains every nilpotent element of A.

Conversely, suppose that there is some $a \in A$ belonging to every maximal ideal of A. Then $1 - a$ is invertible; otherwise $1 - a$ would belong to an ideal distinct from A, and hence to a maximal ideal I, and as $a \in I$, it would follow that $1 \in I$, whence our assertion. But the intersection $\mathfrak{r}$ (*radical* of A) of all maximal ideals of A is an ideal of A. Consequently, *if $a \in \mathfrak{r}$, then $1 - ax$ is invertible for all $x \in A$.* In particular, $1 - a^n$ is invertible for all n.

But the ideals $Aa, Aa^2, Aa^3, \cdots$ form a decreasing sequence, which is therefore stationary (these are finite-dimensional vector spaces like A). Hence there exists an n such that $Aa^n = Aa^{n+1}$, and so there exists an $x \in A$ such that $a^n = xa^{n+1}$. Then $a^n(1 - ax) = 0$, and as $1 - ax$ is invertible, $a^n = 0$ follows.

In conclusion, *if A has no non-trivial nilpotent element, then the intersection of maximal ideals of A is trivial.*

When $k = \mathbb{R}$, this latter result holds for the algebra $K_{\mathbb{R}}$ of a number field. Indeed, if some $a \in K_{\mathbb{R}}$ is nilpotent, then the endomorphism $u_a : x \mapsto ax$ of the vector space $K_{\mathbb{R}}$ is nilpotent and more generally so are $u_a u_b = u_{ab}$ for all $b \in K_{\mathbb{R}}$ (the commutativity of K is required here), and so $0 = Tr(u_a u_b) = Tr(ab)$. However, as seen previously, the bilinear form $Tr(xy)$ on $K_{\mathbb{R}}$ is non-degenerate; therefore $a = 0$.

Let us return to the algebra A over the field k which was briefly set aside. It only has *finitely* many maximal ideals. Indeed, in finite dimension, it is well known that when the intersection of a family of vector subspaces is trivial, a finite family can be extracted whose intersection is trivial. Hence there are maximal ideals $I_1, \cdots, I_r$ such that $I_1 \cap \cdots \cap I_r = \{0\}$. As a consequence, every maximal ideal I contains this intersection. It therefore contains one of the I_k (*Cours d'Algèbre*, §8, exercise 11), and so is one of them, proving the desired result.

Next consider the various maximal ideals $I_1, \cdots, I_r$ of A, the quotient fields $K_p = A/I_p$ and the canonical homomorphisms $j_p : A \mapsto K_p$. There is a homomorphism

$$j : x \mapsto (j_1(x), \cdots, j_r(x))$$

from the algebra A to the product algebra $K_1 \times \cdots \times K_r$. We show it is bijective. To start with, $\ker(j) = \bigcap \ker(j_p) = \bigcap I_p = \{0\}$ by assumption. To show that j is surjective, it then remains to prove that, for all $a_1, \cdots, a_r \in A$, there exists an $x \in A$ such that $x \equiv a_p \pmod{I_p}$ for all p; but this is the famous "Chinese Theorem" (*Cours d'Algèbre*, §8, exercise 10, question f).

In conclusion, we have proved that *if a finite-dimensional commutative algebra over a field k does not have any nilpotent elements, it is isomorphic to a finite product of finite degree extensions over k.*

Let us return to the case of interest to us: $k = \mathbb{R}$ and $A = K_{\mathbb{R}}$. Our purpose is to determine the "residual fields" $K_1, \cdots, K_r$. These are finite degree extensions over $\mathbb{R}$. But every finite degree extension over $\mathbb{R}$ is isomorphic either to $\mathbb{R}$ or to $\mathbb{C}$. Indeed, if such an extension L has dimension at least 2 over $\mathbb{R}$, choosing some $x \in L$ but not in $\mathbb{R}$, the powers $1, x, x^2, \cdots$ of x are linearly dependent over $\mathbb{R}$ (because L is finite-dimensional), so that x is the root of a real algebraic equation of degree at least 2 (otherwise we would have $x \in \mathbb{R}$). Suppose it is of minimum degree. Then its leading term is an irreducible polynomial over $\mathbb{R}$. However (*Cours d'Algèbre*, §33, no 5) the only irreducible polynomials over $\mathbb{R}$ are the linear factors $aX + b$ and the second degree polynomials without any real roots. The second case necessarily holds here, from which it can be deduced that x generates an $\mathbb{R}$-subfield isomorphic to $\mathbb{C}$. Thus, L contains $\mathbb{C}$. It is common knowledge that $\mathbb{C}$ is algebraically closed. So $L = \mathbb{C}$.

From all these arguments it follows that if K is a number field, then the extended algebra $K_{\mathbb{R}}$ is isomorphic to a finite product of fields equal to $\mathbb{R}$ or $\mathbb{C}$, in other words to an algebra of the form $\mathbb{R}^p \times \mathbb{C}^q$. Hence, by dimension arguments, $p + 2q = n$, the degree of K over $\mathbb{Q}$. Moreover, if $x \mapsto (j_1(x), \cdots, j_{p+q}(x))$ denotes the isomorphism in question, then $j_1, \cdots, j_p$ are homomorphisms of $K_{\mathbb{R}}$ *onto* $\mathbb{R}$, whereas $j_{p+1}, \cdots, j_{p+q}$ are homomorphisms *onto* $\mathbb{C}$. Restricting the first p homomorphisms to K defines real embeddings of K, the following q complex embeddings of K (complex because if any of the j_h maps K to $\mathbb{R}$, it also maps $K_{\mathbb{R}}$ to $\mathbb{R}$ since every element of $K_{\mathbb{R}}$ is a real linear combination of elements of K). Restrictions to K of these $p + q$

homomorphisms are mutually distinct, for if $j_k(x) = j_h(x)$ for all $x \in K$, the relation obviously continues to hold for all $x \in K_{\mathbb{R}}$. Finally, let j be an arbitrary embedding of K into $\mathbb{R}$ or $\mathbb{C}$. It can be extended to a homomorphism from $K_{\mathbb{R}}$ to $\mathbb{R}$ or $\mathbb{C}$, whose image is a field (namely $\mathbb{R}$ or $\mathbb{C}$), and whose kernel in $K_{\mathbb{R}}$ is therefore a maximal ideal. But the previous constructions clearly show that the only maximal ideals of $K_{\mathbb{R}}$ are the $p + q$ kernels of $j_1, \cdots, j_{p+q}$. Hence, $\ker(j) = \ker(j_k)$ for some k, and so j is obtained by taking the composition of j_k with an automorphism of $\mathrm{Im}(j_k)$ regarded as an $\mathbb{R}$-algebra. But the only automorphism of $\mathbb{R}$ compatible with the vector space structure over $\mathbb{R}$ is the identity, and the only automorphisms of $\mathbb{C}$, considered as an $\mathbb{R}$-algebra, are the identity and the well-known conjugation. As a result, the only embeddings of K into $\mathbb{R}$ are $j_1, \cdots, j_p$, while any complex embedding is one of the embeddings $j_{p+1}, \cdots, j_{p+q}$ or one of their conjugates. Hence $p = r_1$, $q = r_2$, and all is done. The rest can notably be found in S. Lang, *Algebraic Number Theory* (Addison-Wesley, 1970 or Springer, 1986).

Chapter 2
Simply Connected Spaces and Groups

2.1 The Czechist Viewpoint

Let[1] Γ be a discrete group acting properly and freely on a locally compact space X, so that the quotient $B = \Gamma \backslash X$ is separable (and locally compact). As seen on p. 19, in this case, the canonical map p from X onto B satisfies the property that, for any sufficiently small open subset U of B, the inverse image $p^{-1}(U)$ is the disjoint union of open sets mapped homeomorphically onto U under p. In practice, the space B is *locally connected*, i.e. every point of B admits a fundamental system of connected neighbourhoods. In the above, we can then restrict ourselves to open subsets U that are also connected, and the "components" of $p^{-1}(U)$ that are homeomorphically mapped onto U under p must then be the *connected components* of the open set $p^{-1}(U)$.

The ensuing definition of a *covering* of a locally connected space B follows from this example: it is a pair (X, p) consisting of a locally connected space X and of a continuous map $p : X \to B$ such that every point of B has an open connected neighbourhood U satisfying the following condition: p induces a homeomorphism from each connected component of the open set $p^{-1}(U)$ onto U itself.

Note that neither B nor X is assumed to be connected, but this degree of generality is deceptive (albeit practical). First of all, for every open subset U of B (and even for every locally connected subspace U of B), the pair consisting of $p^{-1}(U)$ and of the map induced by p is a covering of U, which will be called the *restriction* to U of the given covering of B. In particular, every covering of B defines a covering of each connected component of B and proving the converse is not difficult...

Besides, the map p from X onto B is clearly open since each $x \in X$ has a neighbourhood mapped homeomorphically by p onto a neighbourhood of the point $p(x)$. The image under p of a connected component Y of X is thus open in B. It is also closed. Indeed, let b be an adherent point of $p(Y)$ and U a connected open neighbourhood of b in B such that every connected component of $p^{-1}(U)$ is

---

[1]The material in this chapter will not really be needed before Chap. 6. See MA X, no 3.

© Springer International Publishing AG 2017
R. Godement, *Introduction to the Theory of Lie Groups*,
Universitext, DOI 10.1007/978-3-319-54375-8_2

homeomorphically mapped onto U under p. There exists a $b' \in p(Y) \cap U$ and if $y \in Y$ is chosen so that $b' = p(y)$, then the connected component of y in $p^{-1}(U)$ meets Y, and so is fully in Y, whence $b \in p(Y)$ as claimed. It obviously follows that if B is connected, then each connected component of X is also a connected covering space of B.

If (X, p) and (X', p') are coverings of B, a *homomorphism* from (X, p) to (X', p') is a continuous map $f : X \to X'$ such that $p' \circ f = p$. If f is also a homeomorphism, it is said to be an *isomorphism* of covering spaces. We will show how to classify the coverings of a given space B up to isomorphism, subject to some restrictions on the local nature of B.

A covering of B is said to be *trivial* or *decomposable* if it is isomorphic to a covering (X, p) obtained by choosing a discrete space F (i.e. a set equipped with the discrete topology) and setting $X = B \times F$ and $p = pr_1$. An arbitrary covering (X, p) of B is not all that different from a trivial covering. Indeed, consider a sufficiently small connected open subset U of X so that each connected component of $p^{-1}(U)$ is homeomorphically mapped onto U under p. Let F be the set of connected components, and to each $x \in p^{-1}(U)$, associate the pair (b, f), where $b = p(x)$ and $f \in F$ is the connected component of x in $p^{-1}(U)$. This is obviously a bijection from $p^{-1}(U)$ onto $U \times F$ and in fact an isomorphism of $p^{-1}(U)$, regarded as a covering space of U, onto the trivial covering space $U \times F$. In other words, *the restriction of* (X, p) *to every sufficiently small open connected*[2] *subset of B is trivial.*

Moreover, note that if the restrictions of (X, p) to two open subsets U and V are trivial and if U and V meet, then there is a set F such that $p^{-1}(U)$ and $p^{-1}(V)$ are isomorphic to $U \times F$ and $V \times F$. To see this, it suffices to observe that, for any $c \in U \cap V$, the set $p^{-1}(c)$ meets every connected component of $p^{-1}(U)$ or of $p^{-1}(V)$ at exactly one point. This gives a (non-canonical if $U \cap V$ is not connected) method of constructing a bijection between the set of connected components of $p^{-1}(U)$ and the analogous set with respect to V.

We show that if B is connected, which by the way has been assumed from the beginning, then the above implies the existence of a set F such that, for every sufficiently small connected open subset U of X, the restriction $p^{-1}(U)$ of (X, p) to U is isomorphic to $U \times F$. This is equivalent to showing that *the cardinality of the set* $p^{-1}(b)$ *is independent of the point $b \in B$*. But let us choose a cover of B consisting of non-empty connected open sets U_i, $i \in I$, such that $p^{-1}(U_i)$ is trivial for all i, and hence isomorphic to $U_i \times F_i$, where F_i is the set of connected components of $p^{-1}(U_i)$. We saw above that F_i and F_j are equipotent if U_i and U_j meet, hence more generally if there is a finite sequence of indices $i = k_1, \ldots, k_n = j$ in I such that U_{k_p} and $U_{k_{p+1}}$ meet for $1 \leq p < n$. But the existence of such a sequence defines an obvious equivalence relation on I, whence a partition of I into classes I_λ and a corresponding *partition* of the space B into non-empty open sets

---

[2]This assumption is not really necessary since every "sufficiently small" open subset is contained in a connected open subset over which (X, p) is trivial.

$$\bigcup_{i \in I_\lambda} U_i.$$

As B is connected, there is only one class I_λ, from which our assertion readily follows.

If we carry on with this Czechist viewpoint (from the Czech mathematician, Eduard Čech, who was the first to systematically use open covering spaces and their intersection properties to define homology groups of a space), then we must choose an isomorphism ϕ_i from $p^{-1}(U_i)$ onto $U_i \times F$ for all i, and, for every pair of indices i and j, compare ϕ_i and ϕ_j over

$$U_{ij} = U_i \cap U_j. \tag{2.1.1}$$

We obviously get two isomorphisms from $p^{-1}(U_{ij})$ onto $U_{ij} \times F$, which therefore can only differ by an automorphism of the trivial covering $U_{ij} \times F$, i.e. by an automorphism of the topological space $U_{ij} \times F$ compatible with pr_1, hence of the form

$$(u, f) \mapsto (u, \theta_{ij}(u, f)), \tag{2.1.2}$$

where θ_{ij} is a map from $U_{ij} \times F$ to F. For (2.1.2) to be continuous, this must also be the case for θ_{ij}. As F is discrete this means that, for all f, the map $u \mapsto \theta_{ij}(u, f)$ must be locally constant, and hence *constant on every connected component of U_{ij}* (such a component is open since B is assumed to be locally connected). Moreover, (2.1.2) must be bijective, which means that for all $u \in U_{ij}$, the map

$$\theta_{ij}(u) : f \mapsto \theta_{ij}(u, f) \tag{2.1.3}$$

from F to F must be bijective. Conversely, for each $u \in U_{ij}$ and permutation $\theta_{ij}(u)$ of the set F constant on each connected component of U_{ij}, formula (2.1.2) clearly defines an automorphism of the trivial covering $U_{ij} \times F$.

Since, over U_{ij}, (2.1.2) transforms the isomorphism $\phi_i : p^{-1}(U_i) \to U_i \times F$ into the isomorphism $\phi_j : p^{-1}(U_j) \to U_j \times F$, the family of maps θ_{ij} from the various intersections U_{ij} to the group Γ of permutations of the set F has the following properties:

(i) $\theta_{ii}(u) = e$ for all i and $u \in U_i$;
(ii) $\theta_{ij}(u)\theta_{ji}(u) = e$ for all i, j and $u \in U_{ij} = U_{ji}$;
(iii) if i, j and k are three indices such that $U_{ijk} = U_i \cap U_j \cap U_k \neq \emptyset$, then

$$\theta_{ij}(u) = \theta_{ik}(u)\theta_{kj}(u) \quad \text{for all} \quad u \in U_{ijk}. \tag{2.1.4}$$

Conversely, let us take an open cover $(U_i)_{i \in I}$ of B, a set F, and, for every pair (i, j) such that $U_i \# U_j$, a *locally constant* map θ_{ij} from U_{ij} to the group Γ of permutations of F satisfying the above conditions (i), (ii) and (iii). We use them to construct a covering (X, p) of B and, for each i, an isomorphism ϕ_i from $p^{-1}(U_i)$ onto $U_i \times F$. We proceed as follows.

First consider the topological space

$$S = \coprod_{i \in I} U_i \times F = \bigcup_{i \in I} \{i\} \times U_i \times F \qquad (2.1.5)$$

defined as follows: the elements of S are the pairs consisting of an index $i \in I$ and of an element of $U_i \times F$, in other words of the triples (i, u, f) with $i \in I$, $u \in U_i$ and $f \in F$, so that S is the "disjoint sum" of the sets $U_i \times F$ (*no element* of $U_i \times F$ *being allowed* to be identified with an element of $U_j \times F$ for $i \neq j$, contrary to expectations...). Finally, define an open subset of S as a subset whose intersection with $U_i \times F$ is open in $U_i \times F$ for all i. This gives a topology on S. The reader will naturally note that the use of the term "intersection" is a misnomer since $U_i \times F$ is not strictly speaking a subset of S; but there is a canonical injection $(u, f) \mapsto (i, u, f)$ from $U_i \times F$ to S, which justifies the terminology used.

X will be constructed through an identification process in S by means of an equivalence relation R, defined as follows: two elements (i, u, f) and (j, v, g) of S are in the same class if and only if

$$u = v \quad \text{and} \quad f = \theta_{ij}(u)g. \qquad (2.1.6)$$

The condition is empty unless $U_i \# U_j$. As is easily seen, this is an equivalence relation in (2.1.5) if and only if the above conditions (i), (ii) and (iii) hold. It is *open*.[3] Since, by construction, every open subset of S is the union of sets of the form $W \times \{f\} \subset U_i \times F$, where W is an open subset of U_i and f a fixed element of F, it suffices to show that, for all j, the elements $(j, v, g) \in U_j \times F$ whose classes meet $W \times \{f\}$ form an open subset of $U_j \times F$. This is clear since these are elements for which $v \in W \cap U_j$ and $g = \theta_{ij}(v)f$, with $\theta_{ji}(v)$ locally constant. The closure of the graph of relation R follows from similar arguments. Then set $X = S/R$ and define $p : X \mapsto B$ by observing that the map from S onto B which transforms (i, u, f) into u is constant on the classes mod R.

Finally, (X, p) needs to be shown to be a covering of B. Let q be the canonical map from S onto X. By definition (2.1.6) of R, q clearly induces a *bijection* from every subset $\{i\} \times U_i \times F$ of S onto the subset $X_i = p^{-1}(U_i)$ of X. As $U_i \times F$ is open in S and as relation R is open, this bijection from $U_i \times F$ onto X_i is a homeomorphism onto an open subset of X. As the projection $S \to B$ is continuous, so is p. Next, fix a point $a \in B$ and choose i such that $a \in U_i$. Let U be a connected open neighbourhood of a such that $U \subset U_i$. Then consider the homeomorphism from $U_i \times F$ onto $p^{-1}(U_i)$ induced by q. Its restriction to the open subset $U \times F$ of $U_i \times F$ is a homeomorphism onto $p^{-1}(U)$. Therefore, the connected components

---

[3] An equivalence relation R on a space X is said to be open if the R-saturation of every open subset is open. If, moreover, the graph of R is closed and if X is separable, the same holds for X/R.

of $p^{-1}(U)$ are the images under q of the sets $U \times \{f\}$ with $f \in F$. Consequently, (X, p) is a covering of B, as claimed.

We leave it to the reader to check that if, for all i, ϕ_i denotes the inverse isomorphism of q from $p^{-1}(U_i) = X_i$ onto $U_i \times F_r$, then over $U_i \cap U_j$, ϕ_i is transformed into ϕ_j by using the given θ_{ij}.

It is useful to find the conditions under which the above covering is trivial. Let θ be an isomorphism from the trivial covering $B \times F$ onto X. For each i, the composition of θ and the isomorphism $\phi_i : p^{-1}(U_i) \to U_i \times F$ defined above gives an automorphism $\theta_i = \phi_i \circ \theta$ of the trivial covering $U_i \times F$ of U_i, hence of type $(u, f) \mapsto (u, \psi_i(u)f)$, where ψ_i is a locally constant map from U_i to the group Γ of permutations of F. But for $X \in p^{-1}(U_{ij})$, the automorphism $(u, f) \mapsto (u, \theta_{ij}(u)f)$ of $U_{ij} \times F$ transforms $\phi_i(x)$ into $\phi_j(x)$. Hence

$$\psi_i(u) = \theta_{ij}(u) \circ \psi_j(u) \quad \text{in} \quad U_{ij},$$

or else computations in the group Γ imply

$$\theta_{ij}(u) = \psi_i(u)\psi_j(u)^{-1} \text{ in } U_{ij}. \tag{2.1.7}$$

The converse is readily verified; namely, every family of *locally constant* maps $(\psi_i)_{i \in I}$ from the open subsets U_i to the discrete group Γ satisfying (2.1.7) leads to an isomorphism from the trivial covering $B \times F$ onto (X, p). The existence of such maps is therefore a necessary and sufficient condition for the covering defined by the θ_{ij} to be trivial.

Note that if the sets U_i are connected, which is the most common case, the functions ψ_i are necessarily constant, and hence so are the θ_{ij} (although the sets U_{ij} may not be connected). This condition—that the θ_{ij} are constant on each U_{ij}—although necessary for the triviality of the covering (X, p) considered, is not sufficient in general. There is, however, a very particular case where the situation is much simplified, namely the case where the cover $(U_i)_{i \in I}$ consists of only *two* sets U_1 and U_2. It is then a matter of showing that if there is a *constant* map θ_{12} from U_1 to U_2 in Γ, then there are *constant* maps ψ_1 and ψ_2 from U_1 and U_2 to Γ such that $\theta_{12}(u) = \psi_1(u)\psi_2(u)^{-1}$ in $U_1 \cap U_2$—a trivial problem! As θ_{ij} is necessarily constant whenever U_{ij} is connected, it notably follows that *if $B = U \cup V$ with U and V open and $U \cap V$ connected, then every trivial covering of B over U and V is globally trivial*, a result that can be directly proved from the definitions (take x such that $p(x) = c \in U \cap V$, consider the components U' and V' of $p^{-1}(U)$ and $p^{-1}(V)$ containing x, check that $U' \cap V'$ is the component of $p^{-1}(U \cap V)$ containing x, and deduce that p bijectively maps $U' \cup V'$ onto B).

More specifically, let us call a space B *simply connected* if it is connected, locally connected and if it does not admit any non-trivial connected covering (or equivalently if every covering of B is trivial). Then, *if $B = U \cup V$, where U and V are simply connected and open and if $U \cap V$ is connected, then B is necessarily simply connected.* Applying this argument to a sphere of dimension at least two, we will later show that such a sphere is simply connected. To finish the proof, it is best to use the "homotopic" processes set out in the next section. (The Earth can be covered by (i) everything located above latitude 10° south (ii) everything south of latitude 10° north. It remains to check that a space homeomorphic to a ball is simply connected.) Note that the argument no longer holds in dimension one (nor in fact the result: $\mathbb{R} \to \mathbb{T}$)...

2.2 Extensions of Local Homomorphisms of a Simply Connected Group

Since these notes are supposed to address the theory of Lie groups, in this section we give an example to motivate the presentation of these theories in this context. Let G be a connected and locally connected group, U a connected open neighbourhood of e in G and φ a map from U to an "abstract" group F such that

$$\varphi(xy) = \varphi(x)\varphi(y) \text{ whenever } x, y \text{ and } xy \text{ are in } U. \tag{2.2.1}$$

We may ask ourselves whether φ can be extended to a homomorphism from G to F. This is not always the case (were it always possible, the canonical map from $\mathbb{R}$ to $\mathbb{T}$ would have an inverse homomorphism from $\mathbb{T}$ to $\mathbb{R}$). However, the non-existence of such an extension can be expressed in terms of the non-triviality of a covering. In particular, the following result can be proved:

Theorem 1 *Let G be a simply connected topological group. Let U be an open connected neighbourhood of e in G and φ a map from U to a group F such that*

$$\varphi(xy) = \varphi(x)\varphi(y) \text{ whenever } x, y \text{ and } xy \text{ are in } U.$$

Then, φ can be extended in a unique way to a homomorphism from G to F.

The proof rests on constructions that hold in all cases, regardless of whether G is simply connected or not.

Consider the set I of pairs $i = (V, \theta)$ consisting of a sufficiently small connected open set $V \subset G$ so that $V^{-1}V \subset U$ and of a map $\theta : V \to F$ such that

$$\theta(x)^{-1}\theta(y) = \varphi(x^{-1}y) \quad \text{for all } x, y \in V. \tag{2.2.2}$$

A first remark: (2.2.2) has solutions for every open subset V such that $V^{-1}V \subset U$: some of them can be constructed by choosing an element $a \in V$ and setting $\theta(x) = \varphi(a^{-1}x)$. For $x, y \in V$, $a^{-1}y = a^{-1}x.x^{-1}y$, and assumption (2.2.1) then shows that $\theta(y) = \theta(x)\varphi(x^{-1}y)$ as desired. Next, for $i = (V, \theta) \in I$, let us set $V = U_i$ and $\theta = \theta_i$, and consider indices i and j such that $U_i \# U_j$. Clearly,

$$\theta_i(x)^{-1}\theta_i(y) = \theta_j(x)^{-1}\theta_j(y) \quad \text{for all } x, y \in U_{ij} = U_i \cap U_j. \tag{2.2.3}$$

As a consequence there is a *constant* map θ_{ij} from U_{ij} to F such that

$$\theta_i(x) = \theta_{ij}(x)\theta_j(x) \quad \text{for all } x \in U_{ij}. \tag{2.2.4}$$

If the group F is identified (under left translations) with a subset of the group Γ of permutations of the set F, the family of maps θ_{ij} from U_{ij} to Γ satisfy conditions (i), (ii) and (iii) given on page 35. *Hence specifying U and φ enables us to canonically construct a covering of the topological space G.*

Suppose that this covering is trivial. Then, for each $i \in I$, there is a locally constant map ψ_i from U_i to the group Γ of permutations of the set F such that, for all $x \in U_{ij}$, the permutation $\psi_j(x)$ is the composition of the permutation $\psi_j(x)$ and the left translation by $\theta_{ij}(x)$. Denoting by $s_i(x) \in F$ the image of the identity of F under $\psi_i(x)$, we thus get locally constant maps (hence *constant* since the U_i are connected) s_i from the subsets U_i to F such that $s_i(x) = \theta_{ij}(x)s_j(x)$ (product in the group F) in U_{ij} whenever U_i and U_j meet. Taking into account (2.2.4), we get

$$s_i(x)^{-1}\theta_i(x) = s_j(x)^{-1}\theta_j(x) \quad \text{in } U_{ij} \tag{2.2.5}$$

whenever U_i and U_j meet. As the U_i cover G, there is one and only one map Φ from G to F such that $s_i(x)^{-1}\theta_i(x) = \Phi(x)$ in U_i for all i. But as (2.2.2) can be rewritten as

$$\theta_i(x)^{-1}\theta_i(y) = \varphi(x^{-1}y) \quad \text{for all } x, y \in U_i, \tag{2.2.6}$$

a trivial calculation shows that $\varphi(x^{-1}y) = \Phi(x)^{-1}\Phi(y)$ for all $i \in I$ and $x, y \in U_i$.

We then choose a connected open neighbourhood V of e such that $V = V^{-1}$, $V.V \subset U$. For all $a \in G$, there exists an i such that $aV = U_i$ since aV is connected and satisfies $(aV)^{-1}aV \subset U$. Then $\varphi(x^{-1}y) = \Phi(x)^{-1}\Phi(y)$ for all $a \in G$ and $x, y \in aV$, whence

$$\Phi(xv) = \Phi(x)\varphi(v) \quad \text{for all } x \in G \text{ and } v \in V \tag{2.2.7}$$

and more generally,

$$\Phi(v_1 \ldots v_n) = \Phi(e)\varphi(v_1) \ldots \varphi(v_n) \tag{2.2.8}$$

for all n and points $v_1, \ldots, v_n \in V$. The connected set G is the union of the sets V^n since $V = V^{-1}$. So (2.2.8) defines Φ on the whole of G and shows that if $\Phi(x)$ is replaced by $\Phi(e)^{-1}\Phi(x)$, the map Φ is a *homomorphism* from G to F coinciding with φ on V.

We still need to show that $\Phi(x) = \varphi(x)$ for all $x \in U$. But for $x \in U$, the elements $v \in V$ such that $xv \in U$ form a neighbourhood of e. For such a v,

$$\Phi(xv)\varphi(xv)^{-1} = \Phi(x)\varphi(v)[\varphi(x)\varphi(v)]^{-1} = \Phi(x)\varphi(x)^{-1} \tag{2.2.9}$$

clearly holds by (2.2.1) and (2.2.7). The function $\Phi(x)\varphi(x)^{-1}$ is therefore locally constant on U. As U is connected, it is constant, and hence everywhere equal to e since this is the case at the origin. As a result, Φ and φ are indeed equal on U as claimed. This completes the proof of Theorem 1 since, if G is simply connected, then the covering associated to φ by the preceding construction is necessarily trivial.

In the general case, if φ extends to a homomorphism from G to F, the associated covering is trivial. Indeed, let φ also denote such an extension. Then (2.2.2) or, if preferred (2.2.6), shows that for each i there exists an $a_i \in F$ such that $\theta_i(x) = a_i\varphi(x)$ in U_i. By (2.2.4), it follows that $\theta_{ij}(x) = a_i a_j^{-1}$ for all i and j and $x \in U_{ij}$. For every $x \in U_i$, denoting by $\psi_i(x)$ the left translation on F defined by a_i obviously gives us solutions of (2.1.7), and, as seen above, the covering is trivial. We leave the rest for Sect. 2.6.

Remark. Instead of using the cover $(U_i)_{i \in I}$ defined at the beginning of the proof, to construct the covering associated to φ it would be simpler, although not canonical, to use the cover $(aV)_{a \in G}$, where V is chosen as above. Then, in each *non-empty* intersection $aV \cap bV$, there is a constant "transition function" $\theta_{ab}(x) = \varphi(a^{-1}b)$ and the family of functions θ_{ab} defines the desired covering just as well. The proof can be easily adapted to this viewpoint.

2.3   Covering Spaces and the Fundamental Group

Applications of the theory of covering spaces nearly always concern *path*-connected spaces. Let us remind the reader of their definition. A *path* in a space B is a continuous map γ from the interval $I = [0, 1]$ to B. The points $\gamma(0)$ and $\gamma(1)$ are respectively the *origin* and *extremity* (or the endpoints, in the plural) of γ. If $\gamma(0) = \gamma(1) = b$, γ is said to be a *loop based at the point b*. A space B is then said to be *path-connected* if any two arbitrary points of B can be connected by a path, and *locally path-connected* when each of its points has a fundamental system of path-connected open neighbourhoods. A covering (X, p) of a locally path-connected space B is clearly of the same kind. Given a path γ in B, a path γ' in X such that $\gamma = p \circ \gamma'$ is called a *lifting* of γ to X.

Lemma 1 *Let (X, p) be a covering of B and γ a path in B. For any $x \in X$ such that $p(x) = \gamma(0)$, there exists a unique lifting γ' of γ such that $\gamma'(0) = x$.*

Existence is readily proved. Covering $\gamma(I)$ with a finite number of open sets over which (X, p) is trivial, well-known arguments give us a sequence of numbers $0 = t_0 < t_1 < \cdots < t_n = 1$ such that, for each k, the covering (X, p) is trivial over some open subset U_k containing $\gamma(I_k)$, where $I_k = [t_k, t_{k+1}]$. Consider first the connected component in $p^{-1}(U_0)$ containing the point x given in the statement. There is evidently a unique continuous map γ' from I_0 to it such that $p \circ \gamma' = \gamma$ on I_0. Once this is done, consider the connected component in $p^{-1}(U_1)$ of the point $\gamma'(t_1)$, which enables us to construct a map, still denoted by γ', from I_1 to this component such that $p \circ \gamma' = \gamma$ on I_1, and so on from index to index, whence the existence of γ'.

To prove the uniqueness of γ', it is preferable to show that more generally *if f_0 and f_1 are two continuous maps from a space Y to X, such that $p \circ f_0 = p \circ f_1$, then the set of $y \in Y$ such that $f_0(y) = f_1(y)$ is both open and closed in Y*, and hence is the whole of Y if Y is connected and if the set in question is not empty. Now, the closure of the set in question follows from the continuity of f_0 and f_1 (remember that in these Notes, we only consider separable spaces, except spaces that may need to be derived from others, by taking the quotient for example...). To show that it is open, let us suppose that $f_0(b) = f_1(b)$ for some $b \in Y$ and let U be a neighbourhood of $f_0(b)$ homeomorphically and hence injectively mapped to B by p. There is a neighbourhood V of b in Y mapped to U by f_0 and f_1. For any $y \in V$, the points $f_0(y)$ and $f_1(y)$ of U have the same projection in B, and hence coincide since p is injective on U; as a result, $f_0(y) = f_1(y)$ in V, qed.

To go further, the notion of *homotopy* between two paths γ_0 and γ_1 in a space B is needed: a homotopy is a continuous map f from the square $I \times I$ to B such that $f(0,t) = \gamma_0(t)$, $f(1,t) = \gamma_1(t)$, $f(s,0) = \gamma_0(0) = \gamma_1(0)$, and finally, $f(s,1) = \gamma_0(1) = \gamma_1(1)$ for all s and t. So the definition assumes that the two paths considered have the same endpoints. When such a homotopy exists, γ_0 and γ_1 are said to be *homotopic* and this is often written $\gamma_0 \sim \gamma_1$.

Lemma 2 *Let (X, p) be a covering of B and γ_0, γ_1 two paths in X. Suppose that $\gamma_0(0) = \gamma_1(0)$ and that the paths $p \circ \gamma_0$ and $p \circ \gamma_1$ are homotopic in B. Let $f : I \times I \to B$ be a homotopy between $p \circ \gamma_0$ and $p \circ \gamma_1$. Then γ_0 and γ_1 are homotopic, and there exists a unique homotopy $g : I \times I \to X$ between γ_0 and γ_1 such that $f = p \circ g$.*

The proof is suggested by the similarity between this statement and the previous one. We start by constructing sequences of points $0 = s_0 < s_1 < \cdots < s_m = 1$ and $0 = t_0 < t_1 < \cdots < t_n = 1$ such that for all k and h, the covering (X, p) is trivial over an open subset U_{kh} of B containing $f(I_k \times J_h)$, where $I_k = [s_k, s_{k+1})$ and $J_h = [t_h, t_{h+1}]$. Considering the connected component of $p^{-1}(U_{00})$ containing the point $\gamma_0(0) = \gamma_1(0)$, the construction of a continuous map g from $I_0 \times J_0$ to this connected component which "lifts" the map f from $I_0 \times J_0$ to U_{00} is immediate. This being done, there is a connected component of $p^{-1}(U_{10})$ containing $g(s_1, t)$ for all $t \in J_0$. This gives a (continuous) map g from $I_1 \times J_0$ which coincides with the previous one on the common border and which "lifts" f over $I_1 \times J_0$. Continuing from index to index, we find a continuous map $g : I \times I \to X$ which lifts f. It is a homotopy since $s \mapsto g(s, 0)$ and $s \mapsto g(s, 1)$ are continuous and hence *constant* maps from I to discrete subsets of X. As $g(0, 0) = \gamma_0(0) = \gamma_1(0)$, the uniqueness part of Lemma 1 shows that g is a homotopy from γ_0 to γ_1. Finally, uniqueness of g follows as in Lemma 1. See also MA X, no 3, (iii).

Note that if we start from a loop γ based at some point of B, the liftings of γ to X *are not* generally loops in X (counterexample: take $B = \mathbb{T}$ and $X = \mathbb{R}$ with the map p given by $p(\theta) = e^{2\pi i\theta}$, and try to lift the loop in $\mathbb{T}$ given by $\gamma(t) = e^{2\pi it}$ for all $t \in I$ to a loop in $\mathbb{R}$...). However, Lemma 2 tells us that the liftings to X of a loop *homotopic to a point* are themselves loops of the same kind.

To make this type of observation systematic and to reach a classification of the coverings of a given space B, the *fundamental group* based at a point of a path-connected space needs to be introduced. It is defined as follows.

Let E be a path-connected space and a a given point of E. The product of two loops γ_0 and γ_1 based at a in E is defined to be the loop given by

$$\gamma_0\gamma_1(t) = \begin{cases} \gamma_0(2t) & \text{if } 0 \leq t \leq \frac{1}{2}, \\ \gamma_1(2t-1) & \text{if } \frac{1}{2} \leq t \leq 1. \end{cases} \tag{2.3.1}$$

(Note that the definition only assumes that the extremity of γ_0 and the origin of γ_1 coincide, and in a more general framework makes it possible to define the product of two paths satisfying this condition; it will be needed later.) Consider the equivalence relation defined by the homotopy. It trivially follows that the homotopy class of the product of two loops based at a only depends on the class of the two factors. Hence there is a composition law in the set $\pi_1(E, a)$ of loops at a. This turns $\pi_1(E, a)$ into a *group*, which is shown as follows. First, if there are three loops, γ_0, γ_1 and γ_2 based at a, we get a homotopy between the loops $\gamma_0(\gamma_1\gamma_2)$ and $(\gamma_0\gamma_1)\gamma_2$ by considering the map $f : I \times I \to E$ given by

$$f(s,t) = \begin{cases} \gamma_0(\frac{4t}{1+s}) & \text{for } 0 \leq t \leq \frac{s+1}{4}, \\ \gamma_1(4t-1-s) & \text{for } \frac{s+1}{4} \leq t \leq \frac{s+2}{4}, \\ \gamma_2(1-4\frac{1-t}{2-s}) & \text{for } \frac{s+2}{4} \leq t \leq 1. \end{cases} \tag{2.3.2}$$

This implies the associativity of the multiplication in $\pi_1(E, a)$. The existence of an identity is clear: it is the class of the loop $t \mapsto a$ reduced to a. Finally, every class admits an inverse which can be seen by associating to each loop γ the loop γ' given by $\gamma'(t) = \gamma(1-t)$, and whose geometric interpretation is obvious.

The *fundamental group* $\pi_1(E, a)$ of E based at a seemingly depends on a, but if a is replaced by another point b of E, fixing a path γ connecting a to b gives an isomorphism from $\pi_1(E, a)$ onto $\pi_1(E, b)$ by associating to a loop σ based at a the loop based at b which consists in going from b to a by following the inverse of γ, then in going along σ, and finally in returning to b via γ (we leave it to the reader to transform this colourful description into formulas). In this sense, we can talk of a fundamental group $\pi_1(E)$ of E; it is defined up to isomorphism.

Now a remark: *every path-connected and locally path-connected space B such that $\pi_1(B) = \{e\}$ is simply connected*. Indeed, let (X, p) be a connected covering of B and consider the points x' and x'' of X such that $p(x') = p(x'') = b$. They can be connected by a path γ', projecting onto a loop $\gamma = p \circ \gamma'$ based at b. It is homotopic to a point so that, as seen after the proof of Lemma 2, the lifting γ' of γ is a loop based at x'. Hence $x' = x''$ and p is bijective, proving the result. The converse will be shown later.

This result obviously shows that every *contractible* space B (i.e. for which there exists a continuous map from $I \times B$ to B which "starts" with the identity map from B to B and "ends" with a map from B onto a point) is simply connected. This is for instance the case of a ball in the space $\mathbb{R}^n$ (as remarked on p. 37, this implies that a sphere is also simply connected in dimension ≥ 2).

Let E and F be two path-connected spaces and f a continuous map from E to F. For all $a \in E$, with $b = f(a)$, f clearly defines a homomorphism from $\pi_1(E, a)$ to $\pi_1(F, b)$ obtained by associating to each loop γ based at a the loop $f \circ \gamma$ based at b. The map (sic) $(E, a) \mapsto \pi_1(E, a)$ is therefore a *covariant functor* defined on the category of pointed topological spaces, and taking values in the category of groups, a pointed topological space being a pair (E, a) consisting of a topological space E and of a point a of E, a morphism from a pointed space (E, a) to a pointed space (F, b) being a continuous map from E to F mapping a onto b, the composition of these maps being defined as usual, etc., etc. Envy the children of the XXIth century who, after a hundred and thirty two educational and syllabus reforms, each more final than the preceding one, will learn about functors as soon as they enter middle school. (False: in 2003, they are learning to calculate 6×7 with the help of a computer).

In particular, consider a connected covering (X, p) of a space B, so that for all $x \in X$, the map p defines a homomorphism from $\pi_1(X, x)$ to $\pi_1(B, p(x))$. By Lemma 2, which enables us to "lift" homotopies to B, this homomorphism is *injective*. If, moreover, x is replaced by another point x' located "over" the same point of B as x, then the images of $\pi_1(X, x)$ and of $\pi_1(X, x')$ in $\pi_1(B, p(x))$ are its *conjugate* subgroups, as is immediately seen by choosing a path in X connecting x to x' and using it as above to transform every loop based at x into a loop based at x'. Besides, as every loop over B based at $p(x)$ can be lifted to a path in X having x as origin, by making x' vary, we get all the conjugates in $\pi_1(B, p(x))$ of the image of $\pi_1(X, x)$ under p. We conclude that, *for each point b of B, the class of conjugate subgroups in $\pi_1(B, b)$ is fully determined by a covering (X, p)*. Furthermore, for any $b, b' \in B$, the classes of subgroups defined by (X, p) in $\pi_1(B, b)$ and $\pi(B, b')$ correspond under any isomorphism from the first group onto the second one obtained by connecting b to b' by a path. The classification of the coverings of B rests on these observations since, as will be seen later, the connected coverings of B are in *bijective* correspondence (up to isomorphism) with the classes of conjugate subgroups in $\pi_1(B)$, at least under some simple local assumptions on B.

2.4   The Simply Connected Covering of a Space

The classification of covering spaces is immediate when B is *simply path-connected*, i.e. when $\pi_1(B, b) = \{e\}$ for some point (and hence for all points) of B, since, as seen above, the space B is then simply connected and only has trivial coverings. In the general case, the classification of covering spaces is particularly easy if B is assumed to be *locally simply path-connected*, i.e. if every $b \in B$ has an open path-connected and simply path-connected neighbourhood. This is the case if every $b \in B$ admits a neighbourhood homeomorphic to a ball in a space $\mathbb{R}^n$. The first stage in the classification of covering spaces consists in proving the following result, which is anyhow fundamental:

Theorem 2 *Let B be a path-connected and a locally simply path-connected space. Then B admits a path-connected and a locally simply path-connected covering, and the latter is unique up to isomorphism.*

The proof of this result is somewhat lengthy, but is not very difficult and we trust the reader will fill in the details.

(a) Suppose the problem has been solved and let (X, p) be a connected and simply connected covering of B (*path-connected—this indication will be omitted from now on*). Fix some $b \in B$ and $a \in X$ such that $p(a) = b$. As X is simply connected, two paths in X having a as origin are homotopic if and only if they have the same extremity x. On the other hand, by Lemma 2, two such paths are homotopic in X if and only if their projections in B are homotopic. Since every path in B with b as origin can be lifted to a path in X with a as origin, there must be a *bijection* between the elements of X and the classes of paths in B with origin b. This leads to a construction of (X, p) starting from B and the point b.

(b) To see this, start with a connected and locally simply connected space B and fix a point b of B. Let X denote the set of *classes of paths in B originating at b* and p the map from X onto B which associates to each class of paths originating at b, its extremity (which only depends on the class considered). All that is required is to endow X with a topology turning (X, p) into a simply connected covering of B.

For this purpose, let U be a *connected and simply connected* open subset of B. Then $p^{-1}(U) \subset X$ is partitioned into classes of paths originating at b and ending in U. Let us define (for fixed U) two paths γ_0 and γ_1 originating at b and ending at points u_0 and u_1 of U as being equivalent if γ_1 is homotopic to the product of γ_0 and a path in U connecting u_0 to u_1. This gives an equivalence relation in the set of homotopy classes of paths originating at b and ending in U. In other words, this defines an equivalence relation on the set $p^{-1}(U)$, which is thus partitioned into classes. If V is one of these classes, p induces a *bijection* from V onto U. Indeed, fix a point x of V, defined by the path γ connecting b to $u = p(x) \in U$. The points of V then correspond to paths homotopic to $\gamma\gamma'$, where γ' is an arbitrary path in U having origin u. It is already the case that p maps V *onto* U since U is path-connected. Moreover, U being simply connected, two paths in U having origin u are homotopic whenever they have the same extremity. Taking their product with γ shows that the map p from V onto U is also injective.

Let the equivalence classes defined above be called the components of $p^{-1}(U)$. They are mutually disjoint and their union is $p^{-1}(U)$. Letting U vary within the stated conditions, the components of the inverse images of $p^{-1}(U)$ are clearly seen to form a covering of X. We are now going to endow X with the following topology: *a subset W of X is open if and only if, for every open connected and simply connected set $U \subset B$ and every component V of $p^{-1}(U)$, the set $p(W \cap V)$ is open in U* (i.e. in B). The axioms of a topology hold for trivial set theoretic reasons.

We are now going to show that, *if V is as before a component of $p^{-1}(U)$ with U connected and simply connected, then V is open in X and p is a homeomorphism from V onto U.* We show that (X, p) is a covering of B.

To do so, let us take a subset W of V such that $p(W)$ is open. As p is a bijection from V onto U, it amounts to showing that W is open in X. For this it suffices to check that if U' is an open connected and simply connected subset of B and if V' is a component of $p^{-1}(U')$, then $p(W \cap V')$ is open in B. But let us consider a point $x \in W \cap V' \subset V \cap V'$, its projection $u = p(x) \in U \cap U'$, and a connected neighbourhood U'' of u contained in the open set $p(W) \cap U'$. The point x is the homotopy class of a path γ in B connecting b to u. As U'' is connected, u can be connected to every $v \in U''$ by a path in U''. Taking its product with γ, we get an element $y \in X$ such that $p(y) = v$, and as the path in U'' connecting u to v is also a path in U or in U', x and y clearly belong to the same component of $p^{-1}(U)$ or $p^{-1}(U')$. As $x \in V \cap V'$, it follows that $y \in V \cap V'$ as well. However, as p maps V bijectively onto U, $v = p(y) \in p(W)$ implies that $y \in W \cap V'$. As a consequence, $p(W \cap V')$ contains every point v of U'', and so is open in U as claimed.

To finish this part of the proof, the *connectedness and simple connectedness* of the *covering* (X, p) of B constructed above remains to be checked. First note that if γ is a path in B originating at b, then for every $s \in I$, let γ_s be the "partial" path defined by $\gamma_s(t) = \gamma(st)$. It defines a point x_s of X. For s' sufficiently near s, $\gamma_{s'}$ is obtained (up to equivalence) by the product of γ_s with a path connecting $\gamma_s(1)$ to $\gamma_{s'}(1)$ in a connected and simply connected neighbourhood of $\gamma_s(1)$. Hence the map $s \mapsto x_s$ from I to X is continuous, and so defines a lifting of the path γ to X. It is obviously *the* lifting of γ having as "origin" the base point of X representing the class of the path consisting solely of b in B. Hence this particular point of X can be connected to any other point by a path. This is why X is path-connected like B.

Denoting by a this "base" point of X, let λ be a loop based at a over X and let us consider its projection $\gamma = p \circ \lambda$. Defining partial paths γ_s as above and denoting the class of γ_s by x_s, the map $s \mapsto x_s$ is a lifting of γ to X originating at a. Since such a lifting is unique, it is necessarily λ. In other words, $x_s = \lambda(s)$ for all $s \in I$, and in particular $x_1 = $ class of $\gamma = \lambda(1) = a$, which means that the path γ and the path reduced to b are homotopic. Equivalently, the image of the homomorphism from $\pi_1(X, a)$ to $\pi_1(B, b)$ induced by p is the identity subgroup. As this homomorphism is injective, it follows that X is simply connected.

So far, we have proved the existence of a simply connected covering of the space B (the one constructed in the proof is called the *universal covering* of B at the base point b; the terminology will become clear in the next section). In the following section, we prove its uniqueness and at the same time show how to classify the other coverings of B.

2.5 Classification of Covering Spaces

We continue to assume that B is connected and locally connected. In this case, as seen above, if (X, p) is a connected covering of B, then, for all $b \in B$, (X, p) canonically defines a class of mutually conjugate subgroups in $\pi_1(B, b)$. We are now going to show that the correspondence thus defined between connected coverings of B (defined up to isomorphism) and classes of subgroups of $\pi_1(B, b)$ is *bijective*. Uniqueness up to isomorphism of the simply connected covering (which corresponds to the trivial subgroup) will notably follow.

(i) Let (X, p) be the universal covering of B at the point b constructed in the previous section, and $\Gamma = \pi_1(B, b)$ be the fundamental group of B based at b. We first show that Γ can be made to act on X in such a way that B is identified with the quotient $\Gamma \backslash X$. For this, it suffices to observe that for $\gamma \in \Gamma$ and $x \in X$, γx admits a natural definition: choose a loop based at b in the homotopy class γ, a path with origin b in the homotopy class x, and take their product. This gives a new path with origin b whose class only depends on classes γ and x; thus γx is well-defined. Similarly, the constructions outlined above to check the group axioms in $\pi_1(B, b)$ show that Γ *acts* on the set X. Obviously, $p(\gamma x) = p(x)$ by construction, and moreover Γ acts *transitively* on each "fibre" $p^{-1}(u)$ in X. Equivalently, if there are two paths in B with origin b and extremity u, the latter is homotopic to the product of the former with a loop based at b, which is obvious. If, in particular, a denotes as above the base point of X, in other words the class of the trivial loop based at b, then γa must be the extremity of the lifting with origin a of any loop based at b belonging to the class γ. Finally, Γ *acts freely on* X. This means that if λ is a path having origin b and γ a loop based at b, and if the path $\gamma \lambda$ is homotopic to the path λ, then the loop γ is homotopic to a point. This follows immediately by considering the "reverse" path λ' of λ. Paths $(\lambda'\gamma)\lambda \simeq \gamma'(\gamma\lambda)$ and $\lambda'\lambda$ are then homotopic. However the latter is a loop homotopic to a point at $b' = \lambda(1)$. Hence so is the former. Moreover, as the map $\gamma \mapsto (\lambda'\gamma)\lambda$ specifically defines (see p. 42) an isomorphism from $\pi_1(B, b)$ onto $\pi_1(B, b')$, the loop γ is homotopic to a point, as claimed. [Note that we have used path composition in a more general setting than that of loops].

These remarks provide a canonical bijection between $\Gamma \backslash X$ and B. It remains to show that the action of Γ on X is compatible with the topology on X. For this, consider a loop σ based at b, a point x of X represented by a path λ connecting b to the point $u = p(x)$, and fix a connected and simply connected neighbourhood U of u in B. Let V and V' be the components of $p^{-1}(U)$ respectively containing x and γx, where $\gamma \in \Gamma$ is the class of σ so that V and V' are open neighbourhoods of x and $x' = \gamma x$ homeomorphically mapped onto U by p. By the construction of components, V is the set of classes of paths consisting of λ followed by a path in U connecting u to any point of U. As x' is the class of the path $\sigma \lambda$, similarly V' is the set of classes of paths obtained by adjoining to $\sigma \lambda$ a path in U originating at u. Since path composition is associative up to homotopy, $V' = \gamma V$. As a result, first of all Γ acts continuously on X, and secondly every point in X has a neighbourhood whose images under the elements of Γ are mutually disjoint. More generally, if x and y are points in X belonging to distinct orbits of Γ, i.e. such that $p(x) \neq p(y)$, then there clearly exist neighbourhoods V and W of x and y in X such that γV and $\gamma' W$ are disjoint for all $\gamma, \gamma' \in \Gamma$.

Therefore, if B is locally compact, in which case so is X, Γ acts properly and freely on X as claimed (Chap. 1, Sect. 1.4), the quotient space $\Gamma \backslash X$ being canonically identified with B in all cases.

(ii) Let us next show how to *construct a covering of B corresponding to a given class of subgroups of* $\pi_1(B, b) = \Gamma$. For this, fix a subgroup Γ' in the given class and make it act on the simply connected covering (X, p) as in (i) above. Replace X by the quotient space $X' = \Gamma' \backslash X$ and p by the map p' from X' onto B obtained by observing that two points of X corresponding under the action of Γ' have the same image in B. It is necessary to check that (X', p') is also a covering of B and that Γ' belongs to the class of subgroups of $\pi_1(B, b)$ defined by (X', p').

First note that because of the way Γ acts on X (see end of (i) above), *the canonical map π from X onto X' is locally a homeomorphism*. In particular, if U is a connected and simply connected open subset of B, in which case $p^{-1}(U)$ is the disjoint union of (connected...) components simply transitively permuted by Γ, then the restriction of π to any one of these components, say V, is a homeomorphism from V onto an open subset of X'. Besides if V and W are two components of $p^{-1}(U)$, the images $\pi(V)$ and $\pi(W)$ are clearly either identical or disjoint depending on whether V and W correspond or not under the action of Γ'. It follows that the inverse image $p'^{-1}(U) = \Gamma' \backslash p^{-1}(U)$ of U in X' is the disjoint union of open sets mapped homeomorphically onto U under p'. The pair (X', p') is therefore a covering of B as expected. A similar argument shows that the pair (X, π) is a simply connected covering of X', and with good reason.

The image in $\pi_1(B, b)$ of the fundamental group of X' based at a point over b remains to be found. For example, the point $a' = \pi(a)$, may be chosen, where a is the base point of X (i.e. the class of the identity loop in B). We show that the image of $\pi_1(X', a')$ in Γ is precisely the subgroup Γ', thereby completing the construction. To do so, let us start with an element γ of Γ' and let σ be a loop based at b representing the homotopy class γ. It can be lifted to a path λ in X with origin a and whose extremity, as seen in part (i) of the proof, is the point γa of X. As $\pi(a) = \pi(\gamma a)$ since $\gamma \in \Gamma'$, $\pi \circ \lambda$ is clearly a *loop* based at a' in X', which is mapped onto $p' \circ \pi \circ \lambda = \sigma$ under p'. This argument shows that the image of $\pi_1(X', a')$ in Γ contains Γ'.

To prove the inverse inclusion, we start with a loop λ' based at a' in X'. Since (X, a) is a covering of X', the loop λ' can be lifted to a path λ in X with origin a. Then the map $p' : X' \to B$ transforms the homotopy class of λ' into that of the loop $\sigma = p' \circ \lambda' = p' \circ \pi \circ \lambda = p \circ \lambda$. Denote by γ the element of $\Gamma = \pi_1(B, b)$ thus obtained. Since λ is a path in X with origin a, the extremity $\lambda(1)$ of λ must be the point γa of X. But as $\pi \circ \lambda$ is a loop, $\pi(\gamma(a)) = \pi(\lambda(1)) = \pi(\lambda(0)) = \pi(a)$, and as π is the canonical map from X onto $\Gamma' \backslash X$, it follows that $\gamma \in \Gamma'$, giving the desired result.

(iii) To finish the proof, it remains to show that the covering of B corresponding to a given class of subgroups of Γ is unique up to isomorphism.

For this, let (X', p') be a connected covering of B and choose an element $a' \in X'$ such that $p'(a') = b$, and let Γ' be the image in Γ of the fundamental group of X' at a'. We need to show that (X', p') can be identified with the quotient of the simply connected quotient of (X, p) by Γ', constructed above.

We first construct a map π from X onto X'. So, let x be a point of X, and σ a path in B with origin b representing x, so that the unique lifting λ of σ to X having origin a connects the base point a of X to the given point x. As (X', p') is a covering of B, the path σ can be lifted in a unique way to a path λ' in X' having origin a'. The path σ is determined up to homotopy by x, and so, by Lemma 2, so is λ'. This shows that the extremity $x' = \lambda'(1)$ of λ' only depends on x and not on the choice of σ. We set $x' = \pi(x)$. Surjectivity of π obviously follows from the path-connectedness of X', and clearly $p = p' \circ \pi$.

As (X', p') is a covering of B, any sufficiently small neighbourhood V' in X' of x' has the following properties: p' induces a homeomorphism from V' onto a neighbourhood U of the point $u = p'(x')$ in B, and moreover (X', p') and (X, p) are trivial over U. Suppose that V' (and hence also U) is connected so that V' is the connected component of x' in $p'^{-1}(U)$. If $x \in X$ is chosen so that $\pi(x) = x'$ and if V denotes the connected component of x in $p^{-1}(U)$, then the map π induces a map from V to V' since $\pi(V)$ is connected, contains x' and is contained in $p'^{-1}(U)$. Moreover, the composition of this map from V to V' and the homeomorphism p' from V' onto U gives the homeomorphism p from V onto U. As a result, π induces a homeomorphism from V onto V', which shows that the pair (X, π) is a covering of X', trivial over V'.

We still need to check that if x and y are two given points of X, then $\pi(x) = \pi(y)$ if and only if there exists a $\gamma \in \Gamma'$ such that $y = \gamma x$. Let σ and τ be paths in B with origin b belonging to the homotopy classes x and y and let us consider their liftings λ and μ to X having origin a, as well as their liftings λ' and μ' to X' having origin a'. By definition, $\pi(x) = \lambda'(1)$ and $\pi(y) = \mu'(1)$. Equality $\pi(x) = \pi(y)$ therefore tells us that $\lambda'(1) = \mu'(1)$, namely that the loop based at b obtained by adjoining to the path σ the reverse of the path τ is the projection on B of a loop in X' based at a', or equivalently that the loop based at b defined previously belongs, up to homotopy, to the image subgroup Γ' of $\pi_1(X', a')$ in $\pi_1(B, b)$. This means that $y \in \Gamma'x$, completing the proof.

Note that in part (ii) of the construction of the covering $X' = \Gamma' \backslash X$ corresponding to the given subgroup Γ' of Γ, it is in general impossible to make Γ act on the quotient X'. This is nonetheless the case when the subgroup Γ' is *normal* in Γ. Trivial considerations show that Γ then acts on the left on X', or more precisely that the quotient group $\Gamma \backslash \Gamma'$ acts *freely* on X' similarly to the action of Γ on X. Then, (X', p'), endowed with these group operations, is said to be a *Galois covering* of B.

2.6 The Simply Connected Covering of a Topological Group

Let G be a topological group satisfying the assumptions of Theorem 2, i.e. path-connected and locally simply path-connected (this will be the case for connected Lie groups since each point of such a group has neighbourhoods homeomorphic to $\mathbb{R}^n$ for some appropriate n). Let us choose e as the base point of G and apply to (G, e) the constructions used for (B, b) in Sect. 2.4. This gives a simply connected canonical covering of G, which will be denoted by $(\widetilde{G}, p)$ and which is therefore defined as follows: the paths of $\widetilde{G}$ are the paths in G with origin e, projection p associates to the class of a path γ its extremity $\gamma(1)$, and the topology on $\widetilde{G}$ is defined as in Sect. 2.4.

But as G is a topological group, if λ and μ are paths in G with origin e, then the map

$$\lambda * \mu : t \mapsto \lambda(t)\mu(t) \qquad (2.6.1)$$

from I to G defines another path with origin e. The homotopy class of $\lambda * \mu$ depends solely on those of λ and μ (as for paths, we get a "composition" or a "convolution" of homotopies $I \times I \to G$ using the given composition law of G). Hence, this gives a composition law $(x, y) \mapsto x * y$ (which from now will be written simply as xy) in $\widetilde{G}$. Its associativity follows trivially. It admits an identity element—the path reduced to e which will be written $\widetilde{e}$. And every element of $\widetilde{G}$ has an inverse, which can be seen by associating to a path λ the path $t \mapsto \lambda(t)^{-1}$. In other words, $\widetilde{G}$, endowed with this composition law, is a *group*, the map $p : \widetilde{G} \to G$ clearly becoming a homomorphism.

The group law and the topology of $\widetilde{G}$ are easily seen to be compatible. As an example, let us show the continuity of the map $(x, y) \mapsto x * y^{-1}$ at the origin. For this, we choose a connected and simply connected open neighbourhood U of e, and set $\widetilde{U}$ to be the component of $p^{-1}(U)$ containing e, so that p induces a homeomorphism from $\widetilde{U}$ onto U. Let V be a sufficiently small connected open neighbourhood of e so that $V \cdot V^{-1} \subset U$, and let $\widetilde{V}$ be the component of $p^{-1}(V)$ containing e, i.e. the inverse image of V in $\widetilde{U}$. The elements of $\widetilde{U}$ (resp. $\widetilde{V}$) are clearly the classes of the paths in U (resp. V) with origin e. But if $x, y \in \widetilde{V}$ are represented by the paths σ and τ in V with origin e, then $x * y^{-1}$ is clearly represented by the path $t \mapsto \sigma(t)\tau(t)^{-1}$ in U. As a consequence, $\widetilde{V} \cdot \widetilde{V}^{-1} \subset \widetilde{U}$ and so, identifying $\widetilde{U}$ and $\widetilde{V}$ with U and V using p, the map $(x, y) \mapsto x * y^{-1}$ from $\widetilde{V} \times \widetilde{V}$ to $\widetilde{U}$ is identified with the similar map from $V \times V$ to U, thereby implying the desired continuity property. The other properties are obtained likewise.

Hence $\widetilde{G}$ can be considered a connected and simply connected topological group and the map p a homomorphism of topological groups. Then G can be clearly identified (as a topological group) with the quotient of $\widetilde{G}$ by $\ker(p)$. This kernel, which equals $p^{-1}(e)$ and which, as a *set* for now, can be canonically identified with $\pi_1(G, e) = \pi_1(G)$, is a discrete subgroup of $\widetilde{G}$ since p is locally a homeomorphism. This discrete subgroup is abelian and is even contained in the centre of $\widetilde{G}$ (*all normal discrete subgroups of a connected group are central*, because the set of conjugates of an element x of such a subgroup is a connected subset, which therefore necessarily reduces to the point x itself). Let us show that it can be identified *as a group* with $\pi_1(G)$. It suffices to show that if σ and τ are loops based at e in G, then the loop $\sigma * \tau$ is homotopic to the product $\sigma\tau$ of the loops σ and τ (product defined for every space even in the absence of a composition law). For this let ε denote the path reduced to the point e and let us replace σ and τ by the loops $\sigma' = \sigma\varepsilon$ and $\tau' = \varepsilon\tau$ given by

$$\sigma'(t) = \begin{cases} \sigma(2t) & \text{if } t \leq \frac{1}{2}, \\ e & \text{if } t \geq \frac{1}{2}, \end{cases} \quad \tau'(t) = \begin{cases} e & \text{if } t \leq \frac{1}{2}, \\ \tau(2t - 1) & \text{if } t \geq \frac{1}{2}. \end{cases} \qquad (2.6.2)$$

Clearly, σ' and τ' are homotopic to σ and τ and hence $\sigma * \tau$ is homotopic to $\sigma' * \tau'$. However, from the definition it is obvious that $\sigma' * \tau' = \sigma\tau$, and so the desired result is immediate.

The previous proof remains valid if σ is a loop based at e and τ an arbitrary path with origin e. As the composition law $(\sigma, \tau) \mapsto \sigma\tau$ defines the action of $\pi_1(G)$ on $\widetilde{G}$ in the sense used in the preceding section, *in the case of a topological group G, the general theory prompts us to endow $\widetilde{G}$ with the structure of a topological group, to identify $\pi_1(G)$ with the kernel of the canonical homomorphism p from $\widetilde{G}$ onto G, and to then make $\pi_1(G)$ act on $\widetilde{G}$ by left translations.* $\widetilde{G}$ is called the *universal covering* of G. Since, according to the previous section, any other connected covering of G is the quotient of the space $\widetilde{G}$ by a subgroup Γ' of $\pi_1(G)$, i.e. by a normal (since central) subgroup of G, *every other connected covering of G is the quotient group of $\widetilde{G}$ by the corresponding subgroup of $\pi_1(G)$* (as an aside, note that since $\pi_1(G)$ is abelian, the coverings of G correspond to the subgroups, and not just to the classes of subgroups, of $\pi_1(G)$).

Theorem 1 on extensions of local homomorphisms can now be completed:

Theorem 3 *Let G be a connected and locally connected group, U a connected open neighbourhood of the identity e in G and ϕ a local homomorphism from U to a group F. Let $(\widetilde{G}, p)$ be the universal covering of G and $\widetilde{U}$ the connected component of the identity in $p^{-1}(U)$. Then there is a uniquely defined homomorphism $\widetilde{\phi}$ from $\widetilde{G}$ to F such that*

$$\widetilde{\phi}(x) = \phi(p(x)) \quad \text{for all} \quad x \in \widetilde{U}. \tag{2.6.3}$$

It suffices to apply Theorem 1 to the local homomorphism $x \mapsto \phi(p(x))$ of the *connected* open subgroup $\widetilde{U}$ of the simply connected group $\widetilde{G}$.

Note that the above statement does not intend to say more than it does. For example, it does not say that p is a bijection from $\widetilde{U}$ onto U (this would be the case if U was assumed to be simply connected—but we do not make this assumption). Neither does it say that equality $\widetilde{\phi}(x) = \phi(p(x))$ holds throughout $p^{-1}(U)$—if that was the case, the homomorphism $\widetilde{\phi}$ would be trivial on $p^{-1}(e)$, and as G is the quotient of $\widetilde{G}$ by this discrete subgroup, it would in fact follow that $\widetilde{\phi}$ is the composition of p and a homomorphism from G to F extending ϕ, a conclusion that would destroy the entire theory of coverings! In fact, $\widetilde{\phi}$ may well be constant on a non-trivial subgroup D of $p^{-1}(e)$, in which case $\widetilde{\phi}$ "passes to the quotient" and defines a homomorphism to F from the "intermediate" covering $\widetilde{G}/D$ of G, but in general, this does not reduce to G.

Exercise. Describe in terms of $\widetilde{G}$ and $\widetilde{\phi}$ the covering of G associated to ϕ by the proof of Theorem 1.

2.7   The Universal Covering of the Group $SL_2(\mathbb{R})$

We now illustrate the construction of the universal covering with an example using the group $SL_2(\mathbb{R})$. Its fundamental group will be shortly shown to be $\mathbb{Z}$. Let us begin with a completely different case, that of the group $SL_2(\mathbb{C})$ consisting of complex matrices $\begin{pmatrix} a & b \\ c & d \end{pmatrix}$ with determinant $ad - bc = 1$.

$SL_2(\mathbb{C})$ is *simply connected*. To see this, consider the subgroup K of unitary matrices in $G = SL_2(\mathbb{C})$, in other words of matrices with $c = -\bar{b}$ and $d = \bar{a}$. So K can be topologically identified with the subset of $\mathbb{C}^2$ defined by the equality $a\bar{a} + b\bar{b} = 1$, i.e. with the unit sphere in $\mathbb{R}^4$. As a consequence, K is simply connected. Consider the subgroup B_+ of triangular matrices of G, namely $\begin{pmatrix} a & b \\ 0 & d \end{pmatrix}$ with $ad = 1$ and $a > 0$. It is easy to check that the map $(k, b) \mapsto kb$ is a homeomorphism from $K \times B_+$ onto G. As B_+, homeomorphic to $\mathbb{R}^*_+ \times \mathbb{C}$, is simply connected, so is G (exercise: compute the fundamental group of a Cartesian product).

There are (MA XII, Chap. 4) similar subgroups in $G = SL_2(\mathbb{R})$, namely the subgroup K of rotations

$$k = \begin{pmatrix} \cos\phi & -\sin\phi \\ \sin\phi & \cos\phi \end{pmatrix} \tag{2.7.1}$$

and the subgroup B_+ of matrices

$$b = \begin{pmatrix} 1/t & u \\ 0 & t \end{pmatrix} \quad \text{with } u \in \mathbb{R} \text{ and } t > 0. \tag{2.7.2}$$

Once again G is homeomorphic to $K \times B_+$ and B_+, homeomorphic to $\mathbb{R}^*_+ \times \mathbb{R}$, is simply connected. Hence,

$$\pi_1(G) = \pi_1(K) = \pi_1(\mathbb{T}) = \mathbb{Z}, \tag{2.7.3}$$

as claimed above.

The universal covering $\widetilde{G}$ of $SL_2(\mathbb{R})$ will therefore be a simply connected group with a central subgroup D isomorphic to $\mathbb{Z}$ and such that the quotient $\widetilde{G}/D$ is isomorphic to G. Searching for a "concrete" description of $\widetilde{G}$ using a group of matrices satisfying appropriate conditions is pointless; indeed, it can be proved that every continuous homomorphism from $\widetilde{G}$ to a group $GL_n(\mathbb{R})$ is trivial on the kernel of $p : \widetilde{G} \to G$, in other words is reduced to a linear representation of the group G itself. The description of $\widetilde{G}$ given below follows by taking into account the argument of the function $\phi : G \to \mathbb{T}$ given by

$$\phi \begin{pmatrix} a & b \\ c & d \end{pmatrix} = \text{Am}(ci + d), \tag{2.7.4}$$

where $\mathrm{Am}(z) = z/|z|$ for every non-trivial complex number. More precisely, we consider the set $\widetilde{G}$ of pairs (g, θ) in $G \times \mathbb{R}$ for which

$$\phi(g) = e^{i\theta} \tag{2.7.5}$$

and turn it into a simply connected covering of G. The construction will then be complete once the composition law of $\widetilde{G}$ is described. Hence this construction consists of several parts.

(a) *The covering of a space B defined by a continuous map ϕ from B to $\mathbb{T}$.* Let B be a topological space and ϕ a continuous map from B to $\mathbb{T}$. Consider the set X of pairs (b, θ) in $B \times \mathbb{R}$ such that $\phi(b) = e^{i\theta}$. It is a closed subset of $B \times \mathbb{R}$ and the map p from X onto B with the obvious definition is continuous. We show that the pair (X, p) is a (not necessarily connected) *covering* of B, at least if B is locally connected, since we will restrict ourselves to this case in these Notes.

Let U be an open subset of B. A *continuous* map α from U to $\mathbb{R}$ satisfying

$$\phi(u) = e^{i\alpha(u)} \quad \text{for all } u \in U \tag{2.7.6}$$

will be said to be a *uniform branch of the argument of ϕ in U*. As will be seen, the choice of such a function makes it possible to "trivialize" (X, p) over U and, more precisely, to construct a homeomorphism from $p^{-1}(U)$ onto $U \times \mathbb{Z}$ which transforms p into the projection $U \times \mathbb{Z} \to U$. For this, observe that if $x = (u, \theta)$ is in $p^{-1}(U)$, i.e. if $u \in U$, then there exists an $n \in \mathbb{Z}$ such that $\theta = \alpha(u) + 2\pi n$ and conversely. Since $2\pi n = \theta - \alpha(u) = \theta - \alpha(p(x))$, the map $x \mapsto (p(x), n)$ from $p^{-1}(U)$ to $U \times \mathbb{Z}$ is continuous and bijective; the inverse map, given by $(u, n) \mapsto (u, \alpha(u) + 2\pi n)$, is also continuous, giving the desired homeomorphism.

The pair (X, p) will therefore be a covering of B if for every sufficiently small open subset U of B, a uniform branch of the argument of ϕ is shown to exist in U. To this end, take any $b \in B$ and consider the set $U = U(b)$ of $u \in B$ such that

$$|\phi(u) - \phi(b)| < 2. \tag{2.7.7}$$

If $\mathbb{T}(b)$ denotes the open subset of $\mathbb{T}$ obtained by omitting the point $-\phi(b)$, then $U(b) = \phi^{-1}(\mathbb{T}(b))$, and so is open in B. Let us show that there is a uniform branch of the argument of ϕ in $U(b)$. To do so, set $\mathbb{T}^* = \mathbb{T} - \{-1\}$ and for $z \in \mathbb{T}^*$, let $\arg(z)$ denote the value of the argument of z strictly contained between $-\pi$ and $+\pi$ ("principal value"); it is a continuous map from $\mathbb{T}^*$ to $\mathbb{R}$ such that

$$z = e^{i \cdot \arg(z)} \text{ for all } z \in \mathbb{T}^* = \mathbb{T} - \{-1\}. \tag{2.7.8}$$

We then choose an arbitrary value θ for the argument of the complex number $\phi(b)$. For $u \in U(b)$, $\phi(u)/\phi(b) \in \mathbb{T}^*$ and so we can define a continuous map α from $U(b)$ to $\mathbb{R}$ by setting

$$\alpha(u) = \theta + \arg(\phi(u)/\phi(b)). \tag{2.7.9}$$

This gives the desired uniform branch.

Hence, in short, *every continuous map ϕ from B to $\mathbb{T}$ canonically defines a covering of B* acted on freely by the discrete group $\mathbb{Z}$ in an obvious way—a good example of a Galois covering in the sense of Sect. 2.5, which the reader will check as an exercise. Besides, the previous arguments show that this covering is (globally) trivial if and only if there exists a uniform branch of the argument of ϕ defined throughout B. As a result, *if ϕ is a continuous map from a simply connected space B to $\mathbb{T}$, then there exists a continuous map α from B to $\mathbb{R}$ such that*

$$\phi(b) = e^{i \cdot \alpha(b)} \text{ for all } b \in B. \tag{2.7.10}$$

This result is frequently used in the classical theory of analytic functions [construction of uniform branches of $\log f(z)$, where f is a holomorphic function without any zeros in a simply connected domain B of $\mathbb{C}$; the choice of a uniform branch α of the argument of the function $z \mapsto f(z)/|f(z)|$ enables us to construct a uniform branch of $\log f(z)$, namely the function $\log |f(z)| + i \cdot \alpha(z)$, where $\log |f(z)| \in \mathbb{R}$; as is well known, the uniform branch $\log |f(z)| + i \cdot \alpha(z)$ is then not only continuous but also holomorphic in B].

Exercise. Let (R, q) be a covering of a space T (the choice of the letters R and T is meant to suggest analogies...) and let ϕ be a continuous map from a space B to T. Consider the set $X \subset B \times R$ of pairs (b, θ) consisting of $b \in B$ and $\theta \in R$ such that $q(\theta) = \phi(b)$. Let p be the projection of X onto B. Show that (X, p) is a covering of B (the "inverse image" of a covering by a continuous map). In what way is this exercise related to the content of the present section? Suppose B is simply connected. Show that there is a continuous map $\psi : B \to R$ such that $\phi = q \circ \psi$. What about the case when B, R, T are groups and ϕ and q homomorphisms?

(b) *Construction of the universal covering of $G = SL_2(\mathbb{R})$.* To obtain a universal covering of $SL_2(\mathbb{R})$, we apply the preceding construction to the map from G to $\mathbb{T}$ given by (2.7.4). The pair (X, p) of (a) is therefore transformed into the set of $(g, \theta) \in G \times \mathbb{R}$ such that $\phi(g) = e^{i\theta}$. With the benefit of hindsight regarding the success of this operation, we will denote this set by $\widetilde{G}$. The map p is then the first projection. We thereby get a closed subset, hence locally compact, of $G \times \mathbb{R}$ and we know that $(\widetilde{G}, p)$ is indeed a covering of the space G. Let us show that it is connected and simply connected. For this, we use the decomposition $G = B_+ \cdot K$, where B_+ and K are the subgroups of G described earlier.

Since $t > 0$, for

$$g = bk = \begin{pmatrix} 1/t & u \\ 0 & t \end{pmatrix} \begin{pmatrix} * & * \\ \sin \omega & \cos \omega \end{pmatrix} = \begin{pmatrix} * & * \\ t \cdot \sin \omega & t \cdot \cos \omega \end{pmatrix}, \tag{2.7.11}$$

we get

$$\phi(bk) = e^{i\omega} = \phi(k). \tag{2.7.12}$$

The map $(b, k) \mapsto bk$ being a homeomorphism, $\widetilde{G}$ is identified (as a set and topo-
logically) with the subset of $B_+ \times K \times \mathbb{R} = B_+ \times \mathbb{T} \times \mathbb{R}$ consisting of triples (b, z, θ)
with $b \in B_+$, $z \in \mathbb{T}$, $\theta \in \mathbb{R}$ and $z = e^{i\theta}$. As $\theta \mapsto (e^{i\theta}, \theta)$ is a homeomorphism
from $\mathbb{R}$ to the set of pairs (z, θ) considered, $\widetilde{G}$ is finally seen to be homeomorphic
to $B_+ \times \mathbb{R}$. The space $\widetilde{G}$ is therefore indeed a universal covering of G up to isomorphism. Its general construction was given in Sect. 2.6. It has a group structure which
we now need to explicitly describe in the context of the above model. We already
know that the map p should be a homomorphism from the *group* $\widetilde{G}$ to G, but this
information is naturally insufficient to compute the composition law of $\widetilde{G}$. To do so,
we first present another realization of G.

(c) *The covering $\widetilde{G}$ as the set of uniform branches of the arguments of functions*
$\mathrm{Am}(cz + d)$. Let $P : \mathrm{Im}(z) > 0$ be the upper half-plane. On $G \times P$ consider the
function

$$J(g, z) = cz + d \quad \text{if} \quad g = \begin{pmatrix} a & b \\ c & d \end{pmatrix}. \tag{2.7.13}$$

It is continuous, without any zeros, and an easy computation shows that[4]

$$J(gg'', z) = J(g', g''(z))J(g'', z), \tag{2.7.14}$$

where G acts on P in the usual way:

$$g(z) = \frac{az + b}{cz + d}. \tag{2.7.15}$$

A formula analogous to (2.7.14) follows if the ($\mathbb{C}^*$-valued) function J is replaced
by the continuous map A from $G \times P$ to $\mathbb{T}$ given by

$$A(g, z) = J(g, z)/|J(g, z)| = \mathrm{Am}(cz + d). \tag{2.7.16}$$

As $G \times P$ is not simply connected, a uniform branch of the argument of the function
A cannot be globally defined, but this becomes feasible if G is replaced by $\widetilde{G}$ since
$\widetilde{G} \times P$ is simply connected. As $A(e, i) = 1$, *there exists a unique continuous map ω
from $\widetilde{G} \times P$ to $\mathbb{R}$ satisfying the conditions*

$$A(p(\gamma), z) = e^{i \cdot \omega(\gamma, z)} \quad \text{for all } \gamma \in \widetilde{G} \text{ and } z \in P, \tag{2.7.17}$$

$$\omega(\widetilde{e}, i) = 0 \tag{2.7.18}$$

where $\widetilde{e}$ is the "base" point of $\widetilde{G}$ defined by

$$\widetilde{e} = \left(\begin{pmatrix} 1 & 0 \\ 0 & 1 \end{pmatrix}, 0 \right). \tag{2.7.19}$$

---

[4]The similarity with equality $d(g'g''z)/dz = d(g'g''z)/d(g''z) \times d(g''z)/dz$ is not entirely a coincidence.

Note that a more general equality

$$\omega(\gamma, i) = \theta \text{ if } \gamma = (g, \theta) \tag{2.7.20}$$

holds because the maps $\gamma \mapsto \omega(\gamma, i)$ and $\gamma = (g, \theta) \mapsto \theta$ from $\widetilde{G}$ to $\mathbb{R}$ are obviously two uniform branches of the argument of the function $(g, \theta) \mapsto A(g, i)$, and as they coincide for $\gamma = \widetilde{e}$, they are equal everywhere.

Formula (2.7.17) also shows that, for a given γ, the map $z \mapsto \omega(\gamma, z)$ from P to $\mathbb{R}$ is a uniform branch of the argument of the map $z \mapsto \mathrm{Am}(cz + d)$ from P to $\mathbb{T}$, where $p(\gamma) = \begin{pmatrix} a & b \\ c & d \end{pmatrix}$. Hence the pair (g, α) consisting of the matrix $g = p(\gamma)$ and the function $\alpha(z) = \omega(\gamma, z)$ can be associated to γ. *We thereby get a bijection from $\widetilde{G}$ onto the set of pairs (g, α) consisting of $g \in G$ and of a uniform branch $z \mapsto \alpha(z)$ of the argument of $z \mapsto A(g, z)$*. Indeed, consider such a pair (g, α); if it is the image of an element γ of $\widetilde{G}$, it is necessarily of the form $\gamma = (g, \theta)$, where θ is a solution of $e^{i\theta} = \phi(g) = \mathrm{Am}(ci + d) = w(g, i)$. Hence, it is a matter of showing that there is a unique solution for which $\omega(\gamma, z) = \alpha(z)$ for all z. But the other two factors, being uniform branches of the argument of the same function—namely $z \mapsto \mathrm{Am}(cz + d)$, only differ by a constant multiple of 2π, so that to check that they are equal for all z, it suffices to do so for $z = i$. Equality (2.7.20) then shows that we only need to choose $\theta = \alpha(i)$, which indeed determines the pair (g, θ) unambiguously.

Let γ' and γ'' then be elements of $\widetilde{G}$, and consider the pairs (g', α') and (g'', α'') corresponding to them in the previous construction. The equalities

$$A(g', z) = e^{i \cdot \alpha'(z)}, \quad A(g'', z) = e^{i \cdot \alpha''(z)} \tag{2.7.21}$$

and the equality analogous to (2.7.13) for the function A then imply that

$$A(g'g'', z) = e^{i \cdot \alpha(z),} \text{ where } \alpha(z) = \alpha'(g''(z)) + \alpha''(z). \tag{2.7.22}$$

The new function α is obviously a uniform branch of the argument of the function $z \mapsto A(g'g'', z)$. As a consequence, the pair $(g'g'', \alpha)$ corresponds to an element γ of $\widetilde{G}$. Hence if $\widetilde{G}$ is identified with the set of pairs (g, α), it is possible to define a composition law on $\widetilde{G}$ by setting

$$(g', \alpha')(g'', \alpha'') = (g'g'', \alpha) \quad \text{where} \quad \alpha(z) = \alpha'(g''(z)) + \alpha''(z). \tag{2.7.23}$$

The last part of the construction consists in showing that this *composition law is indeed the one following from the general constructions of Sect.* 2.6, a result that, on the face of it, is not at all obvious since it is defined using *paths in G* that have not yet been mentioned.

Exercise. A group G acts on a set P. Let F be the set of maps from P to a (not necessarily abelian) group R. Define a composition law on $G \times F$ by setting

$(g', \alpha')(g'', \alpha'') = (g'g'', \alpha)$, where $\alpha(z) = \alpha'(g''(z)) \cdot \alpha''(z)$ for all $z \in P$. Show that $G \times F$ is a group.

(d) *Compatibility with path multiplication.* Formula (2.7.23) enabled us to transform the set $\widetilde{G}$ into the set whose identity element is obviously $\widetilde{e}$. Defining the composition law (2.7.23) by involving the function ω considered above is both useful and easy. Indeed, if (g', α') and (g'', α'') correspond to elements γ' and γ'' of $\widetilde{G}$, then, as seen, by construction,

$$g' = p(\gamma'), \quad g'' = p(\gamma''), \quad \alpha'(z) = \omega(\gamma', z), \quad \alpha''(z) = \omega(\gamma'', z). \qquad (2.7.24)$$

Denote the product given in (2.7.23) by $\gamma'\gamma'' = \gamma$, so that γ corresponds to the pair $(g'g'', \alpha)$ given by the formula in question. Therefore

$$\omega(\gamma, z) = \alpha(z) = \omega(\gamma', g''(z)) + \omega(\gamma'', z). \qquad (2.7.25)$$

Hence if $\widetilde{G}$ is made to act on P by setting

$$\gamma(z) = g(z) \quad \text{where } g = p(\gamma), \qquad (2.7.26)$$

multiplication (2.7.23) in $\widetilde{G}$ is such that the function ω satisfies the identity

$$\omega(\gamma'\gamma'', z) = \omega(\gamma', \gamma''(z)) + \omega(\gamma'', z), \qquad (2.7.27)$$

whose likeness with (2.7.14) is clear. Besides, (2.7.27) fully determines $\gamma'\gamma''$, given that, in addition,

$$p(\gamma'\gamma'') = p(\gamma')p(\gamma''). \qquad (2.7.28)$$

If indeed we return to the definition of $\widetilde{G}$ as a subset of $G \times \mathbb{R}$, and if we set $\gamma' = (g', \theta')$ and $\gamma'' = (g'', \theta'')$, so that $\theta' = \omega(\gamma', i)$ and $\theta'' = \omega(\gamma'', i)$, conditions (2.7.27) and (2.7.28) show that $\gamma'\gamma'' = (g'g'', \theta)$, with

$$\theta = \omega(\gamma'\gamma'', i) = \omega(\gamma', g''(i)) + \omega(\gamma'', i) = \omega(\gamma', g''(i)) + \theta''. \qquad (2.7.29)$$

Note that $\omega(\gamma', g''(i))$ is the value at $z'' = g''(i)$ of the uniform branch of the argument of the function $z \mapsto \text{Am}(c'z + d')$, where c' and d' are the obvious entries of the matrix g', which takes value θ' at $z = i$.

A comparison with the composition law following from Sect. 2.6 requires the latter to be explicitly formulated. Let us temporarily denote it by $(\gamma', \gamma'') \mapsto \gamma' * \gamma''$ so as to avoid confusion with (2.7.23). It is obtained in the following manner, by setting $\widetilde{e}$ to be its identity.

Let γ' and γ'' be two elements of $\widetilde{G}$. Connect them to $\widetilde{e}$ by the paths

$$\gamma'(t) = (g'(t), \theta'(t)) \quad \text{and} \quad \gamma''(t) = (g''(t), \theta''(t)), \qquad (2.7.30)$$

so that $t \mapsto g'(t)$ and $t \mapsto g''(t)$ are their projections in G. Then there is a unique path $t \mapsto \gamma(t)$ in $\widetilde{G}$ with origin $\widetilde{e}$ whose projection follows the path $t \mapsto g'(t)g''(t)$ (product in the group G) and the extremity $\gamma(1)$ of this lifting is the desired element $\gamma' * \gamma''$ of $\widetilde{G}$.

As (2.7.28) and (2.7.29) determine $\gamma'\gamma''$ and as

$$p(\gamma' * \gamma'') = p(\gamma')p(\gamma''), \tag{2.7.31}$$

to prove $\gamma'\gamma'' = \gamma' * \gamma''$, all that needs to be shown is that

$$\omega(\gamma' * \gamma'', i) = \omega(\gamma', g''(i)) + \omega(\gamma', i). \tag{2.7.32}$$

But let $\gamma(t) = (g'(t)g''(t), \theta(t))$ be the lifting with origin e of the path $g'(t)g''(t)$, so that $\gamma(1) = \gamma' * \gamma''$. More generally, $\gamma(t) = \gamma'(t) * \gamma''(t)$ for all $t \in I$, and hence, instead of proving (2.7.32), we need to show that

$$\omega(\gamma(t), i) = \omega(\gamma'(t), g''(t)(i)) + \omega(\gamma''(t), i). \tag{2.7.33}$$

By (2.7.20) this can be rewritten as

$$\theta(t) = \omega(\gamma'(t), g''(t)(i)) + \theta''(t). \tag{2.7.34}$$

Now, the right-hand side is a *continuous* map from I to $\mathbb{R}$ since every function appearing in it—ω, $t \mapsto \gamma'(t)$, $g \mapsto g(i)$, etc...—are continuous and this map vanishes at $t = 0$ since $g'(0) = g''(0) = e$ and $\theta''(0) = 0$. If

$$(g'(t)g''(t), \omega(\gamma'(t), g''(t)(i)) + \theta''(t)) \in \widetilde{G} \tag{2.7.35}$$

is shown to hold for all $t \in I$, then the continuous map

$$t \mapsto (g'(t)g''(t), \omega(\gamma'(t), g''(t)(i)) + \theta''(t)) \tag{2.7.36}$$

from I to $\widetilde{G}$ will be a path with origin $\widetilde{e}$ lifting the path $t \mapsto g'(t)g''(t)$. As its lifting with origin $\widetilde{e}$ is unique, (2.7.36) will then be the lifting $(g'(t)g''(t), \theta(t))$, and Equality (2.7.34) will then follow. Hence, all that is needed is to check (2.7.35). For this, according to the definition of $\widetilde{G}$ as a subset of $G \times \mathbb{R}$ given in subsection (b), it suffices to show that

$$e^{i[\omega(\gamma'(t), g''(t)(i)) + \theta''(t)]} = \phi[g'(t)g''(t)]. \tag{2.7.37}$$

Now, generally speaking $\phi(g) = \text{Am}(ci + d) = A(g, i)$, where we have inserted function (2.7.16). Then, taking into account identity (2.7.14) for A, it follows that

$$\phi[g'(t)g''(t)] = A[g'(t)g''(t), i] = A[g'(t), g''(t)(i)] \cdot A[g''(t), i]. \tag{2.7.38}$$

So it suffices to separately prove the equalities

$$e^{i \cdot \omega(\gamma'(t), g''(t)(i))} = A[g'(t), g''(t)(i)],$$
$$e^{i \cdot \theta''(t)} = A[g''(t), i]. \tag{2.7.39}$$

But these follow readily from the formula

$$A(p(\gamma), z) = e^{i \cdot \omega(\gamma, z)} \tag{2.7.40}$$

applied to (2.7.39), with $\gamma = \gamma'(t)$ and $z = g''(t)(i)$ and to (2.7.40), with $\gamma = \gamma''(t)$ and $z = i$, completing the proof.

(e) *Conclusions.* As these constructions are not very obvious,[5] it may be useful to summarize the essential points obtained along the way:

(i) As a *set*, $\widetilde{G}$ consists of pairs (g, α), where $g \in G$, and $\alpha : P \mapsto \mathbb{R}$ is a uniform branch of the argument of the function $z \mapsto cz + d$, $g = \begin{pmatrix} a & b \\ c & d \end{pmatrix}$; the map $p : \widetilde{G} \to G$ is given by $p(g, \alpha) = g$;

(ii) as a *group*, $\widetilde{G}$ is endowed with composition law (2.7.22):

$$(g', \alpha')(g'', \alpha'') = (g'g'', \alpha), \quad \text{where} \quad \alpha(z) = \alpha'(g''(z)) + \alpha''(z);$$

(iii) as a *topological space*, $\widetilde{G}$ is identified with a closed subspace of $G \times \mathbb{R}$ under the map

$$(g, \alpha) \mapsto (g, \alpha(i));$$

the image of $\widetilde{G}$ under this map consists of all pairs (g, θ) such that

$$ci + d = |ci + d| e^{i\theta} \quad \text{if} \quad g = \begin{pmatrix} a & b \\ c & d \end{pmatrix}.$$

These data fully define $\widetilde{G}$ and its structures, and the rest can be forgotten...

---

[5] They do not appear to be in the published literature. The construction of $\widetilde{G}$ as the set of pairs (g, α) with composition law (2.7.23) has been known to experts for a long time, but for experts, of which I am supposed to be one, this construction obviously leads to a genuine *covering* of the group G. It assumes that a topology is established on the set of these pairs (g, α), that its compatibility with the composition law (2.7.23) holds, and finally that so does the "local triviality" condition of the coverings for the map p. Hence, these "trivial verifications" are usually omitted as it is reckoned that, if a central extension of G by $\mathbb{Z}$ is (abstractly) constructed, Providence will provide the topology that will turn it into a universal covering of G. This expectation is of course fully justified in hindsight, which explains why the detailed construction is little more than a tedious exercise for "beginners".

Exercise. Show that the inverse image $\widetilde{K}$ of K in $\widetilde{G}$ is isomorphic to $\mathbb{R}$, that the subgroup B_+ is isomorphically embedded in $\widetilde{G}$ by $b \mapsto (b, 0)$, where 0 is the map $z \mapsto 0$ from P to $\mathbb{R}$, and that the obvious map from $\widetilde{K} \times B_+$ to $\widetilde{G}$ is a homeomorphism.[6]

Exercise. Let n be a positive integer. Consider the set $\widetilde{G}_n$ of pairs (g, ρ) consisting of $g \in G$ and of a continuous (hence holomorphic) map $\rho : P \mapsto \mathbb{C}^*$ such that

$$\rho(z)^n = cz + d \quad \text{if} \quad g = \begin{pmatrix} a & b \\ c & d \end{pmatrix} \tag{2.7.41}$$

(we know that a holomorphic function with no zeros in a simply connected domain has nth roots for all n). Turn $\widetilde{G}_n$ into a group using the exercise of pages 55–56 and consider the map $\widetilde{G}$ to $\widetilde{G}_n$ (justify this!) which associates to the pair (g, α) the pair (g, ρ) given by

$$\rho(z) = |cz + d|^{1/n} \cdot e^{i \cdot \alpha(z)/n}, \tag{2.7.42}$$

where $|cz + d|^{1/n}$ denotes the *positive* nth root of $|cz + d|$. Show that this map is a surjective homomorphism and that $\widetilde{G}_n$ is identified with the quotient of G by the subgroup $D = \mathbb{Z}$ consisting of the multiples of n (where D is the kernel of $\widetilde{G} \mapsto G$). Deduce that $\widetilde{G}_n$ can be endowed with a topology with respect to which $\widetilde{G}_n$ is a covering of G of order n (which is in fact Galois—we have $G = \widetilde{G}_n/D_n$ where D_n is a cyclic central subgroup of order n).

Exercise. Let r be a real number and Γ a discrete subgroup of G. A (for example) continuous (but not necessarily holomorphic) map f from P to $\mathbb{C}$ is said to be of weight r with respect to Γ if, for all $\gamma = \begin{pmatrix} a & b \\ c & d \end{pmatrix} \in \Gamma$, there is a holomorphic function $M_\gamma(z)$ on P satisfying $|M_\gamma(z)| = |cz + d|^r$ and such that

$$f(\gamma(z)) = M_\gamma(z) f(z) \quad \text{for all } z \in P. \tag{2.7.43}$$

The equality

$$M_{\gamma'\gamma''}(z) = M_{\gamma'}(\gamma''(z)) M_{\gamma''}(z) \tag{2.7.44}$$

must obviously be assumed if we want functions f that do not vanish everywhere. As $M_\gamma(z)$ is holomorphic,

$$M_\gamma(z) = |cz + d|^r e^{ir \cdot \lambda(z)}, \tag{2.7.45}$$

where λ is a uniform branch on P of the argument of $cz + d$. Note that this formula determines (2.7.45) up to multiplication by a factor of type $e^{2\pi irn}$, $n \in \mathbb{Z}$.

---

[6]This result could have been proved beforehand since B_+ is simply connected and it would seem that it could have then easily provided a direct construction of $\widetilde{G}$. Unfortunately, we would have needed to express the composition law of $\widetilde{G}$ in terms of the decomposition considered...

Associate to f a function ϕ on $\widetilde{G}$, considered as the set of pairs (g, α), by setting

$$\phi(g, \alpha) = |ci + d|^{-r} e^{-ir \cdot \alpha(i)} \cdot f(g(i)) \quad \text{for all } (g, \alpha) \in \widetilde{G}, \tag{2.7.46}$$

where $g = \begin{pmatrix} a & b \\ c & d \end{pmatrix}$. Show that the function ϕ is multiplied by a constant with absolute value 1 when it undergoes left translation by an element of the discrete subgroup $p^{-1}(\Gamma)$ of $\widetilde{G}$.

Suppose that $r = p/q$ with p, q integers, $q > 0$. Show that, by passing to the quotient, ϕ can be defined on the group $\widetilde{G}_q$ of the previous exercise. In particular, take $r = 1/2$, and for Γ the subgroup of $SL_2(\mathbb{Z})$ generated by $\begin{pmatrix} 1 & 2 \\ 0 & 1 \end{pmatrix}$ and $\begin{pmatrix} 0 & -1 \\ 1 & 0 \end{pmatrix}$, and for f the Jacobi function

$$\theta(z) = \sum e^{\pi i n^2 z}, \tag{2.7.47}$$

for which

$$\theta(z + 2) = \theta(z), \quad \theta(-1/z) = \left(\frac{z}{i}\right)^{\frac{1}{2}} \theta(i) \tag{2.7.48}$$

is known to hold, $(z/i)^{1/2}$ being the uniform branch of the square root of the function z/i on P which has value 1 at $z = i$. Define directly the function corresponding to θ on the "double" covering $\widetilde{G}_2$ of G, and show that when it undergoes a left translation by an element of the inverse image of Γ in $\widetilde{G}_2$, the function obtained is multiplied by a constant equal to $+1$ or -1 (and whose full computation is one of the hardest exercises in the classical theory of modular functions).

Chapter 3
Analytic Properties of Linear Groups

The aim of this chapter is to introduce a number of fundamental techniques of the theory of Lie groups and to do so in the context of matrix groups. In particular, it aims to show how a "Lie algebra" can be associated to any closed subgroup of a group $GL_n(\mathbb{R})$ in such a way that the connected component can be reconstructed and the given group endowed with what will turn out to be the structure of a Lie group.

The content of this chapter is directly inspired by J. von Neumann's article *Über die analytischen Eigenschaften von Gruppen linearer Transformationen und ihrer Darstellungen* (Mathematische Zeitschrift, 30 (1929), p. 3–42; see also J. von Neumann, Collected Works, vol. I, p. 509–548). Except for its language (von Neumann being Hungarian, it would be wrong to criticize his German, whose importance at the time was almost equal to that of English, or rather American, today), this article is probably the best possible introduction to Lie groups, on the one hand, because of its extraordinary modernity—it is the first "proper" presentation of the topic—, and on the other, because it contains direct proofs of some of the least obvious results of the general theory of Lie groups.

Sections 3.1, 3.2, 3.3, 3.4, 3.6, 3.7 and 3.8 of this chapter present von Neumann's work. The reader can start with these, leaving aside Sect. 3.5 (which extends the results to groups "locally isomorphic" to a linear group, i.e. to the category of Lie groups, as can be proved once the entire topic has been covered), Sects. 3.9 and 3.10, which are a preparation for Chap. 6.

3.1 Exponential and Logarithm

There are several possible ways in which the *norm* of a matrix $X = (x_{pq})_{1 \leq p,q \leq n}$ with real or complex entries can be defined so as to satisfy the inequality

$$\|XY\| \leq \|X\| \cdot \|Y\|, \tag{3.1.1}$$

© Springer International Publishing AG 2017
R. Godement, *Introduction to the Theory of Lie Groups*,
Universitext, DOI 10.1007/978-3-319-54375-8_3

which is very well adapted to the study of power series of the form

$$f(X) = \sum a_p X^p. \tag{3.1.2}$$

Among the norms satisfying (3.1.1), let us mention the Frobenius function

$$X \mapsto \left(\sum |x_{pq}|^2\right)^{\frac{1}{2}}, \tag{3.1.3}$$

the function

$$\|X\| = n \cdot \sup |x_{pq}|, \tag{3.1.4}$$

for which

$$|x_{pq}| \le \frac{1}{n}\|X\| \tag{3.1.5}$$

always holds, and finally any of the norms obtained by identifying a real $n \times n$ matrix with a linear map on $\mathbb{R}^n$ by endowing $\mathbb{R}^n$ with an arbitrary norm and hence by defining, as is done for every normed vector space, the norm of a (continuous—but this condition is superfluous in finite dimensions) linear map. In what follows, we will adopt definition (3.1.4); it simplifies many arguments as will be shortly seen. Von Neumann uses Frobenius' definition.

As remarked by him at the beginning of his article, and by others before him, relation (3.1.1) immediately implies absolute convergence of the following power series[1]:

$$\exp(X) = \sum X^n/n!, \qquad\qquad X \in M_n(\mathbb{R}) \tag{3.1.6}$$
$$\log(g) = \sum (-1)^{n-1}(g-1)^n/n, \qquad \|g-1\| < 1. \tag{3.1.7}$$

The reader will have no trouble checking not only that these functions are continuous on their domain of definition (uniform continuity on every compact domain), but also finding more precise inequalities. For example, for continuity at $X = 0$ (resp. $g = 1$) of the exponential and of the logarithm, we have

$$\|\exp(X) - 1\| \le e^{\|X\|} - 1, \quad \|\log(g)\| \le \log \frac{1}{1 - \|g-1\|} \text{ if } \|g-1\| < 1; \tag{3.1.8}$$

more generally,

$$\|\exp(A) - \exp(B)\| \le e^\rho \|A - B\| \quad \text{if} \quad \|A\|, \|B\| \le \rho, \tag{3.1.9}$$

---

[1] The letters X and g are commonly used in the theory of Lie groups. The aim is to suggest or recall the fact that the logarithm is defined in the neighbourhood of the identity element of a Lie group—notably $GL_n(\mathbb{R})$—whereas the exponential is defined everywhere on its Lie *algebra*. The reason for this notation will become clearer in Chaps. 5 and 6.

$$\| \log(x) - \log(y) \| \le \frac{1}{1-q} \|x - y\| \quad \text{if} \quad \|x - 1\|, \|y - 1\| \le q < 1. \quad (3.1.10)$$

These inequalities follow readily from

$$\|A^p - B^p\| \le pR^{p-1}\|A - B\|, \quad (3.1.11)$$

which holds for $\|A\|, \|B\| \le R$, and which can be proved by induction on p. The functions exp and log (with values in the finite-dimensional vector space $M_n(\mathbb{R})$ over $\mathbb{R}$) are next shown to be *analytic* on their domains of definition—a point which is not made clear by von Neumann (or it might have seemed too obvious to him to state explicitly).

First recall that a function f defined on an open subset U of $\mathbb{R}^p$ and with values in a finite-dimensional real vector space is said to be *analytic* on U if, for all $a \in U$, there is an absolutely convergent multiple power series expansion

$$f(a + x) = \sum c_\alpha(a)x^\alpha, \quad \text{where } x^\alpha = x_1^{\alpha_1} \cdots x_p^{\alpha_p} \text{ if } \alpha = (\alpha_1 \cdots, \alpha_p),$$

which holds whenever the coordinates x_i of x are sufficiently small. An analytic function is obviously C^∞. It is equally obvious that the converse is false. Similarly to the classical case of a single variable, the following properties can be easily proved: if a power series $\sum c_\alpha x^\alpha$ converges absolutely on a set of type $|x_i| < \rho_i$, its sum is analytic on this domain. If there are open sets $U \subset \mathbb{R}^q$ and $V \subset \mathbb{R}^q$ and analytic maps $f : U \to V$ and $g : V \to \mathbb{R}^n$, then the map $g \circ f$ from U to $\mathbb{R}^n$ is also analytic (see J. Dieudonnée, *Eléments d'Analyse*, Sect. IX, Chap. 1 to 3). There are similar definitions and results over $\mathbb{C}$, the method of proof (multiple power series bounded above by geometric series) remaining valid for all complete valuation fields.

Therefore, when the space $\mathbb{R}^p$ considered is the algebra $M_n(\mathbb{R})$ of $n \times n$ matrices, the analyticity of a function $f(X)$ defined on an open subset appears to suppose that $f(X)$ has not just a series expansion like (3.1.2), but a power series expansion with respect to the variables x_{pq}, the entries of the matrix X. Indeed, series (3.1.2) is evidently obtained from the corresponding series in x_{pq} by grouping together the terms which could, on the face of it, turn a multiple series that to begin with was not absolutely convergent, into one that is. Fortunately, this is not the case:

Lemma *If the matrix power series* $f(X) = \sum a_p X^p$ *converges absolutely for all* $X \in M_n(\mathbb{R})$ *such that* $\|X\| < \rho$, *its sum is an analytic function on this open subset.*

The entries of $f(X)$ are indeed given by the multiple power series

$$f(x)_{ij} = \sum a_p x_{ik_1} x_{k_1 k_2} \cdots x_{k_{p-1} j} \quad (3.1.12)$$

where the sum is over $k_1, \ldots k_{p-1}$ and p. Inequality (3.1.5) shows that the general term of the preceding multiple series is bounded above by $|a_p| \|X\|^p / n^p$. On the other hand, there are n^{p-1} terms of degree p in (3.1.12). To prove that (3.1.12) is absolute convergent, it therefore suffices to check that so is the series $\sum n^{p-1}|a_p| \cdot \|X\|^p/n^p$, i.e. the given series $\sum a_p X^p$. As the open set $\|x\| < \rho$ is the cube $|x_{pq}| < \rho/n$, the lemma follows immediately.

That the functions exp and log are analytic on the open sets on which they are clearly defined follows from the previous lemma.

Let us next show that, in some sense, the maps exp and log are mutual inverses:

Theorem 1

$$\boxed{\begin{aligned} \exp(\log(g)) &= g \quad for \quad \|g - 1\| < 1, \\ \log(\exp(X)) &= X \quad for \quad \|X\| < \log 2. \end{aligned}}$$

$$\text{(3.1.13)}$$
$$\text{(3.1.14)}$$

The first formula is self-explanatory since exp is defined everywhere. To make the second one well-defined, we take inequality (3.1.8) into account. It shows that $\| \exp(X) - 1\| < 1$ for $\|X\| < \log 2$, which makes the log of $\exp(X)$ well-defined. For the proof of Theorem 1, we merely reproduce von Neumann's:

The expressions $\exp(\log(X)) - X$ and $\log(\exp(X)) - X$ need to be shown to be trivial. As the power series of these functions converge (according to the assumptions made on X), these expressions are the limits of

$$\sum_{\mu=0}^{r} \frac{1}{\mu!}\left(\sum_{\nu=1}^{s} \frac{(-1)^{\nu-1}}{\nu}(X-1)^{\nu}\right)^{\mu} - X \qquad (3.1.15)$$

resp.

$$\sum_{\mu=1}^{r} \frac{(-1)^{\mu-1}}{\mu}\left(\sum_{\nu=1}^{s} \frac{1}{\nu!}X^{\nu}\right)^{\mu} - X \qquad (3.1.16)$$

when s and then r are made to approach infinity. Hence in both cases it suffices to show that these expressions are arbitrarily small for sufficiently large r and s, i.e. that for all $\varepsilon > 0$, there exists $t = t(\varepsilon)$ such that these expressions are $< \varepsilon$ whenever $r \geq t$ and $s \geq t$. But these expressions are polynomials of degree rs in $X - 1$ (resp. X) and when $r \geq t$ and $s \geq t$, the coefficients of the terms of degree $\leq t$ in these expressions are the same as those in the expansions of the functions $e^{\log(1+x)} - 1 - x$ (resp. $\log(e^x) - x$), where x is a scalar. They are therefore zero. Moreover, in absolute value, all the coefficients of these terms are less than the corresponding coefficients of the power series [replace X by $x + 1$ in (3.1.15)]

$$\sum_{\mu=0}^{\infty} \frac{1}{\mu!}\left(\sum_{\nu=1}^{\infty} \frac{1}{\nu}x^{\nu}\right)^{\mu} + 1 + x = e^{\log\frac{1}{1-x}} + 1 + x = 1 + x + \frac{1}{1-x}$$

resp.

$$\sum_{\mu=1}^{\infty} \frac{1}{\mu} \left(\sum_{\nu=0}^{\infty} \frac{1}{\nu!} x^\nu \right)^\mu + x = \log \frac{1}{1 - (e^x - 1)} + x = x + \log \frac{1}{2 - e^x}.$$

As these functions are holomorphic for $|x| < 1$ (resp. $|x| < \log 2$), their power series converge for $|x| < 1$ (resp. $|x| < \log 2$). For every a such that $0 < a < 1$ (resp. $0 < a < \log 2$), the coefficient of x^p in each of them is therefore $\leq c_a/a^p$, where c_a only depends on a. (Moreover, almost all the coefficients in the first one $= 1$.) This is especially true for those coefficients of interest to us.

"So for $\|X - 1\| = a' < a < 1$ (resp. $\|X\| = a' < a < \log 2$), the absolute value of each of the expressions of interest to us is

$$\sum_{p=t+1}^{rs} \frac{c_a}{a^p} a'^p = c_a \sum_{p=t+1}^{rs} \left(\frac{a'}{a} \right)^p \leq \frac{a c_a}{a - a'} \left(\frac{a'}{a} \right)^{t+1}$$

whenever $r, s \geq t$. For sufficiently large t this is arbitrarily small, proving our claim. It therefore holds whenever such a number a exists, i.e. for all X with $\|x - 1\| < 1$ (resp. $\|X\| < \log 2$)."

Corollary of Theorem 1 *Let U be the ball $\|X\| < \log 2$ in $M_n(\mathbb{R})$ and V its image under the exponential map. Then V is an open neighbourhood of e in $GL_n(\mathbb{R})$, the exponential map is an analytic homeomorphism from U onto V, and the inverse homeomorphism is the logarithmic map from V onto U.*

Indeed, V is contained in the open set $\|g - 1\| < 1$ on which log is defined, and Theorem 1 shows that V is the inverse image of U under this map, and so is open. The other assertions are trivial.

Let us now prove the "well-known" identities.

By formal calculations that essentially reduce to the binomial formula,

$$\boxed{\exp(X + Y) = \exp(X)\exp(Y) \text{ if } XY = YX} \tag{3.1.17}$$

follows from the multiplication formula of absolutely convergent series, which applies to vector or matrix series as well as to numerical series. We next show that

$$\boxed{\log(xy) = \log(x) + \log(y)} \tag{3.1.18}$$

if

$$xy = yx \text{ and } \|x - 1\|, \|y - 1\|, \|xy - 1\| < 1.$$

This is more tricky. We first show that (3.1.18) holds in the particular case when $\|x - 1\|, \|y - 1\| < 1 - 1/\sqrt{2}$. Indeed, by (3.1.8), we then have

$$\| \log(x) \|, \| \log(y) \| < \log \frac{1}{1 - (1 - 1/\sqrt{2})} = \frac{1}{2} \log 2 \tag{3.1.19}$$

and so $\| \log(x) + \log(y) \| < \log 2$. Hence, by (3.1.14),

$$\log(x) + \log(y) = \log[\exp(\log(x) + \log(y))], \tag{3.1.20}$$

and since $\log(x)$ and $\log(y)$ commute as limits of polynomials in x and y, the right-hand side clearly equals $\log[\exp(\log(x)) \exp(\log(y))] = \log(xy)$ by once more appealing to (3.1.14).

To go from here to the general case, first note that if x and y satisfy the assumptions of (3.1.18), then the same is true for matrices $x' = 1 + t(x - 1)$ and $y' = 1 + t(y - 1)$ for any scalar t such that $0 \le t \le 1$. Inequalities $\|x' - 1\| < 1$ and $\|y' - 1\| < 1$ are indeed obvious, and moreover,

$$\begin{aligned}
\|x'y' - 1\| &= \|t(x - 1) + t(y - 1) + t^2(x - 1)(y - 1)\| \\
&= \|t(xy - 1) - (t - t^2)(x - 1)(y - 1)\| \\
&< t + (t - t^2) = 1 - (1 - t)^2 \le 1
\end{aligned} \tag{3.1.21}$$

(strict inequality supposes $0 < t \le 1$, but the case $t = 0$ is directly obvious). Hence the function

$$\begin{aligned}
F(t) = \log([1 + t(x - 1)][1 + t(y - 1)]) &- \log(1 + t(x - 1)) \\
&- \log(1 + t(y - 1))
\end{aligned} \tag{3.1.22}$$

can be defined for $0 \le t \le 1$, and it is clearly continuous in this interval. Let us show it is analytic for $0 < t < 1$. This is clearly the case for the last two terms on the right-hand side, whose power series expansions with respect to t are given by (3.1.7). For the first log, observe that it consists of the analytic map

$$t \mapsto [1 + t(x - 1)][1 + t(y - 1)] \tag{3.1.23}$$

from $0 < t < 1$ to the open set $\|z - 1\| < 1$ and of log, defined and analytic on this open set. The well-known substitution theorem[2] then implies our claim, namely that it is analytic on the interval $0 < t < 1$.

But since (3.1.18) has been proved to hold when $\|x - 1\|$ and $\|y - 1\|$ are sufficiently small, it is clear that $F(t) = 0$ for $0 \le t \le r$ where $r > 0$ is an appropriately chosen number. As F is analytic on $0 < t < 1$, it follows that $F(t) = 0$ for $0 \le t < 1$, and

---

[2]Concerning this matter, von N seems to believe that by substituting (3.1.23) into (3.1.7) gives a power series convergent in $0 \le t \le 1$. However, this amounts to the substitution, in a convergent power series, of a second degree polynomial of type $At + Bt^2$ without constant term. If the given series converges for $\|x\| < 1$ (for example), the power series in t formally resulting from this substitution only converges for

$$\|At\| + \|Bt^2\| < 1,$$

but not necessarily for $\|At + Bt^2\| < 1$. The substitution of one (or many) power series in another one supposes the convergence of the series obtained by replacing *all* the coefficients of *every* series considered by their absolute values so as to avoid miraculous convergence due to accidental term corrections...

as F is continuous in $0 \leq t \leq 1$, we deduce that $F(1) = 0$, which is precisely the desired relation (3.1.18).

Relation (3.1.17) obviously shows that exp *maps* $M_n(\mathbb{R})$ *to the group* $GL_n(\mathbb{R})$, with $\exp(X)^{-1} = \exp(-X)$. Note that, the neighbourhood of 1 on which the log is defined, namely $\|g - 1\| < 1$, is in fact itself contained in $GL_n(\mathbb{R})$ thanks to the geometric series which provides the inverse of $1 - Y$ for $\|Y\| < 1$. The same results hold if $\mathbb{R}$ is replaced by $\mathbb{C}$ (or by any complete valuation field).

The equalities

$$\exp(gXg^{-1}) = g \cdot \exp(X) \cdot g^{-1}, \quad X \in M_n(\mathbb{R}), g \in GL_n(\mathbb{R}) \qquad (3.1.24)$$

$$\log(gxg^{-1}) = g \cdot \log(X) \cdot g^{-1} \quad \text{if } \|x - 1\| < 1 \text{ and } \|gxg^{-1} - 1\| < 1 \quad (3.1.25)$$

are also frequently used. They follow readily from the power series expansions. Note the strange conditions under which (3.1.25) holds. Indeed, if $\|gxg^{-1} - 1\| < 1$, then the series $\sum (-1)^{p-1}(gxg^{-1} - 1)^p/p$ converges absolutely. As it can be derived by applying the automorphism $x \mapsto gxg^{-1}$ to the series $\sum (-1)^{p-1}(x - 1)^p/p$ defining $\log(x)$, (as the limit of its partial sums), it converges even without the assumption $\|x - 1\| < 1$. This obviously suggests that the condition $\|x - 1\| < 1$ used to define log is far too restrictive and that it should be possible—as we have already done for $n = 1$—to define the logarithm of a matrix under more relaxed conditions.

Before moving on from generalities about the exponential function, following von Neumann, note that

$$\log(\exp(X)\exp(Y)) = X + Y + O(\|X\| \cdot \|Y\|) \qquad (3.1.26)$$

when X and Y approach 0. This would be clear if we assumed $XY = XY$, but we naturally do not make such a supposition.

First of all, by (3.1.4)

$$\log(\exp(X)\exp(Y)) - X - Y = \log(\exp(X)\exp(Y)) - \log(\exp(X + Y)) \quad (3.1.27)$$

clearly holds provided X and Y are sufficiently near 0. Moreover, for any number q such that $0 < q < 1$,

$$\|\exp(X)\exp(Y) - 1\| < q, \quad \|\exp(X + Y) - 1\| < q \qquad (3.1.28)$$

also holds provided that here too, X and Y are sufficiently near zero. By (3.1.10),

$$\|\log(\exp(X)\exp(Y)) - \log(\exp(X + Y))\| \\ = O(\|\exp(X)\exp(Y) - \exp(X + Y)\|) \qquad (3.1.29)$$

when X and Y approach 0. Taking into account (3.1.27), we need to show that

$$\| \exp(X) \exp(Y) - \exp(X + Y) \| = O(\|X\| \cdot \|Y\|) \qquad (3.1.30)$$

as X and Y approach 0. However, $\exp(X) \exp(Y) - \exp(X + Y)$ is the limit of the expression

$$\sum_{p=0}^{r} \frac{1}{p!} X^p \sum_{q=0}^{r} \frac{1}{q!} Y^q - \sum_{n=0}^{r} \frac{1}{n!} (X + Y)^n \qquad (3.1.31)$$

as r increases indefinitely. Once all the products have been done (without permuting the terms!), the terms in X^p or in Y^q clearly disappear, and all the remaining monomials involve X and Y. Group together all the monomials containing X k times and Y h times (in any order). In the product of the first two sums appearing in (3.1.31), there is at least one such monomial, namely $X^k Y^h$, with coefficient $1/k!h!$ For reasons of degrees, in $\sum (X + Y)^n / n!$, the monomials considered necessarily come from the expansion of $(X + Y)^{k+h} / (k + h)!$, and there are obviously $\binom{k + h}{k} = \frac{(k+h)!}{k!h!}$ such terms. On account of the coefficients involved in the calculations, the norm of the sum of the (non-commutative) monomials of degree k in X and h in Y is bounded above by $\|X\|^k \|Y\|^h$ multiplied by the factor

$$\frac{1}{k!h!} + \frac{(k + h)!}{k!h!} \frac{1}{(k + h)!} = 2/k!h! \qquad (3.1.32)$$

The terms with either $k = 0$ or $h = 0$ have been seen to cancel each other. Hence the norm of the difference (3.1.31) is bounded above by

$$2 \sum_{k,h=1}^{r} \|X\|^k \|Y\|^h / k!h! \leq 2(e^{\|X\|} - 1)(e^{\|y\|} - 1), \qquad (3.1.33)$$

and since $e^t - 1 = O(t)$ as t approaches 0, (3.1.30) follows readily.

[In 6 we will give an explicit formula in exponential form for $\exp(X) \cdot \exp(Y)$ even when X and Y do not commute, namely

$$\begin{aligned}
\log(\exp(X) \exp(Y)) = {} & X + Y + \frac{1}{2}[X, Y] + \frac{1}{12}[X, [X, Y]] \\
& + \frac{1}{12}[Y, [Y, X]] + \cdots,
\end{aligned} \qquad (3.1.34)$$

where, $[X, Y]$ is defined to be $XY - YX$. This means that, as X and Y tend to 0, the "principal part" of $\log(\exp(X) \exp(Y)) - X - Y$ is $\frac{1}{2}[X, Y]$, whence (3.1.26)!]

If von N's article is strictly followed, it would now be requisite to show how to associate to each closed subgroup H of $GL_n(\mathbb{R})$, a "Lie subalgebra" of $M_n(\mathbb{R})$ whose image under exp generates the connected component $H°$ of H. We will do so in Sect. 3.4. Beforehand, we prove some properties of "1-parameter subgroups" of $GL_n(\mathbb{R})$. These will anyhow prove to be crucial, and will pave the way towards the general theory of Lie groups.

3.2 One-Parameter Subgroups of $GL_n(\mathbb{R})$

A *one-parameter subgroup* of a topological group G is by definition a continuous homomorphism γ from the additive group $\mathbb{R}$ to G. Hence,

$$\gamma(s+t) = \gamma(s)\gamma(t), \quad \gamma(0) = e, \quad \gamma(t)^{-1} = \gamma(-t). \tag{3.2.1}$$

For example, for any $X \in M_n(\mathbb{R}) = \mathfrak{g}$ (small Gothic g), formula

$$\boxed{\gamma_X(t) = \exp(tX) = 1 + tX + t^2X^2/2! + \cdots} \tag{3.2.2}$$

defines a one-parameter subgroup of $G = GL_n(\mathbb{R})$. The map γ_X is not just continuous, but also analytic, and clearly,

$$\frac{d}{dt}\gamma_X(t)\Big|_{t=0} = X. \tag{3.2.3}$$

Conversely, let γ be a one-parameter subgroup of $GL_n(\mathbb{R})$. For any infinitely differentiable function ϕ on $\mathbb{R}$ with compact support,

$$\int \phi(t-s)\gamma(s)ds = \int \phi(s)\gamma(t-s)ds = \gamma(t)A, \tag{3.2.4}$$

where $A = \int \phi(s)\gamma(s)^{-1}ds$ is a constant operator on $\mathbb{R}^n$. But the left-hand side of (3.2.4) is obviously C^∞ (differentiation under the summation sign). Hence so is the function γ as long as ϕ may be chosen in such a way that the operator A is invertible. To see this, choose $\delta > 0$ such that $\|\gamma(s) - 1\| \leq \frac{1}{2}$ for $|s| \leq \delta$, then for ϕ take a positive function vanishing outside $[-\delta, \delta]$ whose total integral is 1. As the integral of ϕ is 1,

$$A - 1 = A - \int \phi(s)\gamma(0)ds = \int \phi(s)[\gamma(s)^{-1} - 1]ds, \tag{3.2.5}$$

and as ϕ is positive, it follows that

$$\|A - 1\| \leq \int \phi(s)\|\gamma(-s) - 1\|ds = \int_{-\delta}^{+\delta} \cdots \leq \frac{1}{2}\int \phi(s)ds = \frac{1}{2}. \tag{3.2.6}$$

Since $\frac{1}{2} < 1$, A is invertible.

The one-parameter subgroup γ is therefore C^∞ and so differentiation with respect to the first relation (3.2.1) is feasible. Setting $s = 0$ in the result and $\gamma'(0) = X$ clearly gives

$$\gamma'(t) = X \cdot \gamma(t). \tag{3.2.7}$$

Hence

$$[\exp(-tX)\gamma(t)]' = -X \cdot \exp(-tX)\gamma(t) + \exp(-tX)\gamma'(t) = 0. \tag{3.2.8}$$

As a consequence, $\gamma(t) = C \cdot \exp(tX)$ with a constant C equal to 1. This can be seen by taking $t = 0$. In conclusion:

Theorem 2 *Every one-parameter subgroup of $GL_n(\mathbb{R})$ is of type $\gamma(t) = \exp(tX)$ with a well-defined $X \in M_n(\mathbb{R})$ given by the equality*

$$X = \frac{d}{dt}\gamma(t)\Big|_{t=0}. \tag{3.2.9}$$

This notably shows that every *continuous* homomorphism from $\mathbb{R}$ to $GL_n(\mathbb{R})$ is *analytic*, a result which will be generalized later. (Von N, who was not acquainted with formula (3.2.4), gives a much more complicated proof. It will be given further down).

Exercise. There is an open neighbourhood U of e in $GL_n(\mathbb{R})$ with the following property: for every $u \in U$, there exists a unique one-parameter subgroup γ such that $\gamma(1) = u$ and $\gamma(t) \in U$ for $0 \le t \le 1$. [Take $U = \exp(\frac{1}{2}S)$, where S is a convex open neighbourhood of 0 mapped homeomorphically by exp onto an open subset of $GL_n(\mathbb{R})$, take X such that $\exp(tX) \in U$ for $0 \le t \le 1$ and show that $tX \in S$ implies $tX \in \frac{1}{2}S$.]

3.3 Limits of Products and Commutators

In what follows we set

$$[X, Y] = XY - YX \tag{3.3.1}$$

for all $X, Y \in \mathfrak{g}$. This endows $\mathfrak{g}$ with the structure of a *Lie algebra*, in other words, of a vector space (over the given base field, which may not be $\mathbb{R}$) endowed with a bilinear composition law $(X, Y) \mapsto [X, Y]$ satisfying the following identities:

$$[X, Y] = -[Y, X], \quad [X, [Y, Z]] + [Y, [Z, X]] + [Z, [X, Y]] = 0. \tag{3.3.2}$$

The next result connects these operations to the composition law of $GL_n(\mathbb{R})$:

Theorem 3 *Let X and Y be matrices in $M_n(\mathbb{R})$. Then,*

$$\gamma_{X+Y}(t) = \lim \left(\gamma_X\left(\frac{t}{n}\right)\gamma_Y\left(\frac{t}{n}\right) \right)^n, \tag{3.3.3}$$

$$\gamma_{[X,Y]}(t^2) = \lim \left(\gamma_X\left(\frac{t}{n}\right)\gamma_Y\left(\frac{t}{n}\right)\gamma_X\left(\frac{t}{n}\right)^{-1}\gamma_Y\left(\frac{t}{n}\right)^{-1} \right)^{n^2}. \tag{3.3.4}$$

Proof of (3.3.3): It suffices to prove it for $t = 1$ (otherwise multiply X and Y by t). For large n, the exponential series then give the partial expansions

$$\gamma_X(1/n) = 1 + X/n + X^2/2n^2 + O(1/n^3),$$
$$\gamma_Y(1/n) = 1 + Y/n + Y^2/2n^2 + O(1/n^3). \tag{3.3.5}$$

Multiplying term by term,

$$\gamma_X(1/n)\gamma_Y(1/n) = 1 + \frac{X+Y}{n} + \frac{X^2 + 2XY + Y^2}{2n^2} + O(1/n^3) = z_n. \tag{3.3.6}$$

As z_n approaches 1, $\log z_n$ can be defined for large n by the power series. The latter shows that $\log u = u - 1 + O(\|u - 1\|^2)$ as u tends to 1. So

$$\log z_n = \frac{X+Y}{n} + O(1/n^2) + O(\|z_n - 1\|^2) = \frac{X+Y}{n} + O(1/n^2) \tag{3.3.7}$$

since obviously $\|z_n - 1\| = O(1/n)$. Consequently,

$$z_n^n = \exp(\log z_n)^n = \exp(n.\log z_n) = \exp(X + Y + O(1/n)), \tag{3.3.8}$$

and since the result obtained converges to $\exp(X+Y) = \gamma_{X+Y}(1)$, this proves (3.3.3).

Proof of (3.3.4): We start from (3.3.6) and from the analogous formula

$$\gamma_X(1/n)^{-1}\gamma_Y(1/n)^{-1} = 1 - \frac{X+Y}{n} + \frac{X^2 + 2XY + Y^2}{2n^2} + O(1/n^3). \tag{3.3.9}$$

Multiplying (3.3.6) and (3.3.9) term by term, we see that the terms in $1/n$ cancel each other, while the terms in $1/2n^2$ lead to a fraction of the numerator

$$2(X^2 + 2XY + Y^2) - 2(X + Y)^2 = 2(XY - YX). \tag{3.3.10}$$

The commutator u_n of the elements $\gamma_X(1/n)$ and $\gamma_Y(1/n)$ is therefore

$$u_n = 1 + \frac{[X, Y]}{n^2} + O(1/n^3), \tag{3.3.11}$$

and so, as above,

$$\log u_n = \frac{[X, Y]}{n^2} + O(1/n^3). \tag{3.3.12}$$

Thus $u_n^{n^2} = \exp([X, Y] + O(1/n))$, which converges to $\gamma_{[X,Y]}(1)$ and proves (3.3.4).

These formulas are fundamental. They provide a heuristic proof of the fact that addition in $\mathfrak{g}$ is obtained by "differentiating" multiplication in G, and that the bracket operation is obtained by twice "differentiating" the map $(x, y) \mapsto xyx^{-1}y^{-1}$ which in some sense measures the non-commutativity of multiplication in the group $GL_n(\mathbb{R})$.

Finally, the obvious equality

$$g\gamma_X(t)g^{-1} = \gamma_{gXg^{-1}}(t) \qquad (3.3.13)$$

follows readily from (3.1.24) and completes these formulas.

We next consider an easy but equally basic corollary of these computations:

Theorem 4 *Let H be a closed subgroup of $GL_n(\mathbb{R})$ and $\mathfrak{h}$ the set of $X \in M_n(\mathbb{R})$ such that $\exp(tx) \in H$ for all $t \in \mathbb{R}$. Then $\mathfrak{h}$ is a Lie subalgebra of $M_n(\mathbb{R})$ and is invariant under $X \mapsto hXh^{-1}$ for all $h \in H$.*

Formulas (3.3.3), (3.3.4) as well as equality $\gamma_{sX}(t) = \gamma_X(st)$ highlight the fact that $\mathfrak{h}$ is a vector subspace of $M_n(\mathbb{R})$ and that $[X, Y] \in \mathfrak{h}$ for all $X, Y \in \mathfrak{h}$, proving the first statement. The second one is obvious by (3.3.13).

Exercise. Show that $\log(\exp(tX)\exp(tY)) = t(X + Y) + \frac{1}{2}t^2[X, Y] + O(t^3)$ as $t \to 0$. How does this relate to (3.1.26)?

3.4   Von Neumann's Theorem for a Closed Subgroup

We now return to von Neumann's essay and show that, if H is a *closed* subgroup of $GL_n(\mathbb{R})$ with Lie algebra $\mathfrak{h}$ defined as in Theorem 4, then the image of $\mathfrak{h}$ under the exponential map is obviously contained in the connected component $H°$ of H. The proof will also show that $H°$ is open in H, more precisely that the exponential map transforms every sufficiently small neighbourhood of 0 in $\mathfrak{h}$ into a neighbourhood of 1 in H (hence H has path-connected neighbourhoods of 1, from which it will follow that $H°$ is open in H).[3]

Von Neumann's method consists in defining the Lie algebra of H in a different way from ours (but as will be seen the results are identical). More precisely, von Neumann associates to H the set $\mathfrak{h} \subset M_n(\mathbb{R})$ of matrices X for which there exist a sequence (h_p) of elements of H tending to 1, and a sequence (ε_p) of strictly positive numbers, tending to 0, such that

$$X = \lim \frac{1}{\varepsilon_p}(h_p - 1). \qquad (3.4.1)$$

The similarity with the quotient defining a derivative is naturally no coincidence. For example, if there is a one-parameter subgroup γ_X in H, it is not very hard to see that its infinitesimal generator X will belong to the set $\bar{\mathfrak{h}}$ just defined. It is, therefore, seemingly *larger than* the Lie algebra $\mathfrak{h}$ of Theorem 3, but it will be shortly seen (Lemma 3) that $\mathfrak{h} = \bar{\mathfrak{h}}$.

To prove Theorem 5 stated further down, von Neumann uses a number of clever arguments heralding "Hilbert's fifth problem", some twenty years in advance. We

---

[3]The reader who prefers to first study the concrete example given further down can do so.

follow his presentation step by step, and keep his notation, including o and O—also very much in advance of his time since this notation only became popular in the 1950s.

Lemma 1 *Let h_p be a sequence of matrices tending to 1 and ε_p a sequence of numbers tending to 0. Then the sequences with general terms*

$$\frac{1}{\varepsilon_p}(h_p - 1) \ \text{ and } \ \frac{1}{\varepsilon_p}\log(h_p) \tag{3.4.2}$$

are either both convergent or both divergent, and tend towards the same limit in the former case.

Indeed,

$$\log(h_p) = \log(1 - (1 - h_p)) = h_p - 1 + o(\|h_p - 1\|^2). \tag{3.4.3}$$

If the first sequence in (3.4.2) converges to X, then obviously $h_p - 1 = \varepsilon_p X + o(\varepsilon_p)$. Hence

$$\log(h_p) = \varepsilon_p X + o(\varepsilon_p) + O(\varepsilon_p^2) = \varepsilon_p X + o(\varepsilon_p), \tag{3.4.4}$$

so that the second sequence indeed tends to X. The converse can be similarly proved.

Lemma 2 *For every $X \in \bar{\mathfrak{h}}$, there is a map $t \mapsto h(t)$ from $I = [0, 1]$ to H such that $h(0) = 1$ and*

$$X = \lim \frac{h(t) - 1}{t} = \lim \frac{1}{t}\log(h(t)). \tag{3.4.5}$$

The equivalence of the two limits can be proved as in Lemma 1 and like von N, we will adopt the latter point of view. The function $h(t)$ remains to be constructed.

Choose a sequence of points h_p of H such that (3.4.1) holds. For all t such that $0 < t \leq 1$, choose a natural number $p(t)$ such that $\varepsilon_{p(t)} < t^2$. There is a unique integer $q(t)$ satisfying

$$q(t)\varepsilon_p(t) \leq t < (q(t) + 1)\varepsilon_{p(t)}, \tag{3.4.6}$$

namely

$$q(t) = [t/\varepsilon_{p(t)}], \tag{3.4.7}$$

where $[x]$ denotes the *integral part* of a real number x (i.e. the largest integer at most equal to x). As t tends to zero, $p(t)$ increases indefinitely since the ε_p are strictly positive, and since

$$q(t) + 1 > t/\varepsilon_p(t) > 1/t, \tag{3.4.8}$$

the same holds for $q(t)$. Set

$$h(t) = (h_{p(t)})^{q(t)} \text{ for } 0 < t \leq 1 \text{ and } h(0) = e. \tag{3.4.9}$$

This is obviously a map from I to H. Since $\log(h_p)/\varepsilon_p$ converges, the following successively hold: $\log(h_p)/\varepsilon_p = O(1)$, then $\log(h_p) = O(\varepsilon_p)$, then

$$q(t)\log(h_{p(t)}) = q(t) \cdot O(\varepsilon_{p(t)}) = O(q(t)\varepsilon_{p(t)}) = O(t) \qquad (3.4.10)$$

by (3.4.6). Hence there exists a t_0 for which $t < t_0$ implies

$$\|q(t) \cdot \log(h_{p(t)})\| < \log 2 \text{ and so perforce } \| \log(h_{p(t)})\| < \log 2. \qquad (3.4.11)$$

As seen previously,

$$\log(\exp(X)) = X \text{ for } \|X\| < \log 2 \qquad (3.4.12)$$

always holds. So does $\exp(nX) = \exp(X)^n$. Hence, it follows that[4]:

$$
\begin{aligned}
q(t) \cdot \log(h_{p(t)}) &= \log(\exp[q(t) \cdot \log(h_{p(t)})]) = \log(\exp[\log(h_{p(t)})]^{q(t)}) \\
&= \log[h_{p(t)}^{q(t)}] = \log(h(t)).
\end{aligned} \qquad (3.4.13)
$$

Hence,

$$\frac{1}{t}\log(h(t)) = \frac{q(t)}{t}\log(h_{p(t)}) = \frac{q(t)\varepsilon_{p(t)}}{t} \cdot \frac{1}{\varepsilon_{p(t)}}\log(h_{p(t)}). \qquad (3.4.14)$$

By Lemma 1, $\log(h_{p(t)})/\varepsilon_{p(t)}$ tends to the matrix X. By (3.4.6)

$$1 \geq q(t)\varepsilon_{p(t)}/t > 1 - \varepsilon_{p(t)}/t > 1 - 1/t. \qquad (3.4.15)$$

So $\log(h(t))/t$ is the product of a scalar tending to 1 and a matrix tending to X, whence the lemma.

At this point, von Neumann shows by a rather simple argument[5] that $\bar{\mathfrak{h}}$ is a Lie subalgebra. For us, it is more convenient to show that $\bar{\mathfrak{h}} = \mathfrak{h}$. Now, if $X \in \bar{\mathfrak{h}}$, then clearly $sX \in \bar{\mathfrak{h}}$ for any scalar s, for if $\log(h(t))/t$ tends to X as t tends to 0, then

---

[4] As an aside, note that, in general, whenever $\|n \cdot \log(g)\| < \log 2, \log(g^n) = n \cdot \log(g)$ if $\|g - 1\| < 1$, the proof being that of (3.4.13). The same conclusion follows if we suppose that

$$\|g^p - 1\| < 1 \text{ for } 1 \leq p \leq n,$$

as is seen using (3.1.18).

[5] If $X = \lim(h'(t) - 1)/t$ and $Y = \lim(h''(t) - 1)/t$, then

$$h'(t)h''(t) = (1 + tX + o(t))(1 + tY + o(t)) = 1 + t(X + Y) + o(t),$$

and so $X + Y \in \bar{\mathfrak{h}}$. To check $XY - YX \in \bar{\mathfrak{h}}$, von Neumann observes that, since $h'(t)$ and $h''(t)$ tend to 1,

$$
\begin{aligned}
\lim(h'(t)h''(t)h'(t)^{-1}h''(t)^{-1} - 1)/t^2 &= \lim(h'(t)h''(t) - h''(t)h'(t))/t^2 \\
&= \lim[(h'(t) - 1)(h''(t) - 1) - (h'(t) - 1)(h''(t) \cdot 1)]t^2 = XY - YX.
\end{aligned}
$$

$\log(h(st))/t$ will clearly tend to sX [this argument supposes that $s > 0$ so that $t \mapsto h(st)$ is defined on a non-trivial interval originating at 0; if $s < 0$, note that, for obvious reasons, $X \in \bar{\mathfrak{h}}$ implies $-X \in \bar{\mathfrak{h}}$, and hence $sX \in \bar{\mathfrak{h}}$]. Since $\mathfrak{h} \subset \bar{\mathfrak{h}}$ is obvious, to check $\bar{\mathfrak{h}} = \mathfrak{h}$ it suffices to prove

Lemma 3 $\exp(X) \in H$ for all $X \in \bar{\mathfrak{h}}$.

To do so, choose $h(t) \in H$, $0 \leq t \leq 1$, such that $(h(t) - 1)/t$ tends to X and set $h_p = h(1/p)$, so that $X = \lim p(h_p - 1)$, or (Lemma 1) that $X = \lim p \cdot \log(h_p)$. As a result, $\exp(X) = \lim \exp(p \cdot \log(h_p)) = \lim \exp(\log(h_p))^p$; but for sufficiently large p, $\|h_p - 1\| < 1$ and in consequence, $\exp(\log(h_p)) = h_p$. Hence

$$\exp(X) = \lim h_p^p \qquad (3.4.16)$$

and as H is closed and contains all h_p^p, this completes the proof.

Theorem 5 *Let H be a closed subgroup of $GL_n(\mathbb{R})$ and $\mathfrak{h}$ its Lie algebra. Then every sufficiently small open neighbourhood of 0 in $\mathfrak{h}$ is homeomorphically mapped by the exponential function onto an open neighbourhood of 1 in H. The connected component H° of H is open in H, it is locally path-connected and path-connected, and it is the subgroup of H generated by the image of $\mathfrak{h}$ under the exponential function.*

Let us show that if S is the closed ball in $\mathfrak{h}$ centered at 0 and of radius r, then $\exp(S)$ is a neighbourhood of 1 in H. We assume this to be false, and show that it leads to a contradiction. Indeed, H would then contain a sequence (h_p) of elements tending to 1, with $h_p \notin \exp(S)$ for all p. Obviously $h_p \neq 1$, and it may be assumed that $\|h_p - 1\| < 1$ for all p. So it is possible to define

$$\delta_p = \|\log(h_p)\|, \text{ where } \lim \delta_p = 0. \qquad (3.4.17)$$

Since $h_p = \exp(\log(h_p))$ for large p, $\log(h_p) \notin S$ cannot hold since $h_p \notin \exp(S)$ (by the way, this means that $\log(h_p) \notin \mathfrak{h}$—otherwise $\log(h_p)$, which tends to 0 would belong to the ball S of $\mathfrak{h}$ for sufficiently large p). As S is compact, the distance ε_p from $\log(h_p)$ to S is strictly positive; obviously, $\varepsilon_p \leq \delta_p$, and so ε_p tends to 0. Let X_p be the point of S at a distance ε_p from $\log(h_p)$. Then,

$$\|h_p \cdot \exp(-X_p) - 1\| \leq \|h_p - \exp(X_p)\| \cdot \|\exp(-X_p)\|$$
$$= \|h_p - \exp(X_p)\| \cdot O(1) = \|\exp(\log(h_p)) - \exp(X_p)\| \cdot O(1) \qquad (3.4.18)$$
$$= O(\|\log(h_p) - X_p\|) \cdot O(1) = O(\varepsilon_p)$$

by (3.1.9)—forethought is always rewarded. Obviously, $h_p \cdot \exp(-X_p) \neq 1$ since $h_p \notin \exp(S)$, and so, for large p,

$$\|\log(h_p \cdot \exp(-X_p))\| = \eta_p > 0. \qquad (3.4.19)$$

Besides, as seen previously, $\eta_p = O(\varepsilon_p)$. If need be by replacing the sequence h_p by a subsequence, the matrix

$$Y = \lim \frac{1}{\eta_p} \log(h_p \cdot \exp(-Xp)) \tag{3.4.20}$$

may be assumed to exist since the unit sphere of $M_n(\mathbb{R})$ is compact. It is non-trivial—it has norm 1—and belongs to $\mathfrak{h}$. This follows from definition (3.4.1) and Lemma 1 applied to the sequence $h_p \cdot \exp(-X_p)$, which is in H and tends to 1. But (3.4.20) also means that

$$\| \log(h_p \cdot \exp(-X_p)) - \eta_p Y \| = o(\eta_p) \tag{3.4.21}$$

and so, again using (3.1.9),

$$\| h_p \cdot \exp(-Xp) - \exp(\eta_p Y) \| = o(\eta_p) \tag{3.4.22}$$

whenever p is sufficiently large for $h_p \cdot \exp(-X_p)$ to be written as the exp of its log. Since $\exp(X_p)$ tends to 1, it follows that

$$\| h_p - \exp(\eta_p Y) \exp(X_p) \| = o(\eta_p) \tag{3.4.23}$$

and so, using (3.1.10),

$$\| \log(h_p) - \log(\exp(\eta_p Y) \exp(X_p)) \| = o(\eta_p). \tag{3.4.24}$$

But by (3.1.26)

$$\begin{aligned}\log(\exp(\eta_p Y) \exp(X_p)) &= \eta_p Y + X_p + O(\|\eta_p Y\| \cdot \|X_p\|) \\ &= \eta_p Y + X_p + o(\eta_p)\end{aligned} \tag{3.4.25}$$

since $\|X_p\|$ tends to 0. Substituting this into (3.4.24), we get

$$\| \log(h_p) - (\eta_p Y + X_p) \| = o(\eta_p) = o(\varepsilon_p). \tag{3.4.26}$$

But as η_p and X_p tend to 0, $\eta_p Y + X_p \in S$ for large p and so the left-hand side is $\geq \varepsilon_p$. Hence, finally, $\varepsilon_p = o(\varepsilon_p)$, and this absurd equality shows that, as claimed, $\exp(S)$ contains a neighbourhood of e in H.

To complete the proof of Theorem 5, let S be a ball centered at 0 of radius $r < \log 2$. Then, $\log(\exp(X)) = X$ for all $X \in S$, and as the log function is continuous, exp clearly induces a homeomorphism from S onto its image, namely a neighbourhood of e in H, establishing the first statement of the theorem. By the above, the subgroup H° contains a neighbourhood of e in H, and so is open in H. It is locally path-connected since the image of a ball is path-connected. The claimed topological properties of H° then follow by obvious arguments (see end of Sect. 1.2 of Chap. 1).

3.5 Extension of von Neumann's Theorem to Locally Linear Groups

Although, as later examples will show, the category of closed subgroups of a linear group already includes many interesting groups, it is insufficient to cover every Lie group and all the more so to show that, conversely, every Lie subalgebra of $M_n(\mathbb{R})$ corresponds to a subgroup of $GL_n(\mathbb{R})$. We essentially meet with three types of difficulties.

(i) A *non-closed* subgroup H of a group $GL_n(\mathbb{R})$ can sometimes be endowed with a reasonable topology (i.e. locally compact and such that the canonical injection j from H to $GL_n(\mathbb{R})$ is continuous). In Chap. 6 we will show for example that this is the case if H is the subgroup of $GL_n(\mathbb{R})$ generated by $\exp(\mathfrak{h})$, where $\mathfrak{h}$ is an arbitrary Lie subalgebra of $M_n(\mathbb{R})$, the topology of H being such that, in this case, the exponential function induces a homeomorphism from a neighbourhood of 0 in $\mathfrak{h}$ onto a neighbourhood of e in H endowed with the topology in question. Example: for $\mathfrak{h}$, take the line generated by a matrix X and for H the set of matrices $\exp(tX)$. If it is not compact, i.e. if the map $t \mapsto \exp(tX)$ is injective, the topology of $\mathbb{R}$ can be transferred to H and H thereby turned into a connected locally compact group with an injective, continuous (but not bicontinuous) homomorphism to $GL_n(\mathbb{R})$.

As an aside, we remind the reader that when it is possible to endow an "abstract" group H with a topology turning it into a locally compact group *countable at infinity* (for example a connected one), then this topology is *unique*. This follows from Theorem 1 of Chap. 1.

(ii) In case (ii), consider pairs (H, j) consisting of a locally compact group H and an injective continuous homomorphism j from H to a group $GL_n(\mathbb{R})$. But passing to the universal covering $\widetilde{H}$ of H (or to any other covering of H) and taking the composition of j with the canonical projection $p : \widetilde{H} \to H$ gives a pair $(\widetilde{H}, \widetilde{j})$ consisting of a locally compact group and of a continuous homomorphism $\widetilde{j}$ from $\widetilde{H}$ to $GL_n(\mathbb{R})$. Instead of being injective, the latter is in general only *locally injective* since its kernel is a discrete subgroup of $\widetilde{H}$. There is no reason why modifying the choice of the homomorphism considered from $\widetilde{H}$ to $GL_n(\mathbb{R})$ would make it injective; the universal covering of $SL_2(\mathbb{R})$, we repeat, *cannot be* injectively embedded into a linear group.

By the way, note that a continuous homomorphism from a locally compact group H to a group $GL_n(\mathbb{R})$ or $GL_n(\mathbb{C})$ or, "more generally", to the linear group of a finite-dimensional real or complex vector space, is usually called a *linear representation* (finite-dimensional and continuous being implied) of H. It is, or was, said to be *faithful* (resp. *locally faithful*) when its kernel is reduced to the identity element (resp. is discrete). Therefore, case (i) above encompasses the groups having faithful representations, case (ii) the groups having locally faithful representations.

(iii) Let us start with a locally compact group H admitting a locally faithful linear representation, and let us replace it by the quotient H/D where D is a *discrete* central subgroup of H—so that H is a covering of H/D. The linear representations of H/D are obtained by considering the representations of H that are trivial on D. As the following simple example shows, H/D may very well have no locally faithful linear representation.

To see this start from the group H of real matrices of the form

$$\begin{pmatrix} 1 & x & z \\ 0 & 1 & y \\ 0 & 0 & 1 \end{pmatrix}.$$

It is now known as the *Heisenberg group* because a study of its representations is tantamount to searching for solutions of the famous equality

$$PQ - QP = h/2\pi i, \text{ (where } h \text{ is Planck's constant)}$$

introduced by W. Heisenberg in quantum mechanics precisely when von Neumann arrived in Göttingen. The group H, whose centre by the way consists of matrices for which $x = y = 0$, is instantly seen to be generated by three one-parameter subgroups. Writing the multiplication in H using x, y, z, observe that a linear representation M of H to $GL_n(\mathbb{R})$ or $GL_n(\mathbb{C})$ transforms the three obvious one-parameter subgroups of H into one parameter subgroups of $GL_n(\mathbb{R})$, necessarily of the form $x \mapsto \exp(xP)$, $y \mapsto \exp(yQ)$ and $z \mapsto \exp(zR)$. These must satisfy the following conditions:

(a) the matrices $\exp(xP)$ and $\exp(yQ)$ commute with $\exp(zR)$ for all real x, y, z (in other words, P and Q commute with R);

(b) for all real x, y,

$$\exp(xP)\exp(yQ)\exp(xP)^{-1}\exp(yQ)^{-1} = \exp(xyR).$$

Replacing x and y by x/p and y/p, taking the p^2th power of the equality obtained by using formula (3.3.4) of Theorem 3, the matrices P, Q, R are seen to satisfy

$$PQ - QP = R.$$

The desired counterexample is obtained by taking for D the central subgroup $x = y = 0$, $z \in \mathbb{Z}$ of H. To show that H/D does not admit any locally faithful linear representation, it is sufficient to show that every representation of H trivial on D is in fact trivial on the whole centre of H—in other words, with the previous notation, that

$$\exp(R) = 1 \quad \text{implies} \quad R = 0.$$

Lemma *Every matrix $A \in M_n(\mathbb{C})$ for which $\exp(A)$ is semisimple is semisimple.*

Indeed, let us write $A = S+N$, where S is semisimple, N nilpotent and $SN = NS$. Then, $\exp(A) = \exp(S)\exp(N)$ and as $\exp(S)$ and $\exp(N)$ are visibly semisimple, unipotent and commute, this is the Jordan decomposition of the matrix $\exp(A)$. Hence, $\exp(N) = 1$. But for a *nilpotent* matrix N, the identity $N = \log\exp(N)$ holds without any restriction on the norm of N or, more clearly,

$$N = -\sum (1 - \exp(N))^p / p.$$

Indeed, the matrix $\exp(N)$ is unipotent, the matrix $1 - \exp(N)$ is nilpotent, and the series reduces to a polynomial. Replacing N by tN, both sides become polynomials in t and are equal for small $|t|$, hence for all t.

As a consequence, if a nilpotent matrix N satisfies $\exp(N) = 1$, then $N = 0$, proving the lemma.

We can now return to the Heisenberg group. Suppose that $\exp(R) = 1$. Then R is semisimple, and if we work in $V = \mathbb{C}^n$, we get a decomposition $V = \oplus V(\lambda)$ into a direct sum of eigenspaces of V. All operators commuting with R, in particular P and Q, leave the $V(\lambda)$ invariant, and then, in each $V(\lambda)$ we find two operators P and Q satisfying $PQ - QP = \lambda 1$. It follows that the scalar operator $\lambda 1$ in $V(\lambda)$ has zero trace, so that $\lambda = 0$ and finally $R = 0$ as claimed. Therefore the quotient group H/D does not admit any locally faithful linear representation (in finite dimensions).

Exercise. If three matrices $P, Q, R \in M_n(\mathbb{C})$ satisfy $[P, R] = [Q, R] = 0$ and $[P, Q] = R$, then R is nilpotent.

Even in this hopeless case, there is nonetheless an alternative to the notion of a locally faithful linear representation. Let j be a locally faithful linear representation of H. The canonical projection $p : H \to H/D$ (where $D = \mathbb{Z}$ for the Heisenberg group) is injective in every sufficiently small neighbourhood of e. So there is a compact neighbourhood V of e in H in which j and p are injective As p then induces a homeomorphism from V onto a compact neighbourhood $V' = p(V)$ of e in $H/D = H'$, it is possible to define a map j' from V' to $GL_n(\mathbb{R})$ by setting $j'(p(x)) = j(x)$ for all $x \in V$. Hence this gives an *injective, continuous* map (hence a homeomorphism) j' from V' to $GL_n(\mathbb{R})$. We show that if V is sufficiently small then the following equality holds. It was used to define *local homomorphisms* in Chap. 2, Sect. 2.2.

$$j'(x'y') = j'(x')j'(y') \text{ whenever } x', y' \text{ and } x'y' \text{ are in } V'. \tag{5.0}$$

Indeed, x', y' and $z' = x'y'$ are the images under p of well-defined points x, y and z of V, and as $p(z) = p(x)p(y)$ we have $xy \in zD$, i.e. $z^{-1}xy \in D$. Hence $z = xy$ — and so (5.0) holds—if V is chosen to be sufficiently small so that $V^{-1}V^2 \cap D = \{e\}$, which is always possible since D is discrete.

A locally compact group H will be said to be *locally linear* if there is a neighbourhood V of e in H and a *continuous injective* map j from V to a group $GL_n(\mathbb{R})$ such that $j(xy) = j(x)j(y)$ whenever x, y and xy are in V. The category of locally linear groups happens to be identical to that of Lie groups (Ado's Theorem). In this chapter we will show how von N's arguments can be extended to this case and be used to introduce an analytic structure on every locally linear group. The converse is out of reach at this point.

The first step consists in extending the constructions of the previous section, which is what we are now going to do. In the following statement, a continuous injective map j from a neighbourhood of e in H to $GL_n(\mathbb{R})$ such that $j(xy) = j(x)j(y)$ whenever both sides are well-defined will be called a *local embedding* of H in $GL_n(\mathbb{R})$.

Theorem 6 *Let H be a locally compact group and j a local embedding of H into a linear group $GL_n(\mathbb{R})$. Let $\mathfrak{h}$ be the set of matrices $X \in M_n(\mathbb{R})$ having the following property: there exists a sequence (h_p) of elements of H tending to e and a sequence (ε_p) of strictly positive numbers tending to 0 such that*

$$X = \lim \frac{1}{\varepsilon_p}[j(h_p) - 1]. \tag{3.5.1}$$

The following properties then hold:

 (i) *For any $X \in \mathfrak{h}$, there is a unique one-parameter subgroup γ_X in H such that*

$$j \circ \gamma_X(t) = \exp(tX) \text{ for all sufficiently small } t \in \mathbb{R}; \tag{3.5.2}$$

   the map $X \mapsto \gamma_X$ from $\mathfrak{h}$ to the set of one-parameter subgroups of H is bijective.
 (ii) *$\mathfrak{h}$ is a Lie subalgebra of $M_n(\mathbb{R})$ and the Lie algebra structure of $\mathfrak{h}$ is given in terms of the one-parameter subgroups of H by the following formulas:*

$$\gamma_{sX}(t) = \gamma_X(st), \tag{3.5.3}$$

$$\gamma_{X+Y}(t) = \lim[\gamma_X(t/n)\gamma_Y(t/n)]^n, \tag{3.5.4}$$

$$\gamma_{[X,Y]}(t) = \lim[\gamma_X(t/n)\gamma_Y(1/n)\gamma_X(t/n)^{-1}\gamma_Y(1/n)^{-1}]^{n^2}. \tag{3.5.5}$$

(iii) *The function $\exp_H : \mathfrak{h} \to H$ given by*

$$\exp_H(X) = \gamma_X(1) \tag{3.5.6}$$

   is continuous and maps every sufficiently small neighbourhood of 0 in $\mathfrak{h}$ homeomorphically onto a neighbourhood of unity in H. Moreover,

$$\exp_H \circ \log \circ j(h) = h \tag{3.5.7}$$

   for all $h \in H$ sufficiently near the identity element.
 (iv) *The connected component H° is open in H and generated by $\exp_H(\mathfrak{h})$.*

The proof is broken down into a number of lemmas. For reasons of simplicity, we will frequently set

$$j(h) = \bar{h} \tag{3.5.8}$$

for any $h \in W$, a neighbourhood of e where the map j is defined.

It is not worth rephrasing Lemma 1 of the previous section in the present context. The first non-trivial result is then the following:

Lemma 1 *For all $X \in \mathfrak{h}$ and sufficiently small $t \in \mathbb{R}$, $\exp(tX) \in j(W)$.*

The proof of the lemma given below would have been an improvement on von Neumann's somewhat complicated arguments, but it was formulated some twenty years after his version.

Let us start from (3.5.1) and choose a *compact* neighbourhood $V = V^{-1}$ of e in H in which j is defined and injective. As j is continuous, $j(V)$ may be assumed to be in the ball $\|g - 1\| < 1$; this enables us to define $\log j(h) = \log(\bar{h})$ for all $h \in V$. Using Theorem 1 of Sect. 1, we thereby get an injective continuous map (hence a homeomorphism) from V to $M_n(\mathbb{R})$. Without loss of generality, assume that $h_p \in V$ for all p.

Since $j(V)$ is contained in the ball $\|g - 1\| < 1$ and it does not contain any non-trivial subgroup of $GL_n(\mathbb{R})$, neither does V contain any non-trivial subgroup of H. Hence, for all $h \in V$ such that $h \neq e$, there are integers $q > 0$ such that $h^q \notin V$. So for each p, there is an integer $q_p > 0$ such that

$$h_p^q \in V \text{ for } 0 \leq q \leq q_p, \quad h_p^{q_p+1} \notin V \tag{3.5.9}$$

(we assume that all $h_p \neq e$, for otherwise $X = 0$ and the lemma is empty). As V is compact,

$$\lim h_p^{q_p} = h \tag{3.5.10}$$

may be assumed to exist, if necessary by extracting a subsequence. Then $h \in V$, and as h is also the limit of the $h_p^{q_p+1} \notin V$, h belongs to the boundary of V. If H is not discrete—and if it is, Theorem 6, as already seen, is trivial—which we will assume, then $h \neq e$, and so $\log(\bar{h}) \neq 0$. Besides,

$$\log(\bar{h}) = \lim \log(\bar{h}_p^{q_p}) = \lim q_p \cdot \log(\bar{h}_p) = \lim q_p \cdot \log(j(h_p)) \tag{3.5.11}$$

follows from the footnote on page 74, which can be applied here by (3.5.9). Taking Lemma 1 from the previous section into account, comparison with (3.5.1) shows that X and $\log(\bar{h})$ are proportional (when a sequence of vectors u_p tends to a non-trivial limit u, the directions $u_p/\|u_p\|$ of the u_p tend to the direction $u/\|u\|$ of u). Instead of proving Lemma 1 in its given form, it therefore suffices to show that

$$\exp(t \cdot \log(\bar{h})) \in j(V) \tag{3.5.12}$$

for small $|t|$. Besides, it is sufficient to do so for $0 < t < 1$. But then, by (3.5.11),

$$\exp(t \cdot \log \bar{h}) = \lim \exp(tq_p \cdot \log(\bar{h}_p)) = \lim \exp([tq_p] \cdot \log(\bar{h}_p)) \tag{3.5.13}$$

since $tq_p \cdot \log(\bar{h}_p) - [tq_p] \log(\bar{h}_p)$, whose norm is bounded above by $\|\log(\bar{h}_p)\|$, tends to 0. On the other hand,

$$\exp([tq_p].\log(\bar{h}_p)) = \exp(\log(\bar{h}_p))^{[tq_p]} = \bar{h}_p^{[tq_p]}. \tag{3.5.14}$$

Hence, as $0 \leq t \leq 1$ implies $0 \leq [tq_p] \leq q_p$, the result obtained belongs to $j(V)$ by (3.5.9). Since V is compact in H, $j(V)$ is closed in $GL_n(\mathbb{R})$ and indeed, taking the limit, $\exp(t \cdot \log(\bar{h})) \in j(V)$ for $0 < t < 1$, proving the lemma.

Lemma 2 *For every* $X \in \mathfrak{h}$, *there is a unique one-parameter subgroup* γ_X *of* H *such that* $j \circ \gamma_X(t) = \exp(tX)$ *for sufficiently small* $|t|$. *The map* $X \mapsto \gamma_X$ *from* $\mathfrak{h}$ *to the set of one-parameter subgroups of* H *is bijective.*

Let us consider the neighbourhood V of Lemma 1. For all $X \in \mathfrak{h}$, there exists a $c > 0$ such that

$$\exp(tX) \in j(V) \text{ whenever } |t| \leq c. \qquad (3.5.15)$$

As j is injective on V, there is a uniquely defined map, namely $t \mapsto \gamma_X(t)$, which by the way is continuous, from the interval $|t| \leq c$ to V, such that (3.5.2) holds for $|t| < c$. Then suppose that V is sufficiently small for j to be defined and injective on V^2, not only in V. If all three inequalities $|s| < c$, $|t| < c$ and $|s + t| < c$ hold, then

$$
\begin{aligned}
j(\gamma_X(s + t)) &= \exp(sX + tX) = \exp(sX)\exp(tX) \\
&= j(\gamma_X(s))j(\gamma_X(t)) = j(\gamma_X(s)\gamma_X(t)).
\end{aligned}
\qquad (3.5.16)
$$

Hence, since j is injective on V^2,

$$\gamma_X(s + t) = \gamma_X(s)\gamma_X(t) \qquad (3.5.17)$$

whenever both sides are defined. As $\mathbb{R}$ is simply connected,[6] there is a unique continuous homomorphism from $\mathbb{R}$ to H which, in the interval $|t| \leq c$, reduces to γ_X. This proves the existence of the one-parameter subgroup γ_X of H.

Conversely, let γ be a one-parameter subgroup of H. The map $t \mapsto j \circ \gamma(t)$ is defined and continuous for small $|t|$ and is obviously a local homomorphism from $\mathbb{R}$ to $GL_n(\mathbb{R})$, and so can be extended to a one-parameter subgroup of $GL_n(\mathbb{R})$. Thus there is a unique $X \in M_n(\mathbb{R})$—namely the derivative of $t \mapsto j \circ \gamma(t)$ at $t = 0$—such that $j \circ \gamma(t) = \exp(tX)$ in the neighbourhood of $t = 0$. The definition of derivatives as limits immediately shows that $X \in \mathfrak{h}$, and hence there is a one-parameter subgroup in H such that $j \circ \gamma(t) = j \circ \gamma'_X(t)$ for sufficiently small t. But $\gamma(t)$ and $\gamma_X(t)$ belong to V for sufficiently small $|t|$ since a one-parameter subgroup

---

[6]There is also a direct argument. Let I be an interval centered on 0 in $\mathbb{R}$ and f a map from I to a group G such that $f(s + t) = f(s)f(t)$ whenever s, t and $s + t$ are in I. Then, obviously

$$f(t) = f(t/n)^n$$

for any $t \in I$ and integer n. It follows that, for all $t \in \mathbb{R}$, the expression $f(t/n)^n$, defined for sufficiently large n, is independent of the choice of n. For if t/p and t/q are in I, then

$$f(t/p)^p = [f(t/pq)^q]^p = f(t/pq)^{pq} = f(t/q)^q.$$

The extension of f to a homomorphism from $\mathbb{R}$ to G follows immediately.

is a *continuous* homomorphism from $\mathbb{R}$ to H. As j is injective on V, $\gamma(t) = \gamma_X(t)$ for sufficiently small $|t|$, and hence for any t, proving the lemma.

Lemma 3 $\mathfrak{h}$ *is a Lie subalgebra of* $M_n(\mathbb{R})$.

Considering the one-parameter subgroups of type $t \mapsto \gamma_X(st)$, $\mathfrak{h}$ is readily seen to be homothety-invariant for homotheties centered at 0. To show that the sum $X + Y$ of two elements of $\mathfrak{h}$ is also in $\mathfrak{h}$, set $h_p = \gamma_X(1/p)\gamma_Y(1/p)$. Then $\lim h_p = e$ in H, and moreover

$$j(h_p) = \exp(X/p)\exp(Y/p) = 1 + \frac{X+Y}{p} + O(1/p^2), \qquad (3.5.18)$$

so that

$$X + Y = \lim p[j(h_p) - 1], \qquad (3.5.19)$$

and thus $X + Y \in \mathfrak{h}$. $[X, Y] \in \mathfrak{h}$ follows similarly by replacing h_p by the commutator of the elements $\gamma_X(1/p)$ and $\gamma_Y(1/p)$ of H, proving the lemma.

Lemma 4 *The map* $\exp_H : \mathfrak{h} \to H$ *given by* $\exp_H(X) = \gamma_X(1)$ *is continuous at the origin.*

We need to prove that, if X_p is a sequence of elements of $\mathfrak{h}$ tending to 0, and V a sufficiently small neighbourhood of e in H, then for sufficiently large p, $\gamma_{X_p}(1) \in V$. The proof we give is by contradiction. We suppose, which we are allowed to do, that V is compact, that j is defined and injective on V and that $j(V)$ is contained in the ball $\|g - 1\| < 1$.

If the statement to be proved is false, then if need be by extracting a subsequence, we may suppose that

$$\gamma_{X_p}(1) \notin V \text{ for all } p. \qquad (3.5.20)$$

But as the map $t \mapsto \gamma_{X_p}(t)$ from $\mathbb{R}$ to H is continuous, it is clear that $\gamma_{X_p}(t) \in V$ in the neighbourhood of 0. Let $[0, t_p]$ be the largest interval having 0 as origin such that

$$\gamma_{X_p}(t) \in V \text{ for } 0 \leq t \leq t_p. \qquad (3.5.21)$$

This interval is in effect closed since V is assumed to be compact, and moreover $t_p < 1$ by (3.5.20). Obviously $\gamma_{X_p}(t_p)$ is a *boundary* point of V, and if need be by extracting a subsequence, we may suppose that

$$\lim \gamma_{X_p}(t_p) = h \qquad (3.5.22)$$

exists. As H is not discrete and as h belongs to the boundary of V, $h \neq e$. Therefore, $\log(\bar{h}) = \log(j(h)) \neq 0$. But applying $\log \circ j$ to (3.5.22) it follows that

$$\log(\bar{h}) = \lim \log(\exp(t_p X_p)) = \lim t_p X_p = 0 \qquad (3.5.23)$$

since $0 \leq t_p \leq 1$ and since $\log(\exp X) \sim X$ for X tending to 0. This contradiction proves the lemma.

Lemma 5 *The map* $\exp_H$ *induces a homeomorphism from every sufficiently small neighbourhood of 0 in* $\mathfrak{h}$ *onto a neighbourhood of e in* H. *The inverse map is* $\log \circ j$.

Let V be a compact neighbourhood of e in H on which j is defined and injective and let S be a closed ball in $\mathfrak{h}$ with centre 0 and radius $r > 0$ such that $\exp_H(S) \subset V$. The existence of S follows from the preceding lemma. As j induces a homeomorphism from V onto $j(V)$, it amounts to showing that the map $j \circ \exp_H = \exp$ from S to $j(V)$ is a homeomorphism from S onto a neighbourhood of 1 in $j(V)$. In fact, it suffices to show that $\exp(S)$ is a neighbourhood of 1 in $j(V)$. For if $r < \log 2$, the map $\exp : M_n(\mathbb{R}) \to GL_n(\mathbb{R})$ induces a homeomorphism from S onto its image, whose inverse is the log map. Then, since for all $X \in S$,

$$\log \circ j \circ \exp_H(X) = \log \circ \exp(X) = X, \tag{3.5.24}$$

with $\log \circ j$ injective and continuous (hence bicontinuous) on $\exp_H(S)$, which can be topological identified with $\exp(S)$ using j, $\exp_H$ and $\log \circ j$ must be mutually inverse homeomorphisms between S and $\exp_H(S)$.

Hence it is now a matter of showing that $\exp(S)$ is a neighbourhood of 1 in the compact set $j(V) \subset GL_n(\mathbb{R})$. The proof is the same as that of Theorem 5 of the previous section. We only need to be careful and choose the h_p in $j(V)$ and not in H, and it is pointless to restate the proof.

Lemma 6 *The map* $\exp_H : \mathfrak{h} \to H$ *is continuous.*

By Lemma 5, there is a ball S in $\mathfrak{h}$ with centre 0 on which $\exp_H$ is continuous. For all $X \in \mathfrak{h}$, there is an integer p such that $Y/p \in S$ for any $Y \in \mathfrak{h}$ sufficiently near X. Then $\exp_H(Y) = \exp_H(Y/p)^p$, and so $\exp_H$ is continuous at X given that the map $h \mapsto h_p$ is continuous from H to H.

The proof of the theorem will be finished once formulas (3.5.3), (3.5.4) and (3.5.5) will have been shown. The first one is obvious. The second one can be re-written as

$$\exp_H(X + Y) = \lim[\exp_H(X/p)\exp_H(Y/p)]^p. \tag{3.5.25}$$

To begin with, note that the sequence whose general term is $\exp_H(X/p)\exp_H(Y/p) = \gamma_X(1/p)\gamma_Y(1/p)$ tends to e in H. Indeed, the homomorphisms γ_X, γ_Y from $\mathbb{R}$ to H are continuous. Moreover, as seen in the proof of Lemma 3,

$$X + Y = \lim p \cdot \log(j[\exp_X(X/p)\exp_H(Y/p)]). \tag{3.5.26}$$

Lemma 5 shows that on the right-hand side there is a limit of elements of $\mathfrak{h}$, and not only of $M_n(\mathbb{R})$. Hence, the map $\exp_H$ being continuous,

$$\exp_H(X + Y) = \lim \exp_H\{p \cdot \log \circ j[\exp_H(X/p)\exp_H(Y/p)]\}$$
$$= \lim\{\exp_H(\log \circ j[\exp_H(X/p)\exp_H(Y/p)]\}^p \qquad (3.5.27)$$
$$= \lim[\exp_H(X/p)\exp_H(Y/p)]^p$$

since $\log \circ j$ and $\exp_H$ are mutual inverses in the neighbourhood of the origin, proving (3.5.25). The commutator formula, which can also be written as

$$\exp_H([X, Y]) = \lim[\exp_H(X/p)\exp_H(Y/p)\exp_H(X/p)^{-1}\exp_H(X/p)^{-1}]^{p^2},$$
$$(3.5.28)$$

follows from similar arguments.

Finally, statement (iv) of the theorem can be proved in the same way as the corresponding statement on closed subgroups of $GL_n(\mathbb{R})$.

3.6 The Lie Algebra and the Adjoint Representation of a Locally Linear Group

Let H be a locally compact and locally linear group. Fixing a local embedding j of H into a group $GL_n(\mathbb{R})$ makes it possible to associate a Lie subalgebra of $M_n(\mathbb{R})$ to H. Theorem 6 (ii) tells us that its internal structure *does not depend on the choice of j*, and in fact enables us to associate to every locally compact and locally linear group H a *canonical* Lie algebra $\mathfrak{h}$, defined as follows: the elements of $\mathfrak{h}$ are the *one-parameter subgroups* of H, and for all $X \in \mathfrak{h}$, denoting by

$$t \mapsto \gamma_X(t) \qquad (3.6.1)$$

the subgroup with parameter X, the structure of the Lie algebra $\mathfrak{h}$ is then defined by formulas (3.5.3), (3.5.4) and (3.5.5). In this regard, note that it is not at all obvious that these formulas do define one-parameter subgroups, and even less that the composition laws thus defined on $\mathfrak{h}$ turn it into a finite-dimensional Lie algebra. The fact that it is indeed the case can be seen by fixing a local embedding of H into a group $GL_n(\mathbb{R})$ and by considering the isomorphism from $\mathfrak{h}$ onto the Lie subalgebra of $M_n(\mathbb{R})$ associated to this embedding by Theorem 6.

If $H = GL_n(\mathbb{R})$, then the "abstract" Lie algebra $\mathfrak{h}$ can clearly be canonically identified with $M_n(\mathbb{R})$, the map $\exp_H$ being merely the usual exponential function.

If $H = \mathbb{R}$, a group to which all these considerations obviously apply, a one-parameter subgroup is of the form $t \mapsto tX$ for some unique $X \in \mathbb{R}$, and $\mathfrak{h}$ can then clearly be identified with $\mathbb{R}$ endowed with its usual vector space structure and with the Jacobi bracket

$$[X, Y] = 0, \qquad (3.6.2)$$

as is more generally the case for all locally linear abelian groups. The exponential map is then reduced to the *identity* map $X \mapsto X$ from $\mathfrak{h} = \mathbb{R}$ to $H = \mathbb{R}$.

Besides, *the Lie algebra of a locally linear group depends functorially on the latter.* In other words, if there are two locally linear groups G and H and a continuous homomorphism M from G to H, then M canonically defines a homomorphism[7] M' from the Lie algebra $\mathfrak{g}$ of G to the Lie algebra $\mathfrak{h}$ of H, and the correspondence $M \mapsto M'$ agrees with the composition of homomorphisms. The definition of M' rests on the observation that for all $X \in \mathfrak{g}$, the map $t \mapsto M(\gamma_X(t))$ is a one-parameter subgroup of H; so $M'(X)$ is a well-defined element of $\mathfrak{h}$ such that

$$\boxed{\gamma_{M'(X)}(t) = M(\gamma_X(t))} \quad \text{for all } t \in \mathbb{R} \text{ and } X \in \mathfrak{g}. \tag{3.6.3}$$

The fact that M' is a Lie algebra homomorphism—i.e. a linear map satisfying

$$M'([X, Y]) = [M'(X), M'(Y)] \quad \text{for all } X, Y \in \mathfrak{g} \tag{3.6.4}$$

follows readily by applying M to transform formulas (3.5.3) to (3.5.5) defining the operations in $\mathfrak{g}$ and $\mathfrak{h}$. As M is compatible with the composition laws and the topologies of G and H, compatibility of M' with the algebraic structures of $\mathfrak{g}$ and $\mathfrak{h}$ follows immediately.

Note that setting $t = 1$ in (3.6.3) gives

$$\exp_H \circ M' = M \circ \exp_G, \tag{3.6.5}$$

an equality equivalent to (3.6.3).

Exercise. Extend the construction of M' to the case where M is a continuous local homomorphism (Chap. 2, Sect. 2.2) from G to H.

Exercise. Suppose that $H = GL_n(\mathbb{R})$ and identify $\mathfrak{h}$ with $M_n(\mathbb{R})$. Show that $M'(X) = Y$ is characterized by the equality

$$M \circ \gamma_X(t) = \exp(tY). \tag{3.6.6}$$

How does this relate to statement (i) of Theorem 6?

Exercise. Suppose that G is a closed subgroup of H and M the identity map. Show that M' is injective on a Lie subalgebra of $\mathfrak{h}$, namely on the set of $X \in \mathfrak{h}$ for which $\exp_H(tX) \in G$ for all t.

Exercise. Let $M : G \to H$ be a continuous homomorphism. Show that M' is injective if and only if the kernel of M is discrete. More generally, the kernel of M' is the subalgebra of $\mathfrak{g}$ which, according to the previous exercise, corresponds to the kernel of M (as an aside, note that a closed subgroup of a locally linear group is

---

[7]As M' is the tangent linear map of M at e in the sense of Chap. 5, it should be denoted by $M'(e)$ or $dM(e)$.

locally linear). In particular, let $\widetilde{G}$ be the universal covering (or any other covering) of a locally linear group G. Then, $\widetilde{G}$ is locally linear, and the canonical projection $p : \widetilde{G} \to G$ induces an isomorphism of the Lie algebra of $\widetilde{G}$ onto that of G.

The construction of the homomorphism M' associated to a homomorphism $M :$ $G \to H$ of locally linear groups notably applies when $H = GL_n(\mathbb{R})$. In this case, as seen above, M is a linear representation of G, and $\mathfrak{h}$ can be canonically identified with $M_n(\mathbb{R})$. Hence we get a homomorphism $M' : \mathfrak{g} \to M_n(\mathbb{R})$, which is obviously given by

$$M \circ \exp_G = \exp \circ M' \tag{3.6.7}$$

or else, in terms of one-parameter subgroups, by

$$M \circ \gamma_X(t) = \exp(tM'(X)) \text{ for all } X \in \mathfrak{g} \quad \text{and all } t \in \mathbb{R}. \tag{3.6.8}$$

By the way, note that a homomorphism from a Lie algebra $\mathfrak{g}$ to an algebra $M_n(\mathbb{R})$, or "more generally" to the Lie algebra $\mathscr{L}(V)$, where V is a finite-dimensional vector space over $\mathbb{R}$ or $\mathbb{C}$, is called a *linear representation* (understood: finite-dimensional) of $\mathfrak{g}$. Hence, any linear representation of a locally linear group G canonically defines a linear representation of its Lie algebra.

Exercise. (i) Let V be a finite-dimensional real vector space, W a subspace of V and X an endomorphism of V. Show that W is X-invariant if and only if it is $\exp(tX)$-invariant for all t. Similarly, an endomorphism A of V commutes with X if and only if it commutes with $\exp(tX)$ for all t.

(ii) Let G be a locally linear group, M a continuous homomorphism from G to $GL(V)$ and M' the corresponding representation of its Lie algebra. Suppose G is connected. Show that a vector subspace W of V is $M(g)$-invariant for all $g \in G$ if and only if it is $M'(X)$-invariant for all $X \in \mathfrak{h}$. Similarly, an endomorphism A of V commutes with all $M(g)$ if and only if it commutes with all $M'(X)$. [When V has no non-trivial subspace invariant under all $M(g)$, the representation M of G is said to be *irreducible*; a similar definition holds for a Lie algebra. Hence, *if G is connected, a linear representation of G is irreducible if and only if so is the corresponding representation of the Lie algebra of G.*]

On the other hand, it is also possible to pass from a group homomorphism $M :$ $G \to H$ to a Lie algebra homomorphism $M' : \mathfrak{g} \to \mathfrak{h}$ when $G = H$ and M is an interior automorphism $g \mapsto aga^{-1}$ of G. Hence there is an automorphism of $\mathfrak{g}$ which generally depends on a and which we denote by $\operatorname{ad}(a)$; it is therefore defined by the equality

$$\boxed{a \cdot \exp_G(X) \cdot a^{-1} = \exp_G(\operatorname{ad}(a)X)} \quad \text{for all } X \in \mathfrak{g}. \tag{3.6.9}$$

The functorial character of the correspondence between M and M' shows that the map $a \mapsto \operatorname{ad}(a)$ is a *homomorphism* from the group G to the group of automorphisms of the vector space (and even of the Lie algebra) $\mathfrak{g}$. It is known as the *adjoint representation* of G. This homomorphism is easily seen to be continuous, i.e. it defines a linear representation of G in the sense given to this expression in Sect. 3.5. To see this, choose a local embedding j of G into a group $GL_{\ell}(\mathbb{R})$. There is an isomorphism j' from $\mathfrak{g}$ to a Lie subalgebra of $M_n(\mathbb{R})$. Identify $\mathfrak{g}$ with its image under j'. Then, for $X \in \mathfrak{g}$, and $a \in G$, the element $Y = \mathrm{ad}\,(a)X$ of $\mathfrak{g}$ seemingly defined by the equality

$$\gamma_Y(t) = a \cdot \gamma_X(t) \cdot a^{-1}, \tag{3.6.10}$$

is given by the equality

$$\exp(tY) = j(a)\exp(tX)j(a)^{-1} \tag{3.6.11}$$

when a is sufficiently small. So, obviously

$$Y = \mathrm{ad}\,(a)X = j(a) \cdot X \cdot j(a)^{-1}, \tag{3.6.12}$$

where the matrix products are in $M_n(\mathbb{R})$. As the map j is continuous, the continuity of the homomorphism $a \mapsto \mathrm{ad}\,(a)$ at the origin (and hence everywhere) is an immediate consequence.

Exercise. If G is connected, the kernel of the adjoint representation is the centre of G. Corollary: if G is a connected locally linear group whose centre is discrete, then G admits a locally faithful linear representation.

According to the above, the adjoint representation of G in $GL(\mathfrak{g})$ should give rise to a representation of $\mathfrak{g}$ in $\mathscr{L}(\mathfrak{g})$, which ought to be written $X \mapsto \mathrm{ad}\,'(X)$ according to the conventions adopted here. It is easy to compute. Indeed, formula (3.6.8) defining M' in the general case readily gives

$$\boxed{M'(X) = \frac{d}{dt}M(\exp_G(tX))_{t=0}.} \tag{3.6.13}$$

In the case at hand, the operator $\mathrm{ad}\,'(X) = U$ is therefore given by

$$U = \frac{d}{dt}\mathrm{ad}\,(\gamma_X(t))_{t=0}. \tag{3.6.14}$$

Hence, for all $Y \in \mathfrak{g}$, using a local embedding j as before, we get

$$\begin{aligned}
U(Y) &= \frac{d}{dt}j(\gamma_X(t)) \cdot Y \cdot j(\gamma_X(t))_{t=0}^{-1}\\
&= \frac{d}{dt}\exp(tX) \cdot Y \cdot \exp(-tX)_{t=0}\\
&= \frac{d}{dt}\exp(tX)Y_{t=0} + \frac{d}{dt}Y \cdot \exp(-tX)_{t=0}\\
&= XY - YX = [X, Y].
\end{aligned} \tag{3.6.15}$$

The following notation is usual in a Lie algebra $\mathfrak{g}$:

$$\boxed{\text{ad}\,(X)Y = [X, Y]} \quad \text{for all } X, Y \in \mathfrak{g}. \qquad (3.6.16)$$

This gives a linear map ad $: \mathfrak{g} \rightarrow \mathscr{L}(\mathfrak{g})$ from $\mathfrak{g}$ to the Lie algebra $\mathscr{L}(\mathfrak{g})$ of endomorphisms of the vector space $\mathfrak{g}$. The Jacobi identity shows that

$$\text{ad}\,([Y, Z]) = \text{ad}\,(Y) \circ \text{ad}\,(Z) - \text{ad}\,(Z) \circ \text{ad}\,(Y) \qquad (3.6.17)$$

also holds, in other words that the map at hand is a Lie algebra homomorphism; it is called the *adjoint representation* of $\mathfrak{g}$. All these computations will be given again in a much more general (?) setting, in the theory of Lie groups where the Lie algebra of such a group will be defined using its analytic structure instead of the existence of local embeddings into linear groups, which, to all appearances, is not very obvious.

Exercise. Let H be a connected closed subgroup of a locally linear group G. Identify the Lie algebra $\mathfrak{h}$ of H with its image in the Lie algebra $\mathfrak{g}$ of G. H is normal if and only if $\mathfrak{h}$ is ad (g)-invariant for all $g \in G$. If G is connected, it suffices that $[X, Y] \in \mathfrak{h}$ for all $X \in \mathfrak{g}$, $Y \in \mathfrak{h}$ ($\mathfrak{h}$ is then said to be an *ideal* of the Lie algebra $\mathfrak{g}$).

Exercise. Let G be a locally linear group and $\mathfrak{g}$ its Lie algebra. Then,

$$\boxed{\text{ad}\,(\exp_G(X)) = \exp(\text{ad}\,(X)) = \sum \text{ad}\,(X)^p / p!} \qquad (3.6.18)$$

for all $X \in \mathfrak{g}$.

3.7   The Analytic Structure on a Locally Linear Group

The aim of von N's article was to show that, if G is a closed subgroup of a linear group $GL_n(\mathbb{R})$, then it is possible to introduce a "coordinate system" in the neighbourhood of each point of G in such a way that (i) if the coordinates $\xi_1(x), \ldots, \xi_p(x)$ are associated to each point x in a sufficiently small open subset U of G, then the map $x \mapsto (\xi_i(x))$ thus obtained is a *homeomorphism* from U onto an *open* subset of a space $\mathbb{R}^p$ (in other words: *every closed subgroup of a linear group is locally Euclidean*), (ii) if $x, y \in G$ remain sufficiently near two given points a and b, so that the product xy remains near ab, then the coordinates (in the neighbourhood of ab) of xy are *analytic* functions of the coordinates of x (in the neighbourhood of a) and of the point y (in the neighbourhood of b). Statement (ii) *in this particular case* solves Hilbert's fifth problem, which is precisely a matter of showing that, in a locally Euclidean group, the coordinate systems can be so chosen that the group multiplication, written using these coordinates, is expressed by analytic functions and not only by continuous ones. In modern language, von N's article shows that every closed subgroup of a linear group is a Lie group.

To introduce a coordinate system in the neighbourhood of each $a \in G$, von N chooses a basis $(X_i)_{1 \le i \le p}$ of the Lie algebra $\mathfrak{g} \subset M_n(\mathbb{R})$ which corresponds to G. He observes that, if U is a sufficiently small neighbourhood of e for the log map to be defined on U and to induce a homeomorphism from U onto an open neighbourhood S of 0 in $\mathfrak{g}$, then, for all $g \in aU$, i.e. for any g sufficiently near a, there is an equality of the form

$$\log(a^{-1}g) = \sum \xi_i(g)X_i \tag{3.7.1}$$

with well-defined scalars $\xi_i(g)$. As $g \mapsto \log(a^{-1}g)$ induces a homeomorphism from aU onto S, the map $g \mapsto (\xi_1(g), \ldots, \xi_p(g))$ induces a homeomorphism from aU onto an open subset of $\mathbb{R}^p$, proving (i) above.

To show (ii), as an example, let us suppose that $a = b = e$. The coordinates of x, of y and of xy are then the numbers appearing in the formulas

$$\log(x) = \sum u_i X_i, \quad \log(y) = \sum v_i X_i, \quad \log(xy) = \sum w_i X_i, \tag{3.7.2}$$

from which it readily follows that

$$\sum w_i X_i = \log\left[\exp(\sum u_i X_i) \exp(\sum v_i X_i) \right]. \tag{3.7.3}$$

The map $(X, Y) \mapsto \log[\exp(X)\exp(Y)]$ being analytic in the neighbourhood of 0, the coordinates of xy can then be written analytically in terms of those of x and y, and point (ii) follows.

We now set aside this outdated view and show more generally how, using Theorem 6, any locally linear locally compact group G can be provided with an analytic structure compatible with its group structure—in other words (Chap. 4) a Lie group structure. Heuristically, it is a matter of making the notion of an "analytic" function on an open set of G well-defined and to show that if f is such a function, then in turn, the function $f(xy)$ depends analytically on the pair (x, y). The constructions in this section will be better understood in Sect. 3.4, where they will be motivated.

a—Let G be a locally linear locally compact group and $\mathfrak{g}$ its Lie algebra. Let U be an open subset of G. For all $a \in U$, the set $a^{-1}U$ is a neighbourhood of e in G, hence contains $\exp_G(S)$, where S is any sufficiently small ball in $\mathfrak{g}$ centered at 0. For every function f defined on U, the map $X \mapsto f(a \cdot \exp_G(X))$ is therefore defined in the neighbourhood of 0 in $\mathfrak{g}$, and since the map $\exp_G$ induces a homeomorphism from any sufficiently small neighbourhood of 0 in $\mathfrak{g}$ onto a neighbourhood of e in G, f is clearly continuous on U if and only if the function $X \mapsto f(a \cdot \exp_G(X))$ is continuous in the neighbourhood of 0 for all $a \in U$.

The function f is then said to be of class C^r (for r integer or $r = \infty$) or *analytic* on U if, for all $a \in U$, the function $f(a \cdot \exp(X))$ is of class C^r or else analytic in the neighbourhood of 0 in $\mathfrak{g}$. The set of C^r functions on U will be written $C^r(U)$, and the set of analytic functions on U, $C^\omega(U)$. In the following, the symbol r will denote either a natural integer, ∞, or ω.

b—The notion of a C^r function is a *local* one. In other words, if the open set U is the (not necessarily finite) union of open subsets U_i and if f is a function defined on U, then

$$f \in C^r(U) \iff f|U_i \in C^r(U_i) \text{ for all } i. \tag{3.7.4}$$

This follows trivially from the definition and from the fact that, in a Euclidean space such as $\mathfrak{g}$, the restriction to a smaller open set of a C^r function defined on an open set is again C^r. Besides, (7.4) clearly also holds for C^r functions on open subsets of a Euclidean space.

c—Clearly, *if f is of class C^r on U, then, by definition, for all $a \in G$, the function $x \mapsto f(ax)$ is of class C^r on $a^{-1}U$.* Similarly, the function $x \mapsto f(xa)$ is of class C^r on Ua^{-1}. Indeed, let $b \in Ua^{-1}$; for $X \in \mathfrak{h}$ near 0,

$$f(b \cdot \exp_G(X) \cdot a) = f(ba \cdot a^{-1}\exp_G(X)a) = f(ba \cdot \exp_G(\mathrm{ad}\,(a)^{-1}X)). \quad (3.7.5)$$

As $ba \in U$, the function $Y \mapsto f(ba \cdot \exp_G(Y))$ is of class C^r in the neighbourhood of 0; its composition with the linear map $X \mapsto \mathrm{ad}\,(a)^{-1}X$ is therefore also C^r in the neighbourhood of $X = 0$, and the desired result follows. Finally, similar arguments show that *if f is of class C^r on U, then the function $x \mapsto f(x^{-1})$ is of class C^r on U^{-1}.*

d—We next show that *if f is a C^r function on an open subset U of G, then the function $f \circ \exp_G$ is of class C^r on the open set $S \subset \mathfrak{g}$ consisting of all elements X such that $\exp_G(X) \in U$.* This requires showing that, for all $A \in S$, the function $X \mapsto f \circ \exp_G(A + X)$ is of class C^r in the neighbourhood of 0 in $\mathfrak{g}$. By assumption, f satisfies the condition

(i) the map $X \mapsto f(\exp_G(A)\exp_G(X))$ is of class C^r in the neighbourhood of 0 in $\mathfrak{g}$.

To derive the desired result, note that, for X near 0, $\exp_G(A + X)$ is near $\exp_G(A)$. As the map $\exp_G$ induces a homeomorphism from any sufficiently small neighbourhood of 0 in $\mathfrak{g}$ onto a neighbourhood of e in G, for sufficiently small X, it is possible to set $\exp_G(A + X) = \exp_G(A)\exp_G(Y)$, where Y is near 0. It then amounts to showing that Y is an analytic function of X in the neighbourhood of the origin.

Suppose first that there is a faithful locally linear representation j (and not just only a local embedding) of G in a group $GL_n(\mathbb{R})$. Then $\mathfrak{g}$ can be identified with a Lie subalgebra of $M_n(\mathbb{R})$, and applying j to both sides of $\exp_G(A + X) = \exp_G(A)\exp_G(Y)$ gives $\exp(A + X) = \exp(A)\exp(Y)$. Hence, since X and Y are near 0 in $M_n(\mathbb{R})$,

$$Y = \log[\exp(A)^{-1}\exp(A + X)]. \quad (3.7.6)$$

The latter highlights the analyticity of Y as a function of X.

In the general case, when only a local embedding j of G into a group $GL_n(\mathbb{R})$, not defined everywhere, is at our disposal, the previous arguments no longer apply since j may well not be defined at $\exp_G(A) \in G$. Nonetheless, j continues to define an isomorphism from the set $\mathfrak{g}$ of one-parameter subgroups of G onto a Lie subalgebra of $M_n(\mathbb{R})$, and, as will be seen, for sufficiently small X and Y the equality $\exp(A + X) = \exp(A)\exp(Y)$ continues to hold.

Indeed, consider the universal covering $\widetilde{G}$ of G and the canonical homomorphism p from $\widetilde{G}$ to G. The map $j \circ p$, a local homomorphism from $\widetilde{G}$ to $GL_n(\mathbb{R})$, can clearly be extended to a locally faithful (obviously continuous) homomorphism $\widetilde{j}$ from $\widetilde{G}$

to $GL_n(\mathbb{R})$ (Chap. 2, Theorem 1). Denoting by $\mathfrak{g}$ and $\widetilde{\mathfrak{g}}$ the Lie algebras of G and $\widetilde{G}$, we get a commutative diagram

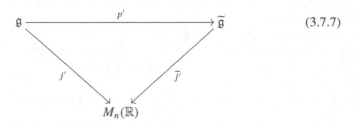

$$(3.7.7)$$

where p' is bijective,[8] whereas j' and $\widetilde{j}$ are injective and enable us to identify $\mathfrak{g}$ and $\widetilde{\mathfrak{g}}$ with the same Lie subalgebra of $M_n(\mathbb{R})$. Identifying elements A, X and Y of $\mathfrak{g}$ with their images in $M_n(\mathbb{R})$ and with the elements of $\widetilde{\mathfrak{g}}$ corresponding to them under p', the proof then consists in showing that, *for sufficiently small X and Y,*

$$\exp_G(A + X) = \exp_G(A)\exp_G(Y) \Longleftrightarrow \exp_{\widetilde{G}}(A + X) = \exp_{\widetilde{G}}(A)\exp_{\widetilde{G}}(Y).$$
$$(3.7.8)$$

For indeed, as seen above, $\widetilde{j}$ will enable us to derive the desired formula (3.7.6) from the second equality.

However, the first equality in (3.7.8) obviously follows from the second one by applying p (this is the functoriality of the exponential map). To show that, conversely, the first equality implies the second, it then suffices to note that if X and Y are sufficiently small, both sides of the second equality belong to a neighbourhood of the point $\exp_{\widetilde{G}}(A)$ in which p is injective.

By the way, note the result obtained during the proof: *for a simply connected group, the existence of a local embedding into a linear group is equivalent to that of a locally faithful linear representation.*

e—We now show that G admits "local charts" in the neighbourhood of each of its points, considering first the neighbourhood of the identity element. The exact statement we prove is the following: *if S is a sufficiently small neighbourhood of 0 in $\mathfrak{g}$ and if $U = \exp_G(S)$, then a function f defined on an open set $V \subset U$ is of class C^r on V if and only if so is the function $f \circ \exp_G$ on the open set corresponding to S.* Taking S to be a sufficiently small open set for the map $\exp_G$ to induce a homeomorphism from S to U, this means that the map $\exp_G$ transforms the analytic structure of U into that of S.

---

[8]As an aside, note that the meaning of the bijectivity of p' is essentially that every one-parameter subgroup of G can be lifted in a unique way to a one-parameter subgroup of $\widetilde{G}$. For a direct proof, see the arguments of Lemma 2, page 82, where $\exp(tX)$ should be replaced by the given one-parameter subgroup of G, j by the homomorphism $p : \widetilde{G} \to G$, and V by a neighbourhood of e in $\widetilde{G}$ in which p is injective. It could also be observed that a one-parameter subgroup of G is a path in G (of infinite length if we may say so) with origin e, and so can be lifted in a unique way to a path with origin e in $\widetilde{G}$.

Exercise. Using a priori arguments, check that the lifting is also a one-parameter subgroup of $\widetilde{G}$.

Since it has already been established that if f is of class C^r, then the same holds for $f \circ \exp_G$ without any further assumptions, only the converse needs to be shown. Hence, we need to prove that, for all $A \in \mathfrak{g}$ *sufficiently near* 0, the statement

(ii) the map $X \mapsto f(\exp_G(A + X))$ is of class C^r in the neighbourhood of 0 in $\mathfrak{g}$ implies statement (i) of p. 91. The proof is "obvious": set $\exp_G(A)\exp_G(X) = \exp_G(A + Y)$ and check that Y is an analytic function of X in the neighbourhood of the origin. Statement (i) then follows by replacing Y by its expression in terms of X in $f(\exp_G(A + Y))$.

To justify this heuristic argument, we first choose an open neighbourhood S of 0 in $\mathfrak{g}$ which the map $\exp_G$ transforms homeomorphically into an open neighbourhood U of e in G. If U, i.e. S, is sufficiently small, then there is a local embedding j of G into a group $GL_n(\mathbb{R})$ defined and injective on the neighbourhood U and mapping it to the ball centered at 1 of radius 1 on which the logarithm is defined. Besides, as seen in the preceding section, j defines an injective homomorphism j' from $\mathfrak{g}$ to the Lie algebra $M_n(\mathbb{R})$, and $j'(S)$ may be assumed to be in the ball with centre 0 and radius $\log 2$ on which the exponential is the inverse of the logarithm. Now, take A in S. For small X, $\exp_G(A)\exp_G(X) \in U$, and so there is a unique $Z \in S$ such that $\exp_G(A)\exp_G(X) = \exp_G(Z)$; Z is obviously a continuous function of X, and tends to A as X tends to 0. Hence $Z = A + Y$, where $Y \in \mathfrak{g}$, is a continuous function of X and tends to 0 with X. Applying j gives $\exp(j'(A))\exp(j'(X)) = \exp(j'(A+X))$, and as $j'(A+Y) \in j'(S)$ for X near zero, $j'(A+Y) = \log[\exp(j'(A))\exp(j'(X))]$ clearly holds. The left-hand side is therefore an analytic function of X in the neighbourhood of 0, and as j' is an injective linear map, it immediately follows that Y is also an analytic function of X in the neighbourhood of 0. This proves the implication (ii) $\Rightarrow$ (i).

f—We have just seen that in the neighbourhood of e, the "analytic structure" of G is that of an open subset of a Cartesian space (namely of an open subset of $\mathfrak{g}$). Let us show that this is also the case in the neighbourhood of any other point a of G, more precisely that *for all $a \in G$, there is an open neighbourhood $U(a)$ of a in G and a homeomorphism ξ from $U(a)$ onto an open subset of a Cartesian space which, for any open subset $V \subset U(a)$, transforms $C^r(V)$ into the set of C^r functions on the open set $\xi(V)$. If $a = e$, for $U(e)$, it suffices to take a sufficiently small neighbourhood U of e, and for ξ the inverse of the map $\exp_G$ (which we could denote by $\log_G$— it is defined in the neighbourhood of e only). In the general case, we take $U(a) = aU$, where U is the open set considered previously, and for ξ the map $x \mapsto \log_G(a^{-1}x)$ from aU to $\mathfrak{g}$. The notion of a C^r function on an open subset of G being defined in such a way that it is invariant under left translations, if the method is successful in the neighbourhood of e, it will clearly be so in the neighbourhood of a. No additional verification is necessary.

As will be seen in the next chapter, these arguments essentially aim to show the existence of *local charts* in the supposed analytic manifold G.

g—Finally, to conclude, observe that *the composition law $(x, y) \mapsto xy$ of G is analytic at the origin* (hence everywhere since it was essentially proved above that the left translations $x \mapsto ax$ or right translations $x \mapsto xb$ are analytic). This means that, if under $\log_G$ a sufficiently small neighbourhood of e in G is identified with a neighbourhood of 0 in $\mathfrak{g}$, in which case the map $(x, y) \mapsto xy$ is visibly transformed into

$$(X, Y) \mapsto \log_G[\exp_G(X)\exp_G(Y)], \tag{3.7.9}$$

then it is analytic at the origin in the usual sense. But if a local embedding j of G into the group $GL_n(\mathbb{R})$ is chosen so that $\mathfrak{g}$ is identified with a Lie subalgebra of $M_n(\mathbb{R})$ under j', then (3.7.9) clearly becomes the restriction to this Lie subalgebra of the map

$$(X, Y) \mapsto \log[\exp(X)\exp(Y)], \tag{3.7.10}$$

defined in the neighbourhood of $(0, 0)$ in $M_n(\mathbb{R}) \times M_n(\mathbb{R})$, and so analyticity at the origin is now obvious...

In fact, the basic mystery or miracle of the theory of Lie groups does not reside in the analyticity of (3.7.9), but in the fact that *if, in* (3.7.10), *X and Y are given values (near zero) in a Lie subalgebra* $\mathfrak{g}$ *of* $M_n(\mathbb{R})$, *then the map* (3.7.10) *is* $\mathfrak{g}$-*valued*. In the case at hand, this is essentially statement (iii) of Theorem 6. The reason why this happens in this context is given by the *Campbell–Hausdorff* formula which will be proved shortly, and which will be greatly (?) generalized in Chap. 6 so as to be applicable to all Lie groups, and not just (sic) to locally linear groups.

Exercise. Applying the previous constructions to $GL_n(\mathbb{R})$, show that for every open subset U of $GL_n(\mathbb{R})$, $C^r(U)$ is the set of C^r functions (in the usual sense) on the open subset U of $M_n(\mathbb{R})$.

Exercise. Apply the constructions of this section to $G = \mathbb{R}$. What do you find? (Observe that if $\mathfrak{g}$ is identified with $\mathbb{R}$, then $a \cdot \exp_G(X) = a + X$ for all $a \in G = \mathbb{R}$ and $X \in \mathfrak{g} = \mathbb{R}$.)

Exercise. Let G be a *closed* subgroup of $GL_n(\mathbb{R})$. Show that for any open subset U of G, the functions $f \in C^r(U)$ are characterized by the following property: for all $a \in U$, there is an open neighbourhood $V(a)$ of a in $GL_n(\mathbb{R})$ and a C^r function on $V(a)$ which coincides with f on $G \cap V(a)$. [Show that the problem amounts to proving that every function defined and analytic in the neighbourhood of 0 in $\mathfrak{g}$ is, in the neighbourhood of 0, the restriction to $\mathfrak{g}$ of a function defined and analytic in the neighbourhood of 0 in $M_n(\mathbb{R})$. The exercise aims to show that a closed subgroup of $GL_n(\mathbb{R})$, together with its analytic structure, is a *closed submanifold* of the analytic manifold $GL_n(\mathbb{R})$ in the sense of the next chapter.] With appropriate modifications, this result can be extended to locally compact groups embedded into $GL_n(\mathbb{R})$, but it would be rather inconvenient to give its precise formulation without the notion of *immersion* which will be introduced in the following chapter.

3.8   Analyticity of Continuous Homomorphisms

The constructions in the preceding section—which in von Neumann are found at least in the neighbourhood of the identity element since in his days the modern notion of manifold had not yet been developed—aim to show that every locally linear group (and in particular the case that von N was interested in, namely that of closed subgroups of some $GL_n(\mathbb{R})$), is canonically endowed with a Lie group structure in the sense of the next chapter.

Von Neumann's other intent, and probably his main intent, was to show that a *continuous* homomorphism M between two closed subgroups G and H of two linear groups is *analytic*. In his time, this property was regarded from the following point of view. Let $\mathfrak{g}$ and $\mathfrak{h}$ be Lie algebras (sets of matrices in this context) of G and H. Choose bases (X_i) and (Y_j) of the vector spaces $\mathfrak{g}$ and $\mathfrak{h}$. By statement (iii) of Theorem 6, each $g \in G$ sufficiently near e can be associated coordinates $t_i(g)$ (called *canonical*, but wrongly so, since they depend on the choice of a basis of $\mathfrak{g}$); namely, as seen at the beginning of the previous section, numbers appearing in the formula

$$\log(g) = \sum t_i(g)X_i. \tag{3.8.1}$$

By the way, the latter is equivalent to $g = \exp(\sum t_i(g)X_i)$ if the $t_i(g)$ are taken to be sufficiently small. Similarly, for $h \in H$ sufficiently near e, "canonical" coordinates $u_i(h)$ can be defined by the equality

$$\log(h) = \sum u_j(h)Y_j. \tag{3.8.2}$$

As the given homomorphism M from G to H is continuous, for g sufficiently near e, equality $h = M(g)$ is conveyed by formulas giving the $u_j(h)$ in terms of the $t_i(g)$.

From this perspective, the analyticity of the continuous homomorphism M from G to H is formulated as follows: for $g \in G$ sufficiently near e, the canonical coordinates $u_j(h)$ of the point $h = M(g)$ of H are analytic functions of the canonical coordinates $t_i(g)$ of the point g. In this form, the result is trivial, for in fact the $u_j(h)$ are even *linear* functions of the $t_i(g)$. Indeed, we know (Sect. 3.6) that there is a homomorphism M' from the Lie algebra $\mathfrak{g}$ to the Lie algebra $\mathfrak{h}$ associated to M such that, notably

$$M \circ \exp_G = \exp_H \circ M'. \tag{3.8.3}$$

By definition of canonical coordinates,

$$g = \exp_G\left(\sum t_i(g)X_i\right), \quad h = \exp_H\left(\sum u_j(h)Y_j\right) \tag{3.8.4}$$

for $g \in G$ and $h \in H$ near e, so that the equality $h = M(g)$ becomes

$$\sum u_j(h)Y_j = M'\left(\sum t_i(g)X_i\right). \tag{3.8.5}$$

Hence the $u_j(h)$ are indeed linear combinations of the $t_i(g)$. These can be computed whenever the matrix of M' is given with respect to the bases of $\mathfrak{g}$ and $\mathfrak{h}$ in question.

From today's perspective, the analyticity of M would be rendered differently. We would say that *if there is an analytic function ψ on an open subset V of H,* *then the composite function* $\psi \circ M$, *defined on the open subset* $U = M^{-1}(V)$ *of* G, *is analytic on it.* This formulation—once again we anticipate Chap. 4—includes the naive view that consists in checking that, for $h = M(g)$, the $u_j(h)$ are analytic functions of the $t_i(g)$. Indeed, writing $h = M(g)$ replaces the function u_j, analytic in the neighbourhood of e in H, by the function $u_j \circ M$. Checking that the latter can be analytically expressed using canonical coordinates of g then obviously amounts to checking that $u_j \circ M$ is analytic on G in the neighbourhood of e in the sense of the previous section.

As for checking that if ψ is analytic—or more generally of class C^r—on an open subset V of H, then $\psi \circ M$ is of class C^r on the corresponding open subset of G, it only requires an application of the definitions. Consider $a \in G$ for which the point $b = M(a)$ is in V. Then

$$\psi \circ M(a \cdot \exp_G(X)) = \psi(b \cdot M(\exp_G(X))) = \psi(b \cdot \exp_H(M'(X))) \qquad (3.8.6)$$

has to be checked to be an analytic function of $X \in \mathfrak{g}$ in the neighbourhood of 0. Now, it is obtained by composing the map $Y \mapsto \psi(b \cdot \exp_H(Y))$, by assumption an analytic function of $Y \in \mathfrak{h}$ in the neighbourhood of 0, with the *linear* map M' from $\mathfrak{g}$ to $\mathfrak{h}$. The analyticity of the resultant map is obvious. There is naturally no difference between this "modern proof" and von Neumann's simple calculation (3.8.6).

On the other hand, the argument just developed quite evidently applies in a more general context, namely for groups G and H which are only locally linear. In other words, the following result holds:

Theorem 7 *Let G and H be two locally linear groups, $\mathfrak{g}$ and $\mathfrak{h}$ their Lie algebras and M a continuous homomorphism from G to H. Then M is analytic and there is a unique homomorphism M' from the Lie algebra $\mathfrak{g}$ to the Lie algebra $\mathfrak{h}$ such that*

$$M \circ \exp_G = \exp_H \circ M'. \qquad (3.8.7)$$

The theorem holds if $H = GL_n(\mathbb{R})$. In this case, M is a linear representation of G, and in matrix form, can be written as

$$M(g) = (M_{ij}(g))_{1 \le i,j \le n} \qquad (3.8.8)$$

by highlighting the entries of $M(g)$. These are real continuous functions on G satisfying

$$M_{ij}(xy) = \sum_k M_{ik}(x)M_{kj}(y). \qquad (3.8.9)$$

Theorem 7 then shows that the functions M_{ij} are analytic on the whole of G in the sense of the previous sections (hence in the usual sense if G itself is a group $GL_n(\mathbb{R})$).

Von N's method for constructing M' is not the one given in Sect. 3.6. Von N never mentions one-parameter subgroups. His definition of $\mathfrak{g}$ (assuming that like H, G is a closed subgroup of a linear group) rests on the existence of an equality

$$X = \lim \frac{1}{\varepsilon_p}(g_p - 1) \tag{3.8.10}$$

as in (3.4.1) or (3.5.1). His task then consists in showing directly that, if the right-hand side converges, so does the sequence having general term

$$\lim \frac{1}{\varepsilon_p}(M(g_p) - 1). \tag{3.8.11}$$

This is obviously the limit of (3.8.11) and is the desired matrix $M'(X)$. Von N's argument is sufficiently amusing to warrant repeating. We start by choosing a constant $c > 0$, for the moment arbitrary, and set $n_p = [c/\varepsilon_p]$, (integral part) so that $n_p \varepsilon_p$ tends to c and that

$$cX = \lim n_p(g_p - 1) = \lim n_p \cdot \log(g_p). \tag{3.8.12}$$

Since M is continuous at the identity, for a strictly positive number $\eta < 1$, there exists a strictly positive number $\delta < 1$ such that

$$\|g - 1\| < \delta \text{ implies } \|M(g) - 1\| < \eta. \tag{3.8.13}$$

Choose c sufficiently small so that $c\|X\| < \log(1 + \delta) < \log 2$. By (3.8.13), $n_p\|\log(g_p)\| < \log(1 + \delta)$ for large p and so perforce

$$r\|\log(g_p)\| < \log(1 + \delta) \quad \text{for} \quad 1 \leq r \leq n_p. \tag{3.8.14}$$

Thus

$$\|g_p^r - 1\| = \|\exp(\log(g_p))^r - 1\|$$
$$= \|\exp(r \cdot \log(g_p)) - 1\| < e^{\log(1+\delta)} - 1 = \delta < 1. \tag{3.8.15}$$

This follows from (3.1.8) and (3.8.13) or from the footnote 4 on page 74. The choice of δ less than 1 is necessary to obtain $\log(1+\delta) < \log 2$ and to justify these calculations. Now, (3.8.15) implies that

$$\|M(g_p)^r - 1\| < \eta < 1 \quad \text{for} \quad 1 \leq r \leq n_p \tag{3.8.16}$$

from which, again using the footnote 4 on page 74, we get

$$n_p \cdot \log(M(g_p)) = \log(M(g_p)^{n_p}) = \log(M(g_p^{n_p})). \tag{3.8.17}$$

But as

$$g_p^{n_p} = \exp(n_p \cdot \log(g_p)) \tag{3.8.18}$$

tends to $\exp(cX)$ by (3.8.12), and as M is continuous, it follows that

$$\lim n_p \cdot \log(M(g_p)) = \log(\exp(cX)) \tag{3.8.19}$$

exists. As $n_p \varepsilon_p$ tends to c, we finally obtain

$$\lim \frac{1}{\varepsilon_p}(M(g_p) - 1) = \lim \frac{1}{\varepsilon_p} \log(M(g_p)) = \frac{1}{c} \log(\exp(cX), \tag{3.8.20}$$

and so the limit of the matrices (3.8.11) exists.

At this point, von N shows that the limit Y of (3.8.11) depends only on the limit X of (3.8.10), and not on the choice of the g_p or ε_p. The argument consists in observing that if there are two equalities

$$X = \lim \frac{1}{\varepsilon'_p}(g'_p - 1) = \lim \frac{1}{\varepsilon''_p}(g''_p - 1), \tag{3.8.21}$$

they can be combined into a single one

$$X = \lim \frac{1}{\varepsilon_p}(g_p - 1) \tag{3.8.22}$$

by constructing a sequence whose odd (resp. even) terms are those of the sequence g'_p (resp. g''_p). Applying M to (3.8.23), we get a *convergent* sequence which is a combination of the sequences obtained by applying M to the two sequences in (3.8.21). Hence, applying M to the two sequences (3.8.21) gives two sequences converging to the same limit, and the desired result follows.

This enables us to define a map M' from $\mathfrak{g}$ to $\mathfrak{h}$, and it remains to show that it is a Lie algebra homomorphism. To do so, von Neumann uses Lemma 2 from Sect. 3.4 and the footnote from 4 page 74: if X' and X'' are in $\mathfrak{g}$, there are sequences (g'_p) and (g''_p) in G tending to e as well as a sequence of numbers $\varepsilon_p > 0$ tending to zero, such that

$$X' = \lim \frac{1}{\varepsilon_p}(g'_p - 1), \quad X'' = \lim \frac{1}{\varepsilon_p}(g''_p - 1). \tag{3.8.23}$$

Then,

$$X' + X'' = \lim \frac{1}{\varepsilon_p}(g'_p g''_p - 1). \tag{3.8.24}$$

So, by definition,

$$M'(X' + X'') = \lim \frac{1}{\varepsilon_p}(M(g'_p g''_p) - 1) = \lim \frac{1}{\varepsilon_p}(M(g'_p) M(g''_p) - 1), \tag{3.8.25}$$

and since likewise

$$M'(X') = \lim \frac{1}{\varepsilon_p}(M(g'_p) - 1), \quad M'(X'') = \lim \frac{1}{\varepsilon_p}(M(g''_p) - 1), \tag{3.8.26}$$

equality $M'(X' + X'') = M'(X') + M'(X'')$ follows. The same proof shows compatibility with commutators. Von N then knows that setting $y = M(x)$, the canonical coordinates of y are linear functions of the canonical coordinates of x, and the analyticity of M is proved!

3.9 The Derivative of the Exponential Map

Let E and F be two finite-dimensional real vector spaces and ϕ a map from an open subset U of E to F; suppose that ϕ is of class C^1 and let a be a point of U. It is well known that[9] there is then a *linear* map $u = \phi'(a)$ from E to F, the "tangent" to ϕ at a, i.e. such that

$$\phi(a + X) = \phi(a) + u(X) + o(\|X\|) \quad \text{as } X \in E \text{ tends to } 0. \qquad (3.9.1)$$

Hence,

$$u(X) = \frac{d}{dt}\phi(a + tX)\Big|_{t=0} \qquad (3.9.2)$$

for all $X \in E$. In this section, we compute u for the exponential map of $M_n(\mathbb{R})$ to itself, and applying *heuristic* methods we give the formula which will be proved in Chap. 6 for "general" Lie groups (the quotation marks alluding to the fact that linear groups are not all that less general than the Lie groups that are veritably so). The result obtained will be used in the next section to explain the basic mystery of the theory, namely that in a Lie subalgebra $\mathfrak{g}$ of $M_n(\mathbb{R})$, $\log(\exp(X)\exp(Y)) \in \mathfrak{g}$ whenever X and Y are sufficiently near zero.

Let u be the tangent map of the exponential at $A \in M_n(\mathbb{R})$. It is a linear map from $M_n(\mathbb{R})$ to $M_n(\mathbb{R})$ given by

$$u(X) = \frac{d}{dt}\exp(A + tX)\Big|_{t=0} = \frac{d}{dt}\sum (A + tX)^r / r!\Big|_{t=0}. \qquad (3.9.3)$$

As

$$(A + tX)^r = (A + tX)\cdots(A + tX), \qquad (3.9.4)$$

$$\frac{d}{dt}\exp(A + tX)^r = \sum_{p+q=r-1}(A + tX)^p X (A + tX)^q \qquad (3.9.5)$$

---

[9] H. Cartan, *Calcul Différentiel* (Hermann, 1971), p. 28 ff., or M. Berger and B. Gostiaux, *Differential Geometry* (Springer, 1988), or J. Dieudonné, *Eléments d'Analyse*, chap. VIII, or MA IX, Chap. 1...

follows immediately. So term by term differentiation of (3.9.3) gives

$$u(X) = \sum_{p,q=0}^{\infty} \frac{A^p X A^q}{(p+q+1)!}. \tag{3.9.6}$$

Term by term differentiation needs to be justified by showing that the series consisting of the derivatives of the functions $t \mapsto (A + tX)^r/r!$ converges normally on every compact subset of $\mathbb{R}$. Estimates given after (3.1.31) will help the reader to resolve this point as an exercise. What interests us here is the formal computation.

Let us consider the linear operators L and M in $M_n(\mathbb{R})$ defined by

$$L(X) = AX, \quad M(X) = XA. \tag{3.9.7}$$

They commute and the general term of (3.9.6) is obtained by applying the operator $L^p M^q/(p+q+1)!$ to X. In other words,

$$\begin{aligned}
u &= \sum L^p M^q/(p+q+1)! = \sum (L^{r-1} + L^{r-2}M + \cdots + M^{r-1})/r! \\
&= \sum \frac{1}{r!}\frac{L^r - M^r}{L - M} = \frac{\exp(L) - \exp(M)}{L - M}
\end{aligned} \tag{3.9.8}$$

in the algebra of endomorphisms of the vector space $M_n(\mathbb{R})$. Division by the operator

$$L - M = \text{ad}\,(A) : X \mapsto AX - XA = [A, X] \tag{3.9.9}$$

is problematic because there is no reason why $L - M$ should be invertible—in fact it is more likely not to be since its kernel clearly contains the element A of the vector space $M_n(\mathbb{R})$ on which $L - M$ acts. But pursuing our formal computation, since L and M commute, we get the equation

$$u = \exp(L)\frac{1 - \exp(M - L)}{L - M} = \exp(L)\frac{1 - \exp(-\text{ad}\,(A))}{\text{ad}\,(A)}, \tag{3.9.10}$$

and as the function

$$\frac{1 - e^{-z}}{z} = 1 - z/2! + z^2/3! - \cdots = \phi(z) \tag{3.9.11}$$

is defined by its power series for all z, y, including at the origin, it can be applied to any continuous operator on any Banach space. So finally we obtain

$$u = \exp(L) \circ \phi(\text{ad}\,(A)). \tag{3.9.12}$$

Next note that $\exp(L)$ is an automorphism of the vector space $M_n(\mathbb{R})$. So to prove (3.9.12), it suffices to check that by multiplying series (3.9.8) by $\exp(-L)$ we recover the series $\phi(\mathrm{ad}\,(A))$, but this amounts to an obvious formal computation.

Since $L(X) = AX$, we get $L^p(X) = A^p X$ for all p and thus, $\exp(L)$ is the automorphism $X \mapsto \exp(A)X$ of the vector space $M_n(\mathbb{R})$. In view of (3.9.11), (3.9.12) can be rewritten as

$$u(X) = \exp(A) \cdot [X - \mathrm{ad}\,(A)X/2! + \mathrm{ad}\,(A)^2 X/3! - \cdots] \qquad (3.9.13)$$

or finally as

$$\boxed{\exp(A)^{-1} \cdot \exp'(A)X = X - \mathrm{ad}\,(A)X/2! + \mathrm{ad}\,(A)^2 X/3! - \cdots} \qquad (3.9.14)$$

where the left-hand side denotes the product (in $M_n(\mathbb{R})$) of the matrix $\exp(A)^{-1}$ and the image of X under the tangent linear map $\exp'(A) = u$ to the exponential map at A.[10] This is the desired result. Note that

$$\mathrm{ad}\,(A)X = [A, X] = AX - XA,$$
$$\mathrm{ad}\,(A)^2 X = [A, [A, X]] = A^2 X - 2AXA + XA^2, \qquad (3.9.15)$$

and in general,

$$\mathrm{ad}\,(A)^r X = \sum_0^r (-1)^p \binom{r}{p} A^{r-p} X A^p \qquad (3.9.16)$$

which can be easily checked, notably by observing that $\mathrm{ad}\,(A) = L - M$, where $LM = ML$.

Formula (3.9.14) enables us to determine the set of matrices A on which the tangent map

$$\exp'(A) : M_n(\mathbb{R}) \to M_n(\mathbb{R}) \qquad (3.9.17)$$

is *bijective*. We know that for any A in this set, the exponential map induces a homeomorphism from an open neighbourhood of U of A to an open neighbourhood V of $\exp(A)$, and that the inverse map from V to U is also analytic (implicit function theorem). The factor $\exp(A)^{-1}$ being invertible, this amounts to finding the matrices A for which the operator $1 - \mathrm{ad}\,(A)/2! + \mathrm{ad}\,(A)^2/3! - \cdots$ is invertible, i.e. does not admit 0 as an eigenvalue. We can obviously work in $M_n(\mathbb{C}) = \mathbb{C} \otimes_{\mathbb{R}} M_n(\mathbb{R})$ and so assume that the matrix of $\mathrm{ad}\,(A)$ is triangular with respect to an appropriate basis of $M_n(\mathbb{C})$. If the eigenvalues of $\mathrm{ad}\,(A)$ are $\lambda_1, \cdots, \lambda_N$ (with $N = n^2$, and possible repeats), the matrices of the powers of $\mathrm{ad}\,(A)$ are also triangular, and hence so is the

---

[10]The notation $\exp'(A)X$ is therefore short for $\exp'(A)(X)$, likewise $\mathrm{ad}\,(A)^p X$ replaces

$$(\mathrm{ad}\,(A)^p)(X).$$

matrix of the desired operator, and its eigenvalues are then clearly numbers $\phi(\lambda_i)$, where ϕ is the function (3.9.11). Now, $\phi(0) = 1 \neq 0$ and for $z \neq 0$, $\phi(z) = 0$ if and only if $e^z = 1$, i.e. $z = 2k\pi i$. Hence, $\exp'(A)$ is invertible if and only if ad (A) does not admit any eigenvalue of type $2k\pi i$, where k is a non-trivial integer. We write down the result using the eigenvalues of the matrix A itself.

Set $A = S + N$ with S semisimple, N nilpotent and $SN = NS$. Going over to complex numbers, it may be assumed that $S = \mathrm{diag}(t_1, \cdots, t_n)$. If $X = (x_{ij})$, then a trivial calculation gives

$$\mathrm{ad}\,(S)X = ((t_i - t_j)x_{ij}) \tag{3.9.18}$$

and so ad (S) is diagonal (with respect to the obvious basis of $M_n(\mathbb{R})$, with respect to which the coordinates of X are the terms x_{ij} of X), its eigenvalues being the numbers $t_i - t_j$. Besides, ad (N) is nilpotent. This can be seen using formula (3.9.15) and observing that if $N^k = 0$, and if (3.9.15) is applied with $r = 2k$, then for each term of (3.9.15), either $r - p \geq k$, or $p \geq k$, whence ad $(N)^{2k} = 0$. Finally, ad (S) and ad (N) clearly commute—furthermore, generally speaking,

$$\mathrm{ad}\,(A)\mathrm{ad}\,(B) - \mathrm{ad}\,(B)\mathrm{ad}\,(A) = \mathrm{ad}\,([A, B]) \tag{3.9.19}$$

follows from the Jacobi identity (3.3.2). As a result, ad (S) *and* ad (N) *are the semi-simple and nilpotent components of* ad (A). Therefore, the eigenvalues of ad (A) are those of ad (S), i.e. the differences $t_i - t_j$ between the eigenvalues of A. In conclusion:

Theorem 8 *The tangent linear map* $\exp'(A)$ *to*

$$\exp : M_n(\mathbb{R}) \to M_n(\mathbb{R})$$

at a point $A \in M_n(\mathbb{R})$ is bijective if and only if the difference between two eigenvalues of A is never an integral multiple of $2\pi i$.

The case $n = 1$ not being of much interest, this result is best illustrated by investigating the case $n = 2$. If the eigenvalues of A are not real, then they are conjugate imaginary numbers, and so the "exceptional" matrices A not belonging to the set on which $\exp'(A)$ is bijective are those that admit two eigenvalues of type $t \pm k\pi i$, where $t \in \mathbb{R}$ and k is a non-trivial integer. Such a matrix is semisimple since its eigenvalues are distinct, and if x denotes an eigenvector (in $\mathbb{C}^2$) associated to the eigenvalue $t + k\pi i$ of A, then the conjugate imaginary vector $\bar{x}$ is an eigenvector associated to the eigenvalue $t - k\pi i$. Considering the *real* vectors

$$u = \frac{x + \bar{x}}{2}, \quad v = \frac{x - \bar{x}}{2i}, \tag{3.9.20}$$

we thus get a basis for $\mathbb{R}^2$ with respect to which

$$Au = tu - k\pi v, \quad Av = tv + k\pi u. \tag{3.9.21}$$

In other words, there is a matrix $g \in GL_2(\mathbb{R})$ such that

$$gAg^{-1} = \begin{pmatrix} t & k\pi \\ -k\pi & t \end{pmatrix} = t1 + \begin{pmatrix} 0 & k\pi \\ -k\pi & 0 \end{pmatrix}. \tag{3.9.22}$$

Since, in general,

$$\exp \begin{pmatrix} 0 & \theta \\ -\theta & 0 \end{pmatrix} = \begin{pmatrix} \cos\theta & \sin\theta \\ -\sin\theta & \cos\theta \end{pmatrix} = \pm 1 \quad \text{if } \theta \in \pi\mathbb{Z}, \tag{3.9.23}$$

the matrices A found also satisfy $\exp(A) \in \mathbb{R} \cdot 1$.

Are they characterized by this property? If $\exp(A) = \lambda \in \mathbb{R}$, then A is semisimple (see the *lemma* of Sect. 3.5) and each eigenvalue z of A must satisfy $e^z = \lambda \in \mathbb{R}$, and so $z = t + k\pi i$, where $t = \log|\lambda|$. As $z - \bar{z} = 2k\pi i$, the answer is yes if $k \neq 0$, i.e. if A *is not* a scalar.

3.10 The Campbell–Hausdorff Formula

In this section, in addition to the everywhere convergent power series $\phi(z)$ of the previous section, we will need to use the series

$$\psi(z) = z \sum (-1)^n (z-1)^n / (n+1) = \frac{z \cdot \log z}{z-1} \tag{3.10.1}$$

converging in the disk $|z - 1| < 1$. This makes it possible to define $\psi(u)$ for every continuous linear operator u on a Banach space provided that $\|u - 1\| < 1$. A trivial calculation, which only needs to be carried out in $\mathbb{C}$, then shows that

$$\phi(A)\psi(\exp(A)) = 1 \quad \text{for} \quad \|A\| < \log 2, \tag{3.10.2}$$

where A is, for example, in $M_n(\mathbb{R})$.

Let us next consider two matrices X and Y in $M_n(\mathbb{R})$ such that

$$\|X\|, \|Y\| < \frac{1}{2}\log 2 \tag{3.10.3}$$

and set

$$F(t) = \log[\exp(X)\exp(tY)] \quad \text{for} \quad 0 \le t \le 1. \tag{3.10.4}$$

The right-hand side is well defined since by (3.1.8), (3.10.3) implies

$$\| \exp(X) \exp(tY) - 1 \| \leq \| \exp(X) - 1 \| \cdot \| \exp(tY) - 1 \|$$
$$+ \| \exp(X) - 1 \| + \| \exp(tY) - 1 \|$$
$$< (\sqrt{2} - 1)^2 + 2(\sqrt{2} - 1) = 1.$$

As the exp and log functions are analytic on the domain of definition, (3.10.4) is clearly a differentiable function of t on $[0, 1]$ (and even a little beyond); we calculate its derivative.

Since

$$\exp(F(t)) = \exp(X) \exp(tY), \tag{3.10.5}$$

the chain rule readily shows that

$$\exp'(F(t)) F'(t) = \exp(X) \cdot \exp'(tY) Y, \tag{3.10.6}$$

where the value of the tangent linear map $\exp'(F(t)) : M_n(\mathbb{R}) \to M_n(\mathbb{R})$ at the matrix $F'(t) \in M_n(\mathbb{R})$ appears on the left-hand side. By (3.9.14),

$$\exp'(tY) Y = \exp(tY)[Y - \mathrm{ad}\,(tY) Y / 2! + \mathrm{ad}\,(tY)^2 Y / 3! - \cdots] = \exp(tY) Y \tag{3.10.7}$$

since obviously $\mathrm{ad}\,(Y) Y = [Y, Y] = 0$. The right-hand side of (3.10.7) is therefore the matrix $\exp(X) \exp(tY) Y = \exp(F(t)) Y$, and so (3.10.7) can also be written as

$$\exp(F(t))^{-1} \cdot \exp'(F(t)) F'(t) = Y, \tag{3.10.8}$$

or, applying (3.9.14) once again, as

$$\phi(\mathrm{ad}\,(F(t)) F'(t) = Y. \tag{3.10.9}$$

For (3.10.2) to be applicable, we need to make sure that $\| \mathrm{ad}\,(F(t)) \| < \log 2$. Since obviously $\| \mathrm{ad}\,(A) \| \leq 2 \| A \|$ for all $A \in M_n(\mathbb{R})$, this amounts to checking that

$$\| F(t) \| < \frac{1}{2} \log 2 \tag{3.10.10}$$

for $0 \leq t \leq 1$, which, as will be seen, is the case if X and Y are sufficiently small. Indeed, the second inequality in (3.1.8) shows that

$$\| F(t) \| \leq \log \frac{1}{1 - \| \exp(X) \exp(tY) - 1 \|}, \tag{3.10.11}$$

whereas

$$\| \exp(X) \exp(tY) - 1 \| < (e^r - 1)^2 + 2(e^r - 1) = e^{2r} - 1 \tag{3.10.12}$$

for $\|X\| < r$ and $\|Y\| < r$. So it is a matter of choosing r so that

$$\log \frac{1}{1 - (e^{2r} - 1)} \le \frac{1}{2} \log 2, \quad \text{i.e. } r \le \frac{1}{2} \log(2 - 1/\sqrt{2}). \tag{3.10.13}$$

Hence, if X and Y are in the ball of radius r, then

$$\begin{aligned}
F'(t) &= \psi(\exp(\text{ad}\,(F(t))))Y = \psi(\text{ad}\,(\exp(F(t))))Y \\
&= \psi(\text{ad}\,[\exp(X)\exp(tY)])Y = \psi[\text{ad}\,(\exp(X))\text{ad}\,(\exp(tY))]Y \quad (3.10.14) \\
&= \psi(e^{\text{ad}\,(X)} e^{t\cdot\text{ad}\,(Y)})Y
\end{aligned}$$

follows by applying (3.6.18). Note that in the previous calculation, the notation ad denotes both the adjoint representation of the group $GL_n(\mathbb{R})$ and the adjoint representation of the Lie algebra $M_n(\mathbb{R})$...

As $F(0) = \log(\exp(X)) = X$, the desired formula follows.

$$F(1) = \log[\exp(X)\exp(Y)] = X + \int_0^1 \psi(e^{\text{ad}\,(X)} e^{t\cdot\text{ad}\,(Y)})Y \, dt, \tag{3.10.15}$$

when X and Y are in the ball in $M_n(\mathbb{R})$ with center 0 and radius $\frac{1}{2}\log(2 - 1/\sqrt{2})$. This makes it possible to show that, as claimed several times, *if $\mathfrak{g}$ is a Lie subalgebra of $M_n(\mathbb{R})$ and if $X, Y \in \mathfrak{g}$ are sufficiently near zero, then $\log[\exp(X)\exp(Y)]$ is also in $\mathfrak{g}$.* For $X, Y \in \mathfrak{g}$, the vector subspace $\mathfrak{g}$ of $M_n(\mathbb{R})$ is indeed invariant under the operators ad (X) and ad (Y), hence under $\exp(\text{ad}\,(X))\exp(t \cdot \text{ad}\,(Y))$ and so also under $\psi(\exp(\text{ad}\,(X))\exp(t \cdot \text{ad}\,(Y)))$. Thus, the integrand on the right-hand side of (3.10.15) is a $\mathfrak{g}$-valued function, and so the result—thus also the left-hand side—is necessarily in $\mathfrak{g}$.

As will be shown in Chap. 6, the authentic Campbell–Hausdorff formula is obtained by replacing the integral on the right-hand side of (3.10.15) by a power series. We stop here our study of general linear (or locally linear) groups, and end this chapter with the computation of some important and simple concrete examples of Lie algebras.

3.11   Examples of Lie Algebras

If H is a closed subgroup of $GL_n(\mathbb{R})$, the set $\mathfrak{h}$ of $X \in M_n(\mathbb{R})$ such that $\exp(tX) \in H$ for all $t \in \mathbb{R}$ is called the *Lie algebra* of H (Theorem 4). Hence, the image of $\mathfrak{h}$ under the exponential map generates H. We next derive some well-known examples of $\mathfrak{h}$.

Example 1 The *group $GL_n(\mathbb{R})$*. Its Lie algebra is obviously the whole of $M_n(\mathbb{R})$. The image of the exponential map contains a neighbourhood of e, hence generates the identity component, namely $GL_n^+(\mathbb{R})$.

Example 2 The *group* $H = SL_n(\mathbb{R})$. The formula

$$\det(\exp(tX)) = e^{t \cdot \mathrm{Tr}(X)} \tag{3.11.1}$$

then shows that $\mathfrak{h}$ is the set of matrices with trace zero in $M_n(\mathbb{R})$. Checking that $\exp(\mathfrak{h})$ contains a neighbourhood of 1 in H is easy. Indeed, $h = \exp(\log h)$ if $\|h - 1\| < 1$ and thus $e^{\mathrm{Tr}(\log h)} = 1$, so that the trace of the matrix $\log h$ is an integral multiple of $2\pi i$. But if h is sufficiently near 1, the matrix $\log h$, and hence also its trace, is sufficiently near 0. Hence, $\mathrm{Tr}(\log h) = 0$ necessarily holds if h is sufficiently near 1 and the desired result follows immediately.

Suppose for example that $n = 2$. Then the three matrices

$$H = \begin{pmatrix} 1 & 0 \\ 0 & -1 \end{pmatrix}, \quad X = \begin{pmatrix} 0 & 1 \\ 0 & 0 \end{pmatrix}, \quad Y = \begin{pmatrix} 0 & 0 \\ 1 & 0 \end{pmatrix} \tag{3.11.2}$$

form a basis of $SL_2(\mathbb{R})$. Easy calculations give the "commutator formulas"

$$[H, X] = 2X, \quad [H, Y] = -2Y, \quad [X, Y] = H. \tag{3.11.3}$$

The one-parameter subgroups generated by X, Y, H are given by

$$\exp(tX) = \begin{pmatrix} 1 & t \\ 0 & 1 \end{pmatrix}, \quad \exp(tY) = \begin{pmatrix} 1 & 0 \\ t & 1 \end{pmatrix}, \quad \exp(tH) = \begin{pmatrix} e^t & 0 \\ 0 & e^{-t} \end{pmatrix}. \tag{3.11.4}$$

Exercise. Show that if the trace of $X = \begin{pmatrix} a & b \\ c & d \end{pmatrix}$ is zero, then

$$\exp(X) = \cosh(\rho)1 + \frac{\sinh(\rho)}{\rho}X, \text{ where } \rho = (a^2 + bc)^{\frac{1}{2}}. \tag{3.11.5}$$

Example 3 For the group H of triangular matrices

$$h = \begin{pmatrix} h_{11} & * & \cdots & * \\ 0 & h_{22} & \cdots & * \\ \vdots & \vdots & \ddots & \vdots \\ 0 & 0 & \cdots & h_{nn} \end{pmatrix}, \quad h_{ii} \neq 0, \tag{3.11.6}$$

$\mathfrak{h}$ consists of all triangular matrices, i.e. of all endomorphisms of $\mathbb{R}^n$ leaving the vector subspace generated by $e_1, \cdots e_i$ invariant for all i. Indeed, if $\exp(tX)$ is triangular for all t, this subspace is invariant under $\exp(tX)$ for all t, hence also under

$$\frac{d}{dt}\exp(tX) = X \cdot \exp(tX),$$

and so equally under X, which must therefore be triangular. The converse is obvious.

The fact that $\exp(\mathfrak{h})$ generates H^0, i.e. the subgroup defined by the inequalities $h_{ii} > 0$, is obvious. Indeed, if $h \in H$ satisfies $\|h - 1\| < 1$, which makes it possible to define $\log(h) = X$, then X is clearly triangular like h and as $h = \exp(\log(h))$ in the neighbourhood of the identity, the desired result follows immediately. We will see later that, in this particular case, $\exp(\mathfrak{h}) = H^0$.

Example 4 Orthogonal and symplectic groups. Let S be an invertible symmetric or skew-symmetric matrix, and H the group of automorphisms $h \in GL_n(\mathbb{R})$ of S defined by

$$h'Sh = S, \tag{3.11.7}$$

where h' is the transpose of h (*orthogonal group* of S when $S' = S$, *symplectic group of S* when $S' = -S$). Now, $X \in \mathfrak{h}$ if and only if

$$S = \exp(tX)'S \cdot \exp(tX) = \exp(tX') \cdot S \cdot \exp(tX) \tag{3.11.8}$$

for all t. Differentiating with respect to t and setting $t = 0$ leads to the condition

$$X'S + SX = 0. \tag{3.11.9}$$

It is in fact sufficient, for differentiating the right-hand side of (3.11.8) gives the expression $\exp(tX')(X'S + SX)\exp(tX)$. Condition (3.11.9) then shows that the function $t \mapsto \exp(tX') \cdot S \cdot \exp(tX)$ is constant, and as it equals S for $t = 0$, our claim follows.

For example, let us take

$$S = \begin{pmatrix} 1_p & 0 \\ 0 & -1_q \end{pmatrix} \text{ with } p + q = n. \tag{3.11.10}$$

It is always possible to reduce a quadratic form on $\mathbb{R}$ to this form. We need to find the matrices of $\mathfrak{h}$ by writing them in the form $X = \begin{pmatrix} a & b \\ c & d \end{pmatrix}$, where $a, \cdots, d$ are matrices of size $p \times p$, $p \times q$, etc. Then (3.11.9) becomes

$$\begin{pmatrix} a' & c' \\ b' & d' \end{pmatrix} \begin{pmatrix} 1 & 0 \\ 0 & -1 \end{pmatrix} + \begin{pmatrix} 1 & 0 \\ 0 & -1 \end{pmatrix} \begin{pmatrix} a & b \\ c & d \end{pmatrix} = 0, \tag{3.11.11}$$

which is equivalent to

$$a + a' = d + d' = 0, \quad b = c'. \tag{3.11.12}$$

So $\mathfrak{h}$ is the set of matrices of type $\begin{pmatrix} a & b \\ b' & d \end{pmatrix}$ with a and d skew-symmetric and b arbitrary. It is easily checked to be a Lie subalgebra of $M_n(\mathbb{R})$, in particular containing the Lie algebra of matrices for which $b = 0$. The latter, consisting of matrices $\begin{pmatrix} a & 0 \\ 0 & d \end{pmatrix}$ with a and d skew-symmetric evidently corresponds to the subgroup of $h \in H$

leaving invariant not just the quadratic form (3.11.10), but also the standard form (corresponding to the identity matrix), i.e. to the subgroup $K = H \cap O(n) \sim O(p) \times O(q)$ of H.

The simplest case is that of $S = 1$, where H is just the *orthogonal group* $O(n)$ of real matrices h such that $h'h = 1$. Its Lie algebra is the set of skew-symmetric matrices X. We show that here the image of the exponential map is the whole of the connected component $H° = SO(n)$. For this we need to show that every orthogonal matrix k having determinant 1 is of the form $\exp(X)$ with $X' + X = 0$. But it is well known that the eigenvalues of a real orthogonal matrix have absolute value 1 and are mutually conjugate, with an even number of eigenvalues being equal to -1 if we are in $SO(n)$, and not just in $O(n)$. If necessary replacing k by uku^{-1}, where u is a real orthogonal matrix, it follows that k may be assumed to be a "diagonal" matrix consisting of a block 1_r, a block -1_s with s even, and a block $\text{diag}(u_1, \cdots, u_m)$ where each u_i belongs to $SO(2)$, and is obtained by grouping together two conjugate imaginary eigenvectors of k belonging to two *non-real* conjugate eigenvalues of k. Obviously, it is then a matter of finding *skew-symmetric real* matrices whose images under exp are respectively equal to 1_r, -1_s and to the u_i. For 1_r, the obvious choice is the null matrix. As s is even, the matrix -1_s can itself be written as $\text{diag}(v_1, \cdots, v_k)$ with all $v_i \in SO(2)$ and equal to -1_2. Hence once every element

$$u = \begin{pmatrix} \cos\theta & -\sin\theta \\ \sin\theta & \cos\theta \end{pmatrix} \qquad (3.11.13)$$

of $SO(2)$ will have been written as $\exp(X)$, where X is a real 2×2 skew-symmetric matrix, we will be done. For this choose

$$X = \begin{pmatrix} 0 & -\theta \\ \theta & 0 \end{pmatrix}, \qquad (3.11.14)$$

notably because the matrices (3.11.13) form a one-parameter group on which Theorem 2 and its proof necessarily apply...

The result obtained about $SO(n)$ can be greatly generalized:

Theorem 9 *(E. Cartan) Let K be a connected compact subgroup of $GL_n(\mathbb{R})$ and $\mathfrak{k}$ its Lie algebra, i.e. the set of matrices $X \in M_n(\mathbb{R})$ such that $\exp(tX) \in K$ for all $t \in \mathbb{R}$. The exponential map from $\mathfrak{k}$ to K is then surjective.*

To prove this result, it is necessary to use everything known about the structure of the compact groups at hand (E. Cartan and H. Weyl's "root" theory) or failing that, albeit no simpler, Hopf and Rinow's theorem which says that in a complete Riemann space (in particular compact), any two arbitrary points can be connected by a geodesic. These results are beyond the level of this presentation.

We will later return to non-compact orthogonal groups, but for now content ourselves with some examples of subgroups of $GL_n(\mathbb{C})$.

Example 5 The group $GL_n(\mathbb{C})$. Theorem 2, which gives the one-parameter real subgroups of $GL_n(\mathbb{R})$, also applies to $GL_n(\mathbb{C})$ and provides the maps

$$t \mapsto \exp(tX), \text{ with } X \in M_n(\mathbb{C}), \tag{3.11.15}$$

from $\mathbb{R}$ to $GL_n(\mathbb{C})$ (assigning complex values to t would give the one-parameter subgroups "over $\mathbb{C}$", i.e. the *complex analytic* homomorphisms from $\mathbb{C}$ to $GL_n(\mathbb{C})$). An alternative equivalent viewpoint consists in observing that $\mathbb{C}^n$ can be canonically identified with $\mathbb{R}^{2n} = \mathbb{R}^n \times \mathbb{R}^n$, multiplication by i becoming the operator $J \in GL_{2n}(\mathbb{R})$ given by $J(x, y) = (-y, x)$. Then, $GL_n(\mathbb{C})$ is the set of $h \in GL_{2n}(\mathbb{R})$ such that $hJ = Jh$, and so, likewise, the Lie algebra $\mathfrak{h} \subset M_{2n}(\mathbb{R})$ of $H = GL_n(\mathbb{C})$ is the set of matrices X such that $XJ = JX$: we thereby recover $M_n(\mathbb{C})$.

Theorem 10 *The map* $\exp : M_n(\mathbb{C}) \to GL_n(\mathbb{C})$ *is surjective.*

Reduction of matrices to the Jordan form indeed shows that (*Cours d'Algèbre*, §35, exer. 14 and 15) every $g \in GL_n(\mathbb{C})$ can be written uniquely as

$$g = g_s g_u \tag{3.11.16}$$

with g_s *semisimple*, g_u *unipotent*, and $g_s g_u = g_u g_s$. Let $\lambda_1, \cdots, \lambda_k$ be the various eigenvalues of g, i.e. of g_s, and $V_1, \cdots, V_k$ the eigenspaces (in $\mathbb{C}^n$) of g_s. For each i, choose $\mu_i \in \mathbb{C}$ such that $\exp(\mu_i) = \lambda_i$, and let H be the operator which, on V_i, reduces to multiplication by μ_i. Clearly, $g_s = \exp(H)$. Moreover, any operator commuting with g_s leaves all V_i invariant, and so commutes with H, and in particular g_u commutes with H. But $g_u = 1 + N$, where N is *nilpotent*, which makes it possible to define

$$\log(1 + N) = N - N^2/2 + N^3/3 - \cdots \tag{3.11.17}$$

without any restrictions—the series reduces to a finite sum. Clearly, formal algebraic computations[11] give

$$g_u = \exp(\log g_u), \tag{3.11.18}$$

and as H commutes with g_u, and so with N, and thus with $\log g_u$, in conclusion,

$$g = \exp(H)\exp(\log g_u) = \exp(H + \log g_u), \tag{3.11.19}$$

---

[11]We remind the reader that it is always possible to substitute a formal series with a formal series *without constant term*, for example the series (3.4.21), which makes the equality

$$\exp(\log(1 + X)) = 1 + X,$$

where X is an indeterminate (and not a matrix!), well defined. Arguing as in Theorem 1—except that there is no longer any need to be concerned about questions of convergence—readily gives the desired identity. When X is replaced by a nilpotent matrix, it reduces to a simple identity between polynomials in X. In other words, $\exp(\log(u)) = u$ for every unipotent matrix u. A similar argument shows that $\log(\exp(N)) = N$ if N is nilpotent (in which case $\exp(N)$ is unipotent, and log becomes well-defined).

qed. (No similar statement holds over $\mathbb{R}$, even for $G_n^+(\mathbb{R})$.)

Example 6   The orthogonal group of a Hermitian form. Writing its matrix $S = S^*$, H is then the subgroup of $h \in GL_n(\mathbb{C})$ such that $h^*Sh = S$. Its Lie algebra $\mathfrak{h}$ is obviously the set of $X \in M_n(\mathbb{C})$ such that $X^*S + SX = 0$. If in particular $S = 1$, in which case H is the *unitary group* $U(n)$, $\mathfrak{h}$ is found to be the set of X such that $X^* = -X$, which says that X is a Hermitian matrix, up to a factor i. In other words, the one-parameter subgroups of $U(n)$ are the maps $t \mapsto \exp(itX)$, where $X = X^*$ is Hermitian.

Élie Cartan's theorem is obvious for $U(n)$. Indeed, any unitary matrix can be diagonalized by applying a unitary transformation. Hence it is a matter of checking that every diagonal unitary matrix is of the form $\exp(iX)$ with $X = X^*$. This amounts to showing that every complex number having absolute value 1 is of the form e^{it} with real t. Note that this argument confirms that $U(n)$ is connected.

3.12   Cartan Decomposition in a Self-Adjoint Group

Although the map $\exp : M_n(\mathbb{C}) \to GL_n^+(\mathbb{R})$ is not surjective,[12] its image contains on the one hand, the orthogonal subgroup $K = SO(n)$, and on the other, the set P^+ of *positive symmetric* matrices in $GL_n^+(\mathbb{R})$. Indeed, diagonalizing a matrix $p = p' \gg 0$, the existence of a *real symmetric matrix* s such that $p = \exp(s)$ follows immediately. It is even unique as is seen by diagonalizing it (the eigenvectors of s and p are the same). It will therefore be denoted by

$$s = \log(p). \tag{3.12.1}$$

For $t \in \mathbb{R}$, set

$$p^t = \exp(t \cdot \log(p)) \tag{3.12.2}$$

so that every $p \in P^+$ can be connected to the identity matrix by a one-parameter subgroup $t \mapsto p^t$ of $GL_n^+(\mathbb{R})$.

We know that every $g \in GL_n^+(\mathbb{R})$ can be expressed uniquely as

$$g = kp \tag{3.12.3}$$

with $k \in K = SO(n)$ and $p \in P^+$; it suffices to take $p = (g'g)^{\frac{1}{2}}$. As k and p are in the image of the exponential map, decomposition (12.3) confirms very clearly that this image generates the connected group $GL_n^+(\mathbb{R})$.

---

[12]If a matrix of the form $\exp(X)$ with $X \in M_n(\mathbb{R})$ has a *negative real* eigenvalue, its multiplicity is *even*, and so this immediately provides counterexamples to the surjectivity of $\exp : M_n(\mathbb{R}) \to GL_n^+(\mathbb{R})$.

The "Cartan decomposition" (3.12.3) in fact applies to many other groups apart from $GL_n^+(\mathbb{R})$.[13] When H is such a subgroup of $GL_n(\mathbb{R})$, then assuredly,

$$g \in H \Longrightarrow K \in H \quad \text{and} \quad p \in H \tag{3.12.4}$$

if (i) $g \in H \Longrightarrow g' \in H$; (ii) $p \in H \cap P^+ \Longrightarrow p^{1/2} \in H$. As will be seen, condition (ii) holds for any *algebraic* subgroup of $GL_n(\mathbb{R})$, i.e. defined by equations of type $q_i(g) = 0$, where the q_i are polynomials in the entries of the matrix g. In fact:

Lemma *Let H be an algebraic subgroup of $GL_n(\mathbb{R})$ and p a semisimple matrix whose eigenvalues are all positive reals. If $p \in H$, then, for all $t \in \mathbb{R}$, $p^t \in H$ as well.*

(As p is semisimple with real eigenvalues, p is diagonalizable in $\mathbb{R}^n$. Define p^t by taking the tth-power of its eigenvalues without modifying its eigenspaces. In this lemma, p need not be assumed to be symmetric.[14])

Since algebraic groups obviously remain algebraic under any basis change, the question reduces to the case where $p = \text{diag}(\lambda_1, \cdots, \lambda_n)$, with strictly positive eigenvalues; hence $p^t = \text{diag}(\lambda_1^t, \cdots, \lambda_n^t)$. As $p^t \in H$ for all $t \in \mathbb{Z}$, it amounts to showing that *if for all $t \in \mathbb{Z}$, p^t is the zero of a polynomial $q(g)$ in g_{ij}, then so is p^t for all $t \in \mathbb{R}$.*

But the value of q at a diagonal matrix $(\lambda_1, \cdots, \lambda_n)$ is obviously of the form

$$\sum a_\alpha \lambda^\alpha, \quad \text{where } \lambda^\alpha = \lambda_1^{\alpha_1} \cdots \lambda_n^{\alpha_n} \tag{3.12.5}$$

where the sum is over all multi-exponents α (with finitely many non-zero coefficients a_α). If the tth-power of each λ_i is taken, t being integral or real, then each monomial in λ_i is of power t. Hence, it is a matter of showing that

$$\sum a_\alpha (\lambda^\alpha)^t = 0 \text{ for all } t \in \mathbb{Z} \Longrightarrow \sum a_\alpha (\lambda^\alpha)^t = 0 \text{ for all } t \in \mathbb{R}. \tag{3.12.6}$$

For all multi-exponents α, the function $f_\alpha(t) = (\lambda^\alpha)^t$ is a homomorphism from the additive group $\mathbb{Z}$ to the multiplicative group of the field $\mathbb{R}$ and the assumption, which can also be stated as $\sum a_\alpha f_\alpha(t) = 0$ for all $t \in \mathbb{Z}$, expresses a linear relation between the f_α. It is, however, well known (*Cours d'Algèbre*, §11, exercise 17), and, moreover easy to see, that *mutually distinct* homomorphisms from a given group to the multiplicative group of a commutative field (or even of a ring) are linearly independent over this field. The equality $\sum a_\alpha f_\alpha(t) = 0$ therefore says that $\sum a_\alpha = 0$, where the sum is over all α such that f_α equals a given f; in other words,

$$\sum_{\lambda^\alpha = \xi} a_\alpha = 0 \text{ for all } \xi \in \mathbb{R}_+^*. \tag{3.12.7}$$

---

[13] And notably to $GL_n(\mathbb{R})$, provided $SO(n)$ is replaced by the non-connected group $O(n)$.

[14] But as p is the conjugate in $GL_n(\mathbb{R})$ of a positive symmetric matrix, this degree of generality is somewhat deceptive.

Then $\sum a_\alpha f_\alpha(t) = 0$ for all $t \in \mathbb{R}$ follows by grouping together terms in an obvious manner, proving the lemma.

Hence, the above arguments finally lead to the following result:

Theorem 11 *Let G be an algebraic subgroup of $GL_n(\mathbb{R})$ such that $G' = G$. For all $g \in G$, the components k and p of the decomposition $g = kp$ ($k \in O(n)$, $p = p' >> 0$) are in G.*

For example, this result applies to the orthogonal group of the quadratic form with matrix

$$S = \begin{pmatrix} 1_p & 0 \\ 0 & -1_q \end{pmatrix}, \quad p + q = n. \qquad (3.12.8)$$

For this group, we obviously have

$$G \cap O(n) = O(p) \times O(q). \qquad (3.12.9)$$

It is also possible to show that it applies to all "semisimple" subgroups of $GL_n(\mathbb{R})$ since such a subgroup is algebraic and invariant under transposition, up to an interior automorphism of $GL_n(\mathbb{R})$.

Note that if $G \subset GL_n(\mathbb{R})$ is a group satisfying the assumptions of Theorem 11, and if $\mathfrak{g} \subset M_n(\mathbb{R})$ denotes its Lie algebra, the anti-automorphism $g \mapsto g' = {}^t g$ of G transforms each one-parameter subgroup $\exp(tX)$ of G into another one, namely $\exp(tX')$. Hence $X \in \mathfrak{g} \implies X' \in \mathfrak{g}$, and so there is a direct sum decomposition

$$\mathfrak{g} = \mathfrak{k} \oplus \mathfrak{p}, \qquad (3.12.10)$$

where $\mathfrak{k}$ is the set of $X \in \mathfrak{g}$ such that $X' = -X$ and $\mathfrak{p}$ the set of $X \in \mathfrak{g}$ such that $X' = X$. Obviously,

$$[\mathfrak{k}, \mathfrak{k}] \subset \mathfrak{k}, \quad [\mathfrak{k}, \mathfrak{p}] \subset \mathfrak{p}, \quad [\mathfrak{p}, \mathfrak{p}] \subset \mathfrak{k}; \qquad (3.12.11)$$

$\mathfrak{p}$ is the Lie subalgebra of the compact subgroup

$$K = G \cap O(n) \qquad (3.12.12)$$

of G, while the elements $X \in \mathfrak{p}$ correspond to the one-parameter subgroups $t \mapsto p^t$ ($p' = p >> 0$) of G. Combining Theorems 9 and 11 finally gives the equality

$$G^0 = \exp(\mathfrak{k}) \cdot \exp(\mathfrak{p}) \qquad (3.12.13)$$

which generalizes (3.12.3). The map $X \mapsto \exp(X)$ being a homeomorphism from $\mathfrak{p}$ (but not from $\mathfrak{k}$!) onto its image, (3.12.13) notably implies that G^0 *is homeomorphic to the Cartesian product of the compact group $G^0 \cap SO(n)$ and a space $\mathbb{R}^N$.*

Chapter 4
Manifolds and Lie Groups

This purpose of this chapter is to present as concisely as possible the basic notions of the theory of manifolds and Lie groups. For a more thorough presentation, the reader is notably referred to M. Berger and B. Gostiaux, *Differential Geometry* (Springer, 1988), M. Spivak, *A Comprehensive Introduction to Differential Geometry, I* (Publish or Perish, Inc., 2d ed., 1979), F.W. Warner, *Foundations of Differentiable Manifolds and Lie Groups* (Scott and Foresman, 1971), J. Dieudonné, *Eléments d'Analyse 3* (Gauthier-Villars, 1970), Chap. XVI (and not Chap. XIX on Lie groups with its totally pointless use of connections) and to go further, especially to V.S. Varadarajan, *Lie Groups, Lie Algebras, and their Representations* (Prentice-Hall, 1974) as well as to A.W. Knapp, *Lie Groups Beyond an Introduction* (Birkhäuser, 2nd ed., 2002), among other possibilities. See also MA IX.

4.1 Manifolds and Morphisms

Throughout this section, r will denote either one of the integers $1, 2, \ldots$, or the symbol ∞, or the symbol ω, and for any open subset U of a finite-dimensional real vector space E, $C^r(U)$ will denote the set of real functions defined on U and of class C^r in the usual sense if $r \neq \omega$, analytic if $r = \omega$. The map $U \mapsto C^r(U)$ is a *sheaf of functions* on E; in other words, given a (not necessarily finite) family of open subsets U_i whose union is U and a function f defined on U,

$$f \in C^r(U) \iff f|U_i \in C^r(U) \text{ for all } i. \tag{4.1.1}$$

An *n-dimensional C^r manifold* is, by definition, an object X consisting of a separable topological space X (first abuse of notation...) and for each open subset U of X, of a set $C^r(U)$ of maps from U to $\mathbb{R}$ satisfying the following conditions:

(V1) *The map $U \mapsto C^r(U)$ is a sheaf of functions*, i.e. for any union U of open sets U_i, condition (4.1.1) characterizes $C^r(U)$;

© Springer International Publishing AG 2017
R. Godement, *Introduction to the Theory of Lie Groups*,
Universitext, DOI 10.1007/978-3-319-54375-8_4

(V2) *for any sufficiently small open subset U of X, there exists a homeomorphism ξ from U onto an open subset Ω of $\mathbb{R}^n$ which, for every open set $U' \subset U$, transforms the set $C^r(U')$ into $C^r(\Omega')$, where $\Omega' = \xi(U')$.*

Condition (V1) states that, for any function defined on an open subset U, the C^r property is a local one; and condition (V2) states that in the neighbourhood of each point of X, the structure is identical to the one obtained by endowing an open subset of $\mathbb{R}^n$ with the sheaf of C^r functions in the usual sense.

Every open subset X of an n-dimensional real vector space E can clearly be considered an n-dimensional C^r manifold (for all r) by endowing it with its C^r functions in the usual sense.

For less trivial examples, consider a C^r manifold X acted on *properly and freely* by a discrete group Γ and respecting the C^r structure of X—this means that, for any open subset U of X and any $f \in C^r(U)$, the function $f \circ \gamma$, where $\gamma \in \Gamma$, is of class C^r on the open set $\gamma^{-1}(U)$ where it is defined. The quotient space X/Γ can then be endowed with a C^r manifold structure as follows: if p is the canonical map from X to X/Γ, then a function f defined on an open subset U of X/Γ is of class C^r if and only if the function $f \circ p$ belongs to $C^r(p^{-1}(U))$. In the neighbourhood of a point $p(a)$ of X/Γ, the C^r structure of X/Γ is obviously identical to that of X in the neighbourhood of a given point (the reader should check this as an exercise).

In particular, this example leads to an analytic structure on $\mathbb{T}^n = \mathbb{R}^n/\mathbb{Z}^n$: take $X = \mathbb{R}^n$, $\Gamma = \mathbb{Z}^n$ and make Γ act on X by integral translations.

Besides, it goes without saying that the constructions of Chap. 3, Sect. 3.7, were aimed at endowing every locally linear group G with the structure of an analytic manifold. The sets of functions $C^r(U)$ were defined in part a of the construction, (V1) was checked in part b, and finally axiom (V2), i.e. the existence of local charts, was checked in parts e and f of the construction.

We will shortly construct more examples using general theorems that will save us from checking in the dark.

Let X be an n-dimensional manifold. A *local chart* (or simply a *chart* when no confusion is possible) of X is defined to be a pair (U, ξ) consisting of an open set $U \subset X$ and of a homeomorphism ξ from U onto an open subset of $\mathbb{R}^n$ satisfying axiom (V2). Setting

$$\xi(x) = (\xi_1(x), \ldots, \xi_n(x)) \quad \text{for all} \quad x \in U, \tag{4.1.2}$$

we then define numerical functions ξ_i on U, namely the *coordinate functions* relative to the chart considered. The chart (U, ξ) will often be written $(U, \xi_1, \ldots, \xi_n)$. For all open sets $U' \subset U$, $C^r(U')$ is fully determined by the functions ξ_i: *a function is of class C^r on U' if and only if it can be expressed in terms of C^r restrictions (in the usual sense) of the coordinate functions ξ_i to U'.* In particular, the ξ_i are of class C^r on U.

Let X and Y be C^r manifolds and f a map from X to Y. Then f is said to be a *morphism* or *of class C^r* if it is continuous and if for any open subset V of Y, and any function $h \in C^r(V)$, the composite function $h \circ f$ is of class C^r on the open set

$f^{-1}(V)$ where it is defined. To state this definition more concretely, let us choose charts $(U, \xi_1, \ldots, \xi_p)$ and $(V, \eta_1, \ldots, \eta_q)$ of X and Y such that $f(U) \subset V$. Then there is a unique map ϕ making the following diagram commutative:

$$
\begin{array}{ccc}
U & \xrightarrow{\quad f \quad} & V \\
{\scriptstyle \xi} \downarrow & & \downarrow {\scriptstyle \eta} \\
\xi(U) & \xrightarrow{\quad \phi \quad} & \eta(V)
\end{array}
\qquad (4.1.3)
$$

This map tells us how to go from the coordinates $x_i = \xi_i(x)$ of a point $x \in U$ to the coordinates $y_j = \eta(y)$ of the corresponding point $y = f(x) \in V$, and it is given by

$$
y_j = \phi_j(x_1 \ldots, x_p), \quad 1 \le j \le q, \qquad (4.1.4)
$$

where the ϕ_j are now numerical functions defined on the open subset $\xi(U)$ of $\mathbb{R}^p$, and such that formulas (4.1.4) define a map from $\xi(U)$ to the open subset $\eta(V)$ of $\mathbb{R}^q$. Besides, note that $y_j = \eta_j(y) = \eta_j \circ f(x)$, so that (4.1.4) can also be written as

$$
\eta_j \circ f(x) = \phi_j(\xi_1(x), \ldots, \xi_p(x)), \quad \text{in short} \quad \eta_j = \phi_j(\xi_1, \ldots, \xi_p). \qquad (4.1.5)
$$

Hence, *a continuous map $f : X \to Y$ is a morphism if and only if, for all charts (U, ξ) and (V, η) such that $f(U) \subset V$, the functions ϕ_j enabling us to compute the map from U to V induced by f are of class C^r in the usual sense (these are functions defined on an open subset of $\mathbb{R}^p$).* This condition is obviously necessary since the left-hand sides of formulas (4.1.5) must be of class C^r on U. Conversely, if it holds, then given that any function h of class C^r on an open set $V' \subset V$ is expressible in terms of the C^r functions η_j, $h \circ f$ is expressible in terms of the C^r functions $\eta_j \circ f$, hence of the ξ_i and so is of class C^r on the open set $U' = f^{-1}(V') \cap U$. Taking account of (V1), it can be immediately deduced that, for any function h of class C^r on an arbitrary subset of Y, the function $h \circ f$ is of class C^r on the corresponding open subset of X.

The composition of the morphisms $f : X \to Y$ and $g : Y \to Z$ clearly results in a morphism $g \circ f : X \to Z$. If a morphism is bijective, and if the converse map is also a morphism, then it is called an *isomorphism* or a *diffeomorphism*. The composition of isomorphisms is again an isomorphism. An isomorphism from X onto X is an *automorphism* of X. Automorphisms of a manifold form a group, etc.

Before we give an example of morphisms, note that if Y is an open subset of a C^r manifold X, then Y can be endowed with an "induced" C^r manifold structure by setting, for any open subset V of Y, the C^r functions on V to be those given by the manifold structure of X. The identity map from Y to X is then clearly a morphism.

In Chap. 3, Sect. 3.7, the construction part also says that, in a locally linear group endowed with its canonical analytic structure, *translations* and the map $x \mapsto x^{-1}$ are automorphisms of the analytic manifold in question. The statement of Theorem 7 of Chap. 3 provides other examples of analytic maps in a similar context.

4.2   The Rank of a Morphism at a Point

Let X and Y be two C^r manifolds and f a C^r map from X to Y; fix $a \in X$ and set $b = f(a)$. Let (U, ξ) and (V, η) be charts of X and Y at a and b such that $f(U) \subset V$. If X and Y are replaced by U and V, identified with open subsets of $\mathbb{R}^p$ and $\mathbb{R}^q$ by ξ and η, then as already seen f may be expressed by formulas $\eta_j = \phi_j(\xi_1, \ldots, \xi_p)$, with $1 \leq j \leq q$. The matrix

$$\left(\frac{\partial \phi_j}{\partial \xi_i}(a) \right), \tag{4.2.1}$$

consisting of the values of the derivatives of ϕ_j at a with respect to the ξ_i, is called the *Jacobian matrix* of ϕ at a relative to the charts considered.

It is dependent on their choice in a way that can be readily determined. Let us indeed replace the charts considered by the charts (U', ξ') and (V', η') satisfying the same conditions as above, and let $\eta'_j = \phi'_j(\xi'_1, \ldots, \xi'_p)$ be the new expressions for f. They indicate how to go from the coordinates $\xi'_i(x)$ of some variable $x \in U'$ to the coordinates $\eta'_j(y)$ of the image $y = f(x) \in V'$, whereas the former formulas told us how to go from the coordinates $\xi_i(x)$ to the coordinates $\eta_j(y)$. In the open set $U \cap U'$, where the charts ξ and ξ' are simultaneously defined, there are two homeomorphisms ξ and ξ' onto open subsets of $\mathbb{R}^p$, differing from each other by an isomorphism from the former open subset onto the latter. In other words, on $U \cap U'$, the ξ'_i are C^r functions of the ξ_i; similarly on $V \cap V'$, the η'_j are C^r functions of the η_j (and conversely). Hence the following operations enable us to go from formulas $\eta_j = \phi_j(\xi)$ to formulas $\eta'_j = \phi'_j(\xi')$:

(i)   write the $\eta'_j(y)$ in terms of the $\eta_j(y)$;
(ii) in the result obtained, replace the η_j by the expressions $\phi_j(\xi)$;
(iii) finally, replace the ξ_j by their expressions *in terms of* the ξ'_i. In other contexts, this is known as a composition of maps...

It is then immediate from the chain rule that formulas such as

$$\frac{\partial \phi'_j}{\partial \xi'_i}(a) = \sum_{k,h} \frac{\partial \eta'_j}{\partial \eta_h}(b) \cdot \frac{\partial \phi_h}{\partial \xi_k}(a) \cdot \frac{\partial \xi_k}{\partial \xi'_i}(a) \tag{4.2.2}$$

hold at a. This means that the Jacobian matrix of f relative to the new charts is obtained from the Jacobian matrix relative to the old charts by multiplying it on the *left* by the Jacobian matrix of the η'_j with respect to the η_h, and on the *right* by the Jacobian matrix of the ξ_k with respect to the ξ'_i.

Note that the obvious formulas

$$\sum_i \frac{\partial \xi_k}{\partial \xi'_i}(a) \cdot \frac{\partial \xi'_i}{\partial \xi_j}(a) = \delta_{kj} \tag{4.2.3}$$

show that the Jacobian matrices arising from a change of chart are *invertible*. Hence if the Jacobian matrix of f at a relative to the given charts is A, its Jacobian matrix relative to any other choice of charts at a will be of the form UAV with U and V invertible. As a consequence, the *rank* of the matrix A is independent of the choice of charts; as a result, it is called the *rank of f at a*. This notion plays a fundamental role in the (initial phase of the) theory.

The rank clearly satisfies

$$\mathrm{rk}_a(f) \leq \inf(\dim(X), \dim(Y)), \tag{4.2.4}$$

and on the other hand, it is a *lower semicontinuous* function at a (in other words, if $\mathrm{rk}_a(f) = r$, then $rk_x(f) \geq r$ for all $x \in X$ sufficiently near a) since if a determinant of order r non-zero at a is extracted from the Jacobian matrix of f, then, by continuity, it will remain non-zero in the neighbourhood of a. The simplest case occurs when

$$\mathrm{rk}_a(f) = \dim X = \dim Y. \tag{4.2.5}$$

Then there is an open set $U \subset X$ containing a and an open set $V \subset Y$ containing b such that f induces an isomorphism from U onto V. The problem being purely local, using charts, X and Y may be assumed to be open subsets of $\mathbb{R}^n$, where $n = \dim X = \dim Y$. The rank of f at a then equals the rank of the usual Jacobian matrix of f at a, and the assumption obviously means that it is invertible (or, what amounts to the same, that its determinant—i.e. the Jacobian of f at a—is non-zero). But then the implicit function theorem shows that there is an open neighbourhood U of a in X and an open neighbourhood V of b in Y such that (i) f induces a homeomorphism from U onto V, (ii) the inverse map from V onto U is of class C^r. Hence the desired result.

The latter is notably useful for the construction of local charts. Let X be an n-dimensional C^r manifold and suppose there are n given C^r functions ξ_i on an open neighbourhood of a given point $a \in X$. These again define a map ξ of class C^r from this neighbourhood to $\mathbb{R}^n$. The restriction of ξ to an open neighbourhood U of a is a chart of U if and only if this restriction is an isomorphism from the manifold U onto an open subset of $\mathbb{R}^n$. By the above, to check the existence of such a neighbourhood of a, it suffices to check that ξ is of rank n at a, i.e. that the Jacobian of the functions ξ_i relative to an arbitrary local chart of X in the neighbourhood of a is non-zero at a. Naturally, from this result, it is not possible to know in advance the open subset U containing a in which the functions ξ_i will prove to be a chart.

4.3 Immersions and Submersions

We now return to a morphism $f : X \to Y$ and investigate the more general case where the rank r of f is *constant in the neighbourhood of a*; then f is said to be a *subimmersion* at a. Two particular cases are especially important: *immersions* at a—that is the case when

$$\text{rk}_a(f) = \dim X \le \dim Y, \tag{4.3.1}$$

and *submersions* at a—that is the case when

$$\text{rk}_a(f) = \dim Y \le \dim X. \tag{4.3.2}$$

The rank is effectively constant in the neighbourhood of a in both cases since it is a lower semicontinuous function of a and attains its *maximum* at a by (4.2.4). For example, the map

$$(x_1, \ldots, x_p) \mapsto (x_1, \ldots, x_p, 0, \ldots, 0) \tag{4.3.3}$$

from $\mathbb{R}^p$ to $\mathbb{R}^q$ with $p \le q$ is an immersion at every point of $\mathbb{R}^p$, and for $p \ge q$, the map

$$(x_1, \ldots, x_p) \mapsto (x_1, \ldots, x_q) \tag{4.3.4}$$

is on the contrary a submersion. More generally, for any fixed integer $r \le \inf(p, q)$, the map

$$(x_1, \ldots, x_p) \mapsto (x_1, \ldots, x_r, 0, \ldots, 0) \tag{4.3.5}$$

from $\mathbb{R}^p$ to $\mathbb{R}^q$ has rank r everywhere; so it is a subimmersion.

   We next show that these three examples are typical of the local structure of immersions, submersions and subimmersions. It suffices to do so in the general case, in other words to prove the following result:

Theorem 1 *Let X and Y be p- and q-dimensional manifolds and f a morphism from X to Y of constant rank r in the neighbourhood of a point a of X. Then there exists a chart (U, ξ) of X at a and a chart (V, η) of Y at $b = f(a)$, with $\xi(a) = \eta(b) = 0$ and $f(U) \subset V$, such that*

$$\eta_1 = \xi_1, \ldots, \eta_r = \xi_r, \eta_{r+1} = \cdots = \eta_q = 0$$

are the equations of f in the charts considered.

   Let (U, ξ) and (V, η) be fixed arbitrary charts of X and Y at a and b. Restricting the domain of the former, we may assume that $f(U) \subset V$. The rank of f may also be assumed to be r at every point of U. Since the rank of a matrix is found by extracting non-zero determinants from it, if necessary by permuting the ξ_i and η_j, we may assume that

$$\det \left(\frac{\partial \phi_j}{\partial \xi_i}(a) \right)_{1 \le i, j \le r} \ne 0 \tag{4.3.6}$$

in the equations $\eta_j = \phi_j(\xi_1, \ldots, \xi_n)$ defining the map from U to V induced by f.

Consider the p functions

$$\xi_1' = \phi_1(\xi), \ldots, \xi_r' = \phi_r(\xi), \xi_{r+1}' = \xi_{r+1}, \ldots, \xi_p' = \xi_p \qquad (4.3.7)$$

on U. Their Jacobian matrix relative to the coordinates $\xi_1, \ldots, \xi_p$ is then invertible at the given point a since it is clearly the matrix

$$\begin{pmatrix}
\partial\phi_1/\partial\xi_1 & \cdots & \partial\phi_r/\partial\xi_1 & 0 & 0 & \cdots & 0 \\
\vdots & & \vdots & \vdots & \vdots & & \vdots \\
\partial\phi_1/\partial\xi_r & \cdots & \partial\phi_r/\partial\xi_r & 0 & 0 & \cdots & 0 \\
* & \cdots & * & 1 & 0 & \cdots & 0 \\
* & \cdots & * & 0 & 1 & \cdots & 0 \\
\vdots & & \vdots & \vdots & \vdots & \ddots & \vdots \\
\cdots & \cdots & * & 0 & 0 & \cdots & 1
\end{pmatrix} \qquad (4.3.8)$$

and since, by assumption, the first r rows and columns form an invertible matrix. Hence, if necessary by replacing U by a smaller open subset, the functions (4.3.7) may be assumed to form a chart in U. Relative to this new chart ξ', f becomes

$$\eta_1 = \xi_1', \ldots, \eta_r = \xi_r', \eta_j = \phi(\xi_1', \ldots, \xi_p') \text{ if } j > r. \qquad (4.3.9)$$

But the rank of the map f is r at every point x of U. Every determinant of order $r+1$ extracted from the Jacobian of f at x is therefore zero. In particular, for $i, j, > r$, the determinant of the matrix consisting of the rows $1, 2, \ldots, r, j$ and the columns $1, \ldots, r, i$ of (4.3.8) is zero; however, taking account of (4.3.9), it now becomes

$$\begin{vmatrix}
1 & 0 & \cdots & 0 & * \\
0 & 1 & \cdots & 0 & * \\
\vdots & \vdots & \ddots & \vdots & \vdots \\
0 & 0 & \cdots & 1 & * \\
0 & 0 & \cdots & 0 & \partial\phi_j'/\partial\xi_i'
\end{vmatrix} = \partial\phi_j'/\partial\xi_i'. \qquad (4.3.10)$$

In other words, $\partial\phi_j'/\partial\xi_i' = 0$ in U for all $i, j > r$. Without any loss of generality, U may be assumed to be connected, and so (4.3.9) becomes

$$\eta_1 = \xi_1', \ldots, \eta_r = \xi_r', \eta_{r+1} = \phi_{r+1}'(\xi_1', \ldots, \xi_r'), \text{ etc.} \qquad (4.3.11)$$

But the Jacobian matrix of the q functions

$$\eta_1, \ldots, \eta_r, \eta_{r+1} - \phi_{r+1}'(\eta_1, \ldots, \eta_r), \ldots, \eta_q - \phi_q'(\eta_1, \ldots, \eta_r) \qquad (4.3.12)$$

of class C^r defined in a neighbourhood of $b = \phi(a)$ of y is clearly invertible at b. Hence, if necessary by replacing V by a smaller open subset (and also U so that condition $\phi(U) \subset V$ continues to hold), the functions (4.3.12) may be assumed to form a chart in V. Denoting this new chart by η', equalities (4.3.9) become

$$\eta'_1 = \xi'_1, \ldots, \eta'_r = \xi'_r, \eta'_{r+1} = \cdots = \eta'_q = 0, \qquad (4.3.13)$$

completing the proof.

We will subsequently return to the particular case of immersions and submersions—in the context of submanifolds and quotient manifolds, but beforehand we present some useful methods for constructing manifolds.

4.4 Gluing Open Submanifolds

As will be experimentally ascertained, we often get the opportunity to endow a topological space X with a manifold structure by endowing every sufficiently small open subset of X with a manifold structure, and by checking that these structures are mutually "compatible". The line of argument is as follows.

As seen above, if X is a C^r manifold, every *open* subset X' of X can be canonically endowed with an "induced" C^r manifold structure, by defining, for any open subset $U \subset X'$, the C^r functions on U, for the manifold X', to be those that are determined by the manifold structure of X. Then if X' and X'' are two open subsets, the manifold structures of X, X' and X'' induce the same manifold structure on the open subset $X' \cap X''$.

Conversely, let us consider a cover of a locally compact space X by open subsets X_i and, for each i, a C^r manifold structure on X_i. If there is a manifold structure on X inducing the given structure on each X_i, then, the induced structures on X_i and X_j induce the same structure on the open subset $X_i \cap X_j$. Conversely, if this condition is satisfied (in which case, the given manifold structures on the X_i are said to be mutually *compatible*), then there is a unique C^r manifold structure on X inducing the given structure on each X_i. For this, define for each open set $U \subset X$ the set $C^r(U)$ of functions f defined on U such that, for all i, the restriction of f to $U \cap X_i$ is of class C^r for the manifold X_i. Checking axiom (V1) is trivial, checking (V2) is almost as trivial (any $a \in X$ is in some X_i, and a chart of X_i in the neighbourhood of a will obviously provide a chart of X in the neighbourhood of a), and finally the manifold structure thereby constructed on X clearly answers the question, and is the only one that does.

For example, let us undertake to endow the projective space

$$X = P^n(\mathbb{R}) = (\mathbb{R}^{n+1} - \{0\})/\mathbb{R}^* \qquad (4.4.1)$$

with a C^r manifold structure. $\mathbb{R}^*$ acts on the non-trivial vectors of $\mathbb{R}^{n+1}$ by homotheties (so that $P^n(\mathbb{R})$ must be the set of 1-dimensional vector subspaces of $\mathbb{R}^{n+1}$). Let p be the canonical map from $\mathbb{R}^{n+1} - \{0\}$ onto $X = P^n(\mathbb{R})$ and let us start by endowing X with the quotient topology: $U \subset X$ is open if so is $p^{-1}(U)$ in $\mathbb{R}^{n+1}$.

Let $x_0, \ldots x_n$ denote the coordinates with respect to the canonical basis of a variable point in $\mathbb{R}^{n+1}$, and X_i ($0 \le i \le n$) the image under p of the set of points of $\mathbb{R}^{n+1} - \{0\}$ such that $x_i \ne 0$. The sets X_i clearly form an *open* cover of X since $p^{-1}(X_i) = \{x_i \ne 0\}$ is open in $\mathbb{R}^{n+1} - \{0\}$ for each i. On the other hand, let H_i be the hyperplane with equation $x_i = 1$ in $\mathbb{R}^{n+1}$. Every line starting at the origin has at most one intersection point with H_i; as a result, p induces a *bijection* from H_i onto X_i. How to endow a hyperplane (or more generally an affine linear manifold in a finite-dimensional vector space over $\mathbb{R}$) with a C^r manifold structure is well known. Hence the map p enables us to "transfer" the manifold structure of H_i to X_i: a function defined on an open subset U of X_i is said to be of class C^r if so is the corresponding function on the corresponding open subset of H_i. The structures thus obtained on the X_i are mutually compatible (whence the desired structure on X) for if we consider the open set $X_{ij} = X_i \cap X_j$ and the corresponding open sets U_{ij} and U_{ji} of H_i and H_j under p, the points $x \in U_{ij}$ and $y \in U_{ji}$ correspond to the same point of X_{ij} if and only if they are collinear with 0. It then remains to check that the map p_{ij} from U_{ij} to U_{ji} associating to each $x \in U_{ij}$ the point $y = p_{ij}(x) \in U_{ji}$ collinear with 0 and x is indeed a manifold isomorphism. But as

$$x = (x_0, \ldots, x_{i-1}, 1, x_{i+1}, \ldots, x_n),$$
$$y = (y_0, \ldots, y_{j-1}, 1, y_{j+1}, \ldots, y_n), \tag{4.4.2}$$

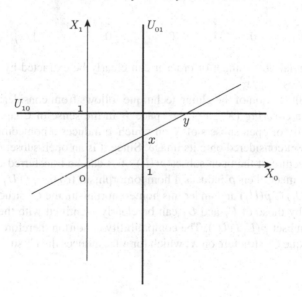

the coordinates of y are calculated in terms of those of x using formulas

$$y_k = x_k / x_j \tag{4.4.3}$$

(note that $x \in U_{ij}$ means precisely that $x_j \ne 0$). Any C^r function of the y_k is therefore a C^r function of the x_k and conversely; hence the desired result.

The manifold structure defined above on $P^n(\mathbb{R})$ is such that the canonical map $p : \mathbb{R}^{n+1} - \{0\} \to P^n(\mathbb{R})$ is a *submersion*. Indeed, consider the hyperplanes H_i and their corresponding open subsets X_i in $X = P^n(\mathbb{R})$. The restriction of p to H_i is a bijection onto X_i, and by construction of the C^r structure on X, it is even an isomorphism from the manifold H_i onto the open subset X_i of the manifold X. As it is also the image under p of the open subset $x_i \neq 0$ of $\mathbb{R}^{n+1} - \{0\}$, the restriction of p to this open subset can be identified with a map from this subset to H_i, namely with the map

$$(x_0, \ldots, x_n) \mapsto (x_0/x_i, \ldots, x_n/x_i). \tag{4.4.4}$$

To check that p is a submersion it suffices to check that so is (4.4.4), which is immediate since the Jacobian matrix of (4.4.4) is

$$\begin{pmatrix} 1/x_i & 0 & \cdots & 0 & 0 & \cdots & 0 \\ 0 & 1/x_i & \cdots & 0 & 0 & \cdots & 0 \\ \vdots & \vdots & \vdots & \vdots & \vdots & \vdots & \vdots \\ 0 & 0 & \cdots & 1/x_i & 0 & \cdots & 0 \\ -x_0/x_i^2 & -x_1/x_i^2 & \cdots & -x_{i-1}/x_i^2 & -x_{i+1}/x_i^2 & \cdots & -x_n/x_i^2 \\ 0 & 0 & \cdots & 0 & 1/x_j & \cdots & 0 \\ 0 & 0 & \cdots & 0 & 0 & \cdots & 0 \\ \vdots & \vdots & \vdots & \vdots & \vdots & \vdots & \vdots \\ 0 & 0 & \cdots & 0 & 0 & \cdots & 1/x_j \end{pmatrix} \tag{4.4.5}$$

and a non-trivial determinant of order n can clearly be extracted by removing the row with entries $-x_k/x_i^2$...

Another illustration of the gluing technique follows from considering a C^r manifold B and a *covering* (X, p) of the space B in the sense of Chap. 2. Denote by (U_i) the family of open subsets of X on which p induces a homeomorphism from the open subset considered onto its image. Since it is an open subset of B, for each i, the C^r structure of the open subset $p(U_i)$ of B can be transferred to U_i using p. If U_i and U_j meet, then p induces a homeomorphism from $U_i \cap U_j$ onto an open subset of $p(U_i) \cap p(U_j)$ and under this homeomorphism, the C^r structures induced on $U_i \cap U_j$ by those of U_i and U_j can be clearly identified with the C^r structure of the open subset $p(U_i \cap U_j)$. The compatibility condition therefore holds, and so there is a unique C^r structure on X, which for all i, induces the C^r structure of $p(U_i)$ on U_i.

4.5   Cartesian Products and Lie Groups

Let X and Y be two C^r manifolds of dimensions p and q respectively. Let (U, ξ) and (V, η) be charts of X and Y, so that U is an open subset of X and ξ an isomorphism from the manifold U onto an open subset of $\mathbb{R}^p$, and with similar conditions holding for V and η. The map $(x, y) \mapsto (\xi(x), \eta(y))$ from $U \times V$ to $\mathbb{R}^p \times \mathbb{R}^q = \mathbb{R}^{p+q}$ is then a bijection (and even a homeomorphism) from the open subset $U \times V$ of the product space $X \times Y$ onto an open subset of $\mathbb{R}^{p+q}$ and enables the transfer of the C^r manifold structure from the open subset $\xi(U) \times \eta(V)$ onto $U \times V$.

Replacing the given charts by two other charts (U', ξ') and $(V'\eta')$ gives a new manifold structure on the open subset $U' \times V'$. We show that it is compatible with the one defined above on $U \times V$, i.e. that the two methods available to endow the open subset $(U \times V) \cap (U' \times V') = (U \cap U') \times (V \cap V')$ of $X \times Y$ with a C^r manifold structure gives the same result. It amounts to showing that for $x \in U \cap U'$ and $y \in V \cap V'$, there is a C^r-isomorphism between two open subsets of $\mathbb{R}^p \times \mathbb{R}^q$ which takes us from the image $(\xi(x), \eta(y))$ of (x, y) in the chart used to define the C^r structure of $U \times V$ to its image $(\xi'(x), \eta'(y))$ in the chart used to define the C^r structure of $U' \times V'$. But there is a C^r-isomorphism from the open subset $\xi(U \cap U')$ of $\mathbb{R}^p$ onto the open subset $\xi'(U \cap U')$ which takes us from $\xi(x)$ to $\xi'(x)$, and similarly there is a C^r-isomorphism from the open subset $\eta(V \cap V')$ of $\mathbb{R}^q$ onto the open subset $\eta'(V \cap V')$ which takes us from $\eta(y)$ to $\eta'(y)$. Hence all we need to show is that if there are two isomorphisms f and g between open subsets of $\mathbb{R}^p$ and $\mathbb{R}^q$ respectively, then the map $(x, y) \mapsto (f(x), g(y))$ is again an isomorphism between open subsets of $\mathbb{R}^{p+q}$; this, we presume, will be obvious to everyone.

The arguments of the previous section then show the existence of a unique C^r manifold structure on the whole of the product $X \times Y$ satisfying the following property: *if (U, ξ) and (V, η) are charts of X and Y, then $(U \times V, \xi \times \eta)$ is a chart of $X \times Y$*, where $\xi \times \eta$ denotes the map $(x, y) \mapsto (\xi(x), \eta(y))$. This gives the product manifold $X \times Y$; by construction, the maps $\mathrm{pr}_1 : X \times Y \to X$ and $\mathrm{pr}_2 : X \times Y \to Y$ are clearly submersions. If there exist real-valued C^r functions f and g defined on the open sets $U \subset X$ and $V \subset Y$, then the function $f \otimes g : (x, y) \mapsto f(x)g(y)$, defined on the open set $U \times V$ is of class C^r on it: it is indeed the product of two C^r functions, namely $f \circ \mathrm{pr}_1$ and $g \circ \mathrm{pr}_2$. Of course it is also possible to use local arguments and reduce the question to open subsets of Cartesian spaces.

The notion of a *group of class C^r*, or of a *Lie group*, follows immediately; it is a group endowed with a manifold structure such that the map $(x, y) \mapsto xy^{-1}$ from $G \times G$ to G is of class C^r. The most obvious example, apart from the additive group $\mathbb{R}^n$, is the group $GL_n(\mathbb{R})$, endowed with the structure obtained by considering it as an open submanifold of the vector space $M_n(\mathbb{R})$. Cramer's formulas for the computation of the inverse of a matrix with a non-zero determinant show that the entries of the matrix xy^{-1} are of the form $p_{ij}(x, y)/\det(y)$, where the p_{ij} are polynomials in the entries of the matrices x and y. These are therefore analytic functions (and so of class C^r for all r) on the open subset $GL_n(\mathbb{R}) \times GL_n(\mathbb{R})$ of the vector space $M_n(\mathbb{R}) \times M_n(\mathbb{R})$, and so the desired result is immediate. We will shortly see that every classical group can be canonically endowed with a Lie group structure. Besides, it might be useful to now indicate that, *if a locally compact group can be endowed with a C^r group structure compatible with its topology, then this structure is unique*. It is also possible to prove that *if a group admits a C^r group structure with $r \neq \omega$, then it necessarily admits an analytic structure* (whose C^r structure in question follows

in an obvious way). In other words, the notion of a C^r group structure is of no value
for $r \neq \omega$...

Note that, on the other hand, to check that a manifold structure on a group G is
compatible with the composition law of G, it suffices to check the following three
conditions:

  (i) left translations $x \mapsto ax$ are of class C^r;
  (ii) the map $x \mapsto x^{-1}$ is of class C^r;
  (iii) the map $(x, y) \mapsto xy$ is of class C^r at the origin.

Combining (i) and (ii) indeed gives the analogue of (i) for right translations, and
combining it with (iii) that $(x, y) \mapsto xy$ is of class C^r everywhere; combining this
with (ii) finally shows that $(x, y) \mapsto xy^{-1}$ is a morphism.

This is what we did in Chap. 3, Sect. 3.7, for $r = \omega$. Checking conditions (i) and
(ii) was the purpose of the construction. As for condition (iii), it was checked in part
g of the construction. The examples of locally linear groups at the end of Chap. 3
are examples of (analytic) Lie groups. We will shortly see that for "classical" or
more generally "algebraic" subgroups of a group $GL_n(\mathbb{R})$, the analytic structure can
be obtained more directly, without having to go through von Neumann's arguments
(which in one form or another remain indispensable for general closed subgroups of
a linear group).

4.6 Definition of Manifolds by Means of Immersions

Let X and Y be two C^r manifolds and $j : Y \to X$ a morphism. Suppose that j is an
immersion at some point b of Y. Set $a = j(b)$ and let p and q be the dimensions of
X and Y, so that $q \leq p$. Then there is a chart $(U, \xi_1, \ldots, \xi_p)$ of X at a and a chart
$(V, \eta_1, \ldots, \eta_q)$ of Y at b such that $j(V) \subset U$ and the equations of j relative to these
charts are given by

$$\xi_1 = \eta_1, \ldots, \xi_q = \eta_q, \, \xi_{q+1} = \cdots = \xi_p = 0. \qquad (4.6.1)$$

To begin with, j is as a consequence injective in the neighbourhood of b. Besides,
if the coordinates are assumed to vanish at a and b, then for all sufficiently small
$r > 0$, there is clearly an open neighbourhood U_r (resp. V_r) of a (resp. b) in U
(resp. V) which the chart ξ (resp. η) maps homeomorphically onto the cube $|\xi_i| < r$
(resp. $|\eta_j| < r$) in $\mathbb{R}^p$ (resp. $\mathbb{R}^q$). $j(V_r) \subset U_r$ necessarily holds, which enables us
to replace U and V by U_r and V_r, in other words to assume that $\xi(U)$ is the cube
$|\xi_i| < r$ and $\eta(V)$ the cube $|\eta_j| < r$. As the cube "of radius r" in $\mathbb{R}^p$ is also the
Cartesian product of the cube of radius r in $\mathbb{R}^q$ and the cube of radius r in $\mathbb{R}^{p-q}$,
it follows that if X and Y are replaced by the open subsets U and V, so that U is
identified with a product $V \times W$, then the map j from V to $U = V \times W$ is identified
with the injection $y \mapsto (y, c)$ for some $c \in W$.

This result—which consists in formulating Theorem 1 in terms of Cartesian products—has important consequences.

Theorem 2 *Let* $j : Y \to X$ *be an immersion (i.e. an immersive morphism at all points of* Y*). Then* j *is locally injective and for all* $b \in Y$*, there exist open neighbourhoods* $U \subset X$ *and* $V \subset Y$ *of* $a = j(b)$ *and* b *such that* $j(V) \subset U$*, and a morphism* $\pi : U \to V$ *such that* $\pi \circ j = \mathrm{id}$ *on* V*. A function* g *defined on an open subset* V *of* Y *is of class* C^r *on* V *if and only if, for all* $b \in V$*, there is a function* f *of class* C^r *defined in the neighbourhood of* $j(b) \in X$ *and such that* $g = f \circ j$ *in the neighbourhood of* b*. A continuous map* h *from a manifold* Z *to* Y *is a morphism if and only if the composite map* $j \circ h$ *from* Z *to* X *is a morphism.*

The first statement is obvious: choose U and V as above so that U can be identified with a product $V \times W$ and j with the injection $y \mapsto (y, c)$ for some $c \in W$; the projection $(y, z) \mapsto y$ then immediately provides the desired morphism π. If, moreover, g is a function defined and of class C^r in the neighbourhood of $b \in Y$, then the function $f = g \circ \pi$ is defined and of class C^r in the neighbourhood of $a = j(b) \in X$, and $f \circ j = g \circ \pi \circ j = g$ in the neighbourhood of b. The claimed characterization of C^r functions on an open subset of Y is then immediate. Finally consider a continuous map h from a manifold Z to Y and suppose that $j \circ h$ is a morphism from Z to X. Consider a point $c \in Z$ and set $b = h(c)$, $a = j(b)$. As above, choose neighbourhoods U and V of a and b and a morphism π from U to V such that $\pi \circ j = \mathrm{id}$ in V. Since h is continuous, there is an open neighbourhood W of c in Z such that $h(W) \subset V$. Then $h = \pi \circ j \circ h$ in W, and as $j \circ h$ and π are of class C^r (on W and U respectively), in conclusion, h is of class C^r on W. Hence the theorem.

The last statement of the theorem will sometimes be referred to as the *universal property of immersions*.

As an aside, note that the functions $g \in C^r(V)$ can also be characterized as follows: since j is continuous and locally injective and since Y is locally compact, j maps every sufficiently small neighbourhood of an element $b \in V$ homeomorphically onto its image in X. So g is of class C^r if, for all $b \in V$ and all sufficiently small neighbourhoods V' of b, the restriction of g to V', considered as a function on $j(V')$, can be extended to a C^r function in the neighbourhood of $a = j(b) \in X$. This raises the question of whether, given a locally compact space Y and a continuous and locally injective map j from Y to a manifold X, it is possible to endow Y with a manifold structure in such a way that j becomes an immersion.

Theorem 3 *Let* X *be a* p*-dimensional* C^r *manifold,* Y *a locally compact space,* j *a continuous locally injective map from* Y *to* X *and* q *an integer less than* p*. The following conditions are equivalent:*

(IMM 1)   For all $b \in Y$*, there exists an open neighbourhood* V *of* b *in* Y *and a chart* (U, ξ) *of* X *at* $a = j(b)$ *such that* j *induces a homeomorphism from* V *onto the subset of* U *defined by equalities*

$$\xi_i(x) = \xi_i(a) \quad for \quad q + 1 \le i \le p; \tag{4.6.2}$$

(IMM 2)   There is a q-dimensional C^r manifold structure on Y such that j is an immersion.

When these conditions hold, there is a unique C^r manifold structure on Y such that j is an immersion.

We only need to show that (IMM 1) implies (IMM 2), the other statements of the theorem having already been proved.

So let us suppose that (IMM 1) holds for some given integer q, and for any open set $V \subset Y$, let us define $C^r(V)$ as follows: the set of functions g on V whose restrictions to any sufficiently small open set $V' \subset V$ can (up to identification by j with a function defined on a subset of X) be extended to a C^r function on an open subset of X. The map $V \mapsto C^r(V)$ is quite clearly a sheaf of continuous maps on Y. The elements of $C^r(V)$ are indeed defined by purely *local* conditions. The validity of condition (V 2) of Sect. 4.1 remains to be checked.

Let V be an open subset of Y. If V is sufficiently small, then according to (IMM 1), there is an open subset U of X and a chart $(\xi_1, \ldots, \xi_p)$ of U such that $0 \in \xi(U)$ and such that j induces a homeomorphism from V onto a subset of U defined by equalities $\xi_{q+1} = \cdots = \xi_p = 0$. Let $\eta_1, \ldots, \eta_q$ be the functions on V which, up to transformation by j, can be identified with the restrictions of the coordinate functions $\xi_1, \ldots, \xi_q$ to $j(V)$. The map

$$\xi : x \mapsto (\xi_1(x), \ldots, \xi_p(x)) \qquad (4.6.3)$$

may be assumed to be a homeomorphism from U onto the cube $|\xi_i| < r$ of $\mathbb{R}^p$; then

$$\eta : y \mapsto (\eta_1(y), \ldots, \eta_q(y)) \qquad (4.6.4)$$

is clearly a homeomorphism from V onto the cube $|\eta_j| < r$ of $\mathbb{R}^q$.

The proof will be complete once we will have shown that for any open set $V' \subset V$, the homeomorphism η from V onto this cube transforms the C^r functions on V' into C^r functions, in the usual sense, on the open subset $\eta(V')$ of $\mathbb{R}^q$. The question being a local one, Y can be replaced by V and X by U, then U and V by their images in $\mathbb{R}^p$ and $\mathbb{R}^q$ respectively; U can then be identified with $V \times W$, where W is a cube in $\mathbb{R}^{p-q}$. The given map j then becomes an injection $j(y) = (y, 0)$ from V to $V \times W = U$. Given the manner in which C^r functions on open subsets of Y have been defined at the beginning of the proof, all that needs to be checked is that, for any open subset V' of V, the C^r functions (in the usual sense) on V' can be *characterized* as follows: a function g defined on V' is of class C^r if and only if, for all $b \in V'$, there is a C^r function f (in the usual sense) defined in a neighbourhood of the point $a = j(b) = (b, 0)$ of U such that $g = f \circ j$ in the neighbourhood of b. But this is obvious, because the projection $\pi : (y, z) \mapsto y$ from U onto V provides a solution to the problem defined on the open subset $\pi^{-1}(V') = V' \times W$ of U, namely the function $f = g \circ \pi$, which obviously is of class C^r on $V' \times W$ if this is also the case for g on V'. The theorem follows.

For example, consider a locally compact group G admitting a locally faithful linear representation $j : G \to GL_n(\mathbb{R})$. *The analytic structure constructed on G in Chap. 3, Sect. 3.7, is then the one which turns j into an immersion.* Indeed, let $\mathfrak{g} \subset M_n(\mathbb{R})$ be the Lie algebra associated to G and j. The analytic structure of Chap. 3, Sect. 3.7, is such that, for all $a \in G$, the map $x \mapsto \log[j(a^{-1}x)]$, which induces a homeomorphism from a neighbourhood of a in G onto a neighbourhood of 0 in $\mathfrak{g}$, is a chart of G in the neighbourhood of a. Similarly, the map $g \mapsto \log[j(a)^{-1}g]$, which induces a homeomorphism from a neighbourhood of $j(a)$ in $GL_n(\mathbb{R})$ onto a neighbourhood of 0 in $M_n(\mathbb{R})$, is a chart of $GL_n(\mathbb{R})$ in the neighbourhood of $j(a)$. Expressing j using these local charts obviously gives a canonical injection (in the neighbourhood of 0) from $\mathfrak{g}$ to $M_n(\mathbb{R})$, which is evidently an immersion since it can be readily formulated as (4.3.3). Hence, the map j from G to $GL_n(\mathbb{R})$ is indeed an immersion, from which it follows that the analytic structure of G is *characterized* by this property.

4.7 Submanifolds

The considerations of the previous sections are especially useful when Y is a (locally compact, i.e. *locally closed*) subspace of X, and j the identity map from Y to X. Condition (IMM 1) can then be reformulated as follows:

(SV) *For any $b \in Y$, there is an open neighbourhood U of b in X and a chart $(\xi_1, \ldots, \xi_p)$ of U such that $U \cap Y$ is a subset of U defined by equations $\xi_{q+1}(x) = \xi_{q+1}(b), \ldots, \xi_p(x) = \xi_p(b)$.*

When this holds for some integer q, Y is said to be a *submanifold* of X. The C^r structure on Y is then obtained by defining $C^r(V)$ for all open subsets V of Y as follows: the set of functions g defined on V and whose restrictions to any sufficiently small open subset of V can be extended to a C^r function on an appropriately chosen open subset of X. Except in the trivial case when Y is open in X, the most important case naturally occurs when Y is a closed subset of X; Y is then said to be a *closed submanifold* of X. It should not be thought that every locally closed, or even closed, submanifold of X satisfies the above condition (SV). In the plane, the graph of the function $y = \sin \frac{1}{x}$ is a locally closed submanifold, but its closure is not a closed submanifold (so that it would be wrong to believe that a locally closed submanifold is always open in a closed *submanifold...*).

In $X = \mathbb{R}^3$, the sphere Y with equation $x^2 + y^2 + z^2 = 1$ is a closed submanifold. To start with, let $(a, b, c) \in Y$ be such that $c \neq 0$. The Jacobian of the functions

$$\xi_1 = x, \quad \xi_2 = y, \quad \xi_3 = x^2 + y^2 + z^2 - 1$$

at the point considered then equals

$$\begin{vmatrix} 1 & 0 & 0 \\ 0 & 1 & 0 \\ 2a & 2b & 2c \end{vmatrix} = 2c \neq 0.$$

So there is an open neighbourhood U of (a, b, c) in $\mathbb{R}^3$ such the restrictions of these functions to U constitute a chart for U. As $U \cap Y$ is defined by the equation $\xi_3 = 0$, condition (SV) holds. The same arguments can be applied to the other points of Y by changing the roles of the variables x, y, and z.

Another method for constructing submanifolds consists in observing that *given C^r submanifolds X and Y and a subimmersion $f : X \to Y$, then $f^{-1}(b)$ is a closed submanifold of X for all $b \in Y$.* Indeed, consider a point $a \in Z = f^{-1}(b)$. There is a chart $(U, \xi_1, \ldots, \xi_p)$ of X at a, a chart $(V, \eta_1, \ldots, \eta_q)$ of Y at b, such that $f(U) \subset V$, and an integer $r \leq \min(p, q)$, such that the restriction of f to U is given by equations $\eta_1 = \xi_1, \ldots, \eta_r = \xi_r, \eta_{r+1} = \cdots = \eta_q = 0$. The ξ_i are obviously assumed to vanish at a and the η_j at b. But then $Z \cap U$ is clearly the subset of U defined by equations $\xi_1 = \cdots = \xi_r = 0$, and the result follows.

In the case of the unit sphere in $\mathbb{R}^3$, we can take $X = \mathbb{R}^3 - \{0\}$ and $Y = \mathbb{R}$, the map f being given by $f(x, y, z) = x^2 + y^2 + z^2$. It is indeed a subimmersion (and even a submersion) on X since the first-order derivatives of f never vanish simultaneously. As a consequence the sphere is a closed subvariety of $\mathbb{R}^3 - \{0\}$, and hence clearly also of $\mathbb{R}^3$.

Similarly, *let X and Y be C^r manifolds and $f : X \to Y$ a submersion* (and not just a subimmersion). Then, *for any submanifold Y' of Y, the inverse image $X' = f^{-1}(Y')$ is a submanifold of X.*

It suffices to show that, for any sufficiently small open subset U of X, $U \cap X'$ is a submanifold of U. But the local structure of submersions, whose universal "model" is the projection of a product manifold onto one of its factors, shows that, setting $V = f(U)$, U may be assumed to be isomorphic to the product of the open subset V of Y with a manifold W. The isomorphism between U and $V \times W$ then transforms f into the projection $\mathrm{pr}_1 : V \times W \to V$. This implies $U \cap X' = U \cap f^{-1}(Y') = (V \cap Y') \times W$. So we finally only need to show another result that is just as indispensable and obvious, namely that *if X and Y are manifolds and X', Y' submanifolds of X and Y, then the product $X' \times Y'$ is again a submanifold of $X \times Y$.* This is an immediate consequence of the definition of products and of submanifolds. Setting out the details of the proof would contribute little to comprehension.

The inverse image of a *submanifold Y'* under a subimmersion $f : X \to Y$ may very well *not be* a submanifold. Take, for example, $X = \mathbb{R}^p$, $Y = \mathbb{R}^{p+q}$ and for the canonical injection f (which in this case is even an immersion). We need to determine whether $Z \cap \mathbb{R}^p$ is a submanifold of $\mathbb{R}^p$ when Z is a submanifold of $\mathbb{R}^{p+q}$. Taking $p = 2, q = 1$, and for Z the hyperboloid of revolution

$$x^2 - y^2 + (z - 1)^2 = 1$$

gives a counterexample. The intersection of Z and the plane $z = 0$ consists of the two lines $y = x$, and $y = -x$, and so *is not* a submanifold because of the "double point" at the origin $x = y = 0$.

Exercise. Show that, in $\mathbb{R}^2$, the equation $x^2 - y^2 = 0$ does not define a submanifold.

The same difficulty arises when we more generally try to find out whether the intersection of two submanifolds Y and Z of a manifold X is also a submanifold of X. Indeed, considering the diagonal map $f : X \to X \times X$ given by $f(x) = (x, x)$, obviously $Y \cap Z = f^{-1}(Y \times Z)$; however, $Y \times Z$ is a submanifold of $X \times X$ and f is clearly an immersion. The answer is yes only if Y and Z cross "transversely" at each of their common points, in a sense that can be specified; see, for example, Dieudonné, Chap. XVI, §8, exercise 9.

4.8 Lie Subgroups

Let G be a Lie group of class C^r and H a subgroup of G. If H is a submanifold of G, then H is said to be a Lie subgroup of G. In this case, H is necessarily closed (a locally compact subgroup of a locally compact group is always closed, Chap. 1, Sect. 1.1). Since, moreover, the map $(x, y) \mapsto xy^{-1}$ from $G \times G$ to G is a morphism, its restriction to $H \times H$ is a morphism from $H \times H$ to H. This can be seen using the following two easy statements:

(i) *if $f : X \to Y$ is a manifold morphism, the restriction of f to a submanifold Z of X is a morphism from Z to Y* (because it is the composition of f and the injection $Z \to X$);

(ii) *if $f : X \to Y$ is a manifold morphism, and if f maps X to a submanifold Z of Y, then f induces a morphism from X to Z* (because the injection $Z \to Y$ is an immersion, the proof is then a consequence of the universal property of Theorem 2).

It follows that if a closed subgroup H of G is also a submanifold of G, the submanifold structure induced on H by that of G is in fact a Lie group structure. As an aside, note that *if a subgroup H of G is closed, it is a submanifold of G*. This has been seen for $G = GL_n(\mathbb{R})$ in Chap. 3, Sect. 3.7, and at the end of Sect. 4.6 of the current chapter, when we proved that, in this case, there exists a unique manifold structure on H such that the canonical injection from H to G is an immersion (equivalently that H is a submanifold of G). The general case, which can be handled likewise, will be explained in Chap. 6 once we will know more about one-parameter subgroups of a Lie group.

The result concerning $GL_n(\mathbb{R})$, namely that its closed subgroups are also sub-manifolds of $GL_n(\mathbb{R})$ and so are Lie groups, applies in particular to all the *classical groups* of Chap. 3. As claimed at the end of Sect. 4.5, we next show how, in this case, this result can be obtained without appealing to von N's arguments.

To start with, consider a Lie group G, a manifold X, and suppose given a morphism from $G \times X$ to X, which we denote $(g, x) \mapsto gx$, and satisfying the obvious algebraic conditions:

$$g'(g''x) = (g'g'')x, \quad ex = x. \tag{4.8.1}$$

The Lie group G is then said to *act* (on the left) on the manifold X or, to avoid confusion with the set theory context, G acts "morphically" or "differentiably" or "analytically" or what have you. A first important remark is that the map $g \mapsto gu$ from G to X is a *subimmersion* for all $u \in X$, i.e. its rank is locally (or even globally) constant. Indeed, taking two points a and b of G, and setting $c = ba^{-1}$, we get a commutative diagram

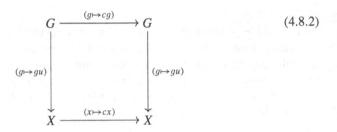

$$\tag{4.8.2}$$

in which the horizontal arrows are manifold automorphisms of G and X. The map $g \mapsto gu$ therefore has the same rank at corresponding points under $g \mapsto cg$, and in particular at a and b, and thus the desired statement follows.

This implies that, for any two points u, v of X, the set of g such that $gu = v$ is a submanifold of G; in particular, *the stabilizer of any point of X is a Lie subgroup of G.*

This result applies notably in the context of a Lie group G and a *linear represen-tation* of G. This is the name given to any morphism $g \mapsto T(g)$ from G to a group $GL_n(\mathbb{R})$, i.e. to any C^r map satisfying the equalities

$$T(e) = 1, \quad T(g'g'') = T(g')T(g''). \tag{4.8.3}$$

In Chap. 6 we will show that, for a map satisfying (4.8.3), continuity implies ana-lyticity. Thanks to such a linear representation T of G, G can be made to act on the space $\mathbb{R}^n$ using the map $(g, x) \mapsto T(g)x$ from $G \times \mathbb{R}^n$ to $\mathbb{R}^n$. It follows that *for every vector $a \in \mathbb{R}^n$, the set of $g \in G$ such that $T(g)a = a$ is a Lie subgroup of G.*

For example, to check that the orthogonal group of an $n \times n$ non-degenerate symmetric matrix S is a Lie group, the previous arguments can be applied to the group $GL_n(\mathbb{R})$ and to its representation T which associates to each $g \in GL_n(\mathbb{R})$ the automorphism $T(g)$ of the vector space of $n \times n$ symmetric matrices given by

$$T(g)X = gXg'; \tag{4.8.4}$$

the orthogonal group of S consists precisely of the elements g such that $T(g)S = S$ and the desired result follows. This argument can easily be modified to apply to other classical groups. For example, to prove that $GL_n(\mathbb{C})$ admits a "natural" Lie group structure, identify $\mathbb{C}^n$ with $\mathbb{R}^{2n}$ and $GL_n(\mathbb{C})$ with a subgroup of $GL_{2n}(\mathbb{R})$, more precisely with the subgroup defined by equation

$$g \circ J = J \circ g, \tag{4.8.5}$$

where J is multiplication by i in $\mathbb{C}^n$, considered as an automorphism of $\mathbb{R}^{2n}$. Then, considering the representation T of $GL_{2n}(\mathbb{R})$ for which $T(g)$ is the automorphism $u \mapsto g \circ u \circ g^{-1}$ of $M_{2n}(\mathbb{R})$, $GL_n(\mathbb{C})$ is seen to be a subgroup of $GL_{2n}(\mathbb{R})$ defined by the equation $T(g)J = J$; hence the desired result.

This kind of reasoning applies to any *real linear algebraic group*, i.e. to any subgroup of the group $GL(E)$ of a finite-dimensional real vector space E definable by finitely many polynomial equations in the entries g_{ij} of the matrix of $g \in GL(E)$ with respect to a given basis of E. To see this, consider first the *rational representations* of $GL(E)$: these are the homomorphisms $g \mapsto T(g)$ from $GL(E)$ to the group $GL(F)$ of a finite-dimensional real vector space F such that, given fixed bases for E and F, the entries of the matrix of $T(g)$ are expressions of type $P(g)/\det(g)^N$, where P is a polynomial function of the entries of g and N a natural integer. Clearly, T is then a morphism from the Lie group $GL(E)$ to the Lie group $GL(F)$. As a consequence, the Lie group $GL(E)$ can be made to act on the manifold F via the map $(g, x) \mapsto T(g)x$. It is also possible to make $GL(E)$ act likewise on the manifold $P(F) = (F - \{0\})/\mathbb{R}^*$ defined in Sect. 4.4 (projective space of F). It follows that, for all $x \in F$, the elements $g \in GL(E)$ such that $T(g)x = x$ form a Lie subgroup (obviously an algebraic one) of $GL(E)$, and similarly, for a line $D \in P(F)$ in F, the elements $g \in GL(E)$ for which D is $T(g)$-invariant form a Lie subgroup (as algebraic as the previous one) of $GL(E)$.

Now, the second method gives *all* the algebraic subgroups H of $G = GL(E)$. Proceed as follows. Choose real polynomial functions P_i $(1 \le i \le N)$ on G such that H is defined by the equations $P_i(g) = 0$. In the vector space of all maps from G to $\mathbb{R}$, consider the subspace V generated by the maps $g \mapsto P_i(ag)$, for all i and all $a \in G$. It is finite-dimensional (because the P_i are *polynomial* functions) and contains the subspace W generated by the functions $P_i(ag)$ with $a \in H$. Left translations

$$L(x) : \{g \mapsto P(g)\} \mapsto \{g \mapsto P(x^{-1}g)\} \tag{4.8.6}$$

by the elements of G define a rational representation $x \mapsto L(x)$ of G in V, and clearly, $x \in H$ if and only if W is $L(x)$-invariant. Letting d be the dimension of W, we then get the desired representation $T : GL(E) \to GL(F)$ by setting[1]

$$F = \wedge^d(V), \quad T(g) = \wedge^d(L(g)), \quad D = \wedge^d(W), \tag{4.8.7}$$

where $\wedge^d(V)$ is the space of skew-symmetric d-linear forms in the dual V^* of V.

4.9 Submersions and Quotient Manifolds

We now state properties of submersions and quotient manifolds corresponding, by some sort of duality, to the properties of immersions and submanifolds proved in Sects. 4.6 and 4.7.

Theorem 4 *Let $\pi : X \to Y$ be a submersion (i.e. a submersive morphism at every point of X). Then π is open and for any $a \in X$, there are open neighbourhoods U and V of a and $b = \pi(a)$ in X and Y, such that $\pi(U) = V$, and a morphism $j : V \to U$ such that $\pi \circ j = $ id in V. A function g defined on an open subset V of $\pi(X)$ is of class C^r on V if and only if the function $g \circ \pi$ is of class C^r on the open subset $\pi^{-1}(V)$ of X. A map h from the open subset $\pi(X)$ of Y to a manifold Z is a morphism if and only if the map $h \circ \pi$ from X to Z is a morphism.*

To construct the neighbourhoods U and V and the morphism j whose existence is asserted by the theorem, it suffices to notice the existence of local charts (U, ξ) and (V, η) of X and Y at a and b, with $\pi(U) \subset V$, in which the equations defining π are

$$(\xi_1, \ldots, \xi_p) \mapsto (\xi_1, \ldots, \xi_q) \tag{4.9.1}$$

if X and Y are respectively of dimensions p and q. It may be assumed that $\xi(a) = \eta(b) = 0$ and that $\xi(U)$ and $\eta(V)$ are the cubes $|\xi_i| < r$ and $|\eta_j| < r$ in $\mathbb{R}^p$ and $\mathbb{R}^q$, respectively. Identifying U and V with these cubes, the question reduces in U to the case where $U = V \times W$ and π is the projection onto the first factor $(y, z) \mapsto y$. The injection $j(y) = (y, 0)$ from V to $V \times W$ then answers the question.

The second and third statements of the theorem follow readily since it is then essentially a matter of checking that, given a product manifold $U = V \times W$, a map from V (or from an open subset V' of V) to $\mathbb{R}$ or more generally to a manifold Z is of class C^r if and only if its composition with the first projection $\mathrm{pr}_1 : U \to V$ is of class C^r on the open subset where it is defined. Simply put: a function of x is, for example, analytic if and only if it is analytic as a function of (x, y).

Theorem 4 shows that if π is *surjective*, then the manifold structure of Y is fully determined by the requirement that π be a submersion: for any open subset V of Y,

---

[1]For a direct proof of the fact that linear algebraic groups are Lie groups, see also Varadarajan, pp. 22–25 and 45.

$C^r(V)$ consists of functions g such that $g \circ \pi \in C^r(\pi^{-1}(V))$. Besides, the requirement that π be a submersion—hence a *continuous open* map—even determines the topology of Y, a subset V of Y being open if and only if so is $\pi^{-1}(V)$ in X. In other words, as a topological space, Y is identified with the quotient of X by the equivalence relation $\pi(x') = \pi(x'')$ determined by π.

This raises the question as to when the quotient X/R of a manifold X by an equivalence relation $R \subset X \times X$ can be endowed with a manifold structure in such a way that $p : X \to X/R$ is a submersion. The answer is well known, but is significantly more difficult than Theorem 3 on the analogous problem for immersions.

Theorem 5 *Let $R \subset X \times X$ be an equivalence relation on a C^r manifold X. There is a C^r manifold structure on X/R such that the projection $p : X \to X/R$ is a submersion if and only if R satisfies the following two conditions:*

(RER 1)   *R is a closed submanifold of $X \times X$;*
(RER 2)   *the map $\mathrm{pr}_2 : R \to X$ is a submersion.*

The manifold structure on X/R is then unique.

The necessity of the conditions is easily seen. Indeed, suppose that the problem has been solved. The graph R is the inverse image of the diagonal of $X/R \times X/R$ under the map $(x, y) \mapsto (p(x), p(y))$ from $X \times X$ onto $X/R \times X/R$. As p is a submersion, then obviously so is this map, and as the diagonal is a closed submanifold of the product, it becomes a question of showing that the inverse image of a closed submanifold under a submersion is always a closed submanifold. But this has already been proved in Sect. 4.7, and hence (RER 1) follows.

To prove the necessity of (RER 2), observe that the nature of the problem is local. Since, locally, a submersion is the projection from a product onto a factor, it may be assumed that $X/R = U$ is a manifold, that $X = U \times V$ with another manifold V, and that $p = \mathrm{pr}_1$. The graph R then consists of the elements $((u_1, v_1), (u_2, v_2)) \in (U \times V) \times (U \times V)$ such that $u_1 = u_2$, so that R can be identified with $U \times V \times V$, the projection pr_2 of (RER 2) then being identified with the map $(u, v_1, v_2) \mapsto (u, v_2)$ from $U \times V \times V$ onto $U \times V$. Indeed, this is clearly a submersion, proving (RER 2).

The fundamental result is that conditions (RER 1) and (RER 2) *characterize* *regular* equivalence relations R, i.e. for which there exists a manifold structure on X/R with the property that p is a submersion. We reproduce the proof given in J.P. Serre, *Lie Algebras and Lie Groups* and which somewhat simplifies an earlier proof by the author.[2] In what follows, we thus start with a relation R satisfying (RER 1) and (RER 2).

Lemma 1 *Let $X = \bigcup X_i$ be a cover of X by open subsets X_i, saturated with respect to R. Suppose that R induces a regular equivalence relation on each X_i. Then, R is regular.*

---

[2]For a somewhat different proof, see Pham Mau Quan, *Introduction à la géométrie des variétés différentiables* (Dunod, 1969).

By assumption, each open subset X_i/R of X/R can be endowed with a C^r structure such that $p : X_i \to X_i/R$ is a submersion. These structures are compatible in the sense of Sect. 4.4; indeed, the C^r structures of X_i/R and X_j/R induce two C^r structures on $(X_i/R) \cap (X_j/R) = (X_i \cap X_j)/R$ such that the restriction of the projection p to $X_i \cap X_j$ is a submersion in both cases. But as the X_i are saturated, the map

$$p : X_i \cap X_j \to (X_i/R) \cap (X_j/R) \tag{4.9.2}$$

is surjective, and hence, as seen above, there can only be one C^r structure on the right-hand set which transforms p into a submersion.

Hence, Sect. 4.4 shows the existence of a C^r structure on X/R inducing the desired structure on each X_i/R, and with respect to this structure, the map $p : X \to X/R$ is a submersion, since the question is of a local nature. Hence the lemma.

Lemma 2 *The map p is open and X/R is locally compact.*

Indeed, if U is an open subset of X, then

$$p^{-1}(p(U)) = \mathrm{pr}_2[(U \times X) \cap R], \tag{4.9.3}$$

and as $\mathrm{pr}_2 : R \to X$ is a submersion, and so is open, the right-hand side is open in X. As a consequence, $p(U)$ is open in X/R, and the map p is indeed open. As R is closed, the quotient space X/R is separable. Indeed, take points $a, b \in X$ such that $p(a) \neq p(b)$. We need to find neighbourhoods U and V of a and b in X such that $p(U)$ and $p(V)$ are disjoint. For obvious reasons at every point of X there is a countable fundamental system of neighbourhoods. Hence, if this were not the case, then X would contain a sequence (a_n) converging to a and a sequence (b_n) converging to b with $p(a_n) = p(b_n)$, i.e. such that $(a_n, b_n) \in R$ for all n. As R is closed, it would follow that at the limit, $(a, b) \in R$, contrary to the fact that $p(a) \neq p(b)$. Finally, the local compactness of X/R follows from the fact that p maps a compact neighbourhood of $x \in X$ onto a compact neighbourhood of $p(x)$.

Lemma 3 *Let U be an open subset of X such that $p(U) = X/R$. If R induces a regular equivalence relation on U, then R is regular on X.*

By assumption, there is a manifold structure on X/R such that $p : U \to X/R$ is a submersion. It is a matter of showing that likewise there is a canonical map from X (and no longer U) onto X/R. To do so, consider the commutative diagram

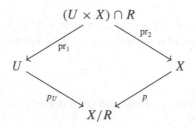

where p_U is the restriction of p to U. As $\text{pr}_2 : R \to X$ is a submersion, then by symmetry so is $\text{pr}_1 : R \to X$. Hence, in turn so is the left pr_1 arrow, and by assumption this is also the case for p_U. As a consequence, $p_U \circ \text{pr}_1 = p \circ \text{pr}_2$ is a submersion, and perforce so is p, since $\text{rk}(AB) \leq \text{rk}(A)$ for all matrices A and B.

Lemma 4 *Let $X = \bigcup X_i$ be an open covering of X; suppose that R induces a regular equivalence relation on each X_i; then R is regular.*

Indeed, Lemma 3 enables us to replace X_i by the R-saturation of X_i, and then Lemma 1 can be applied to this open set.

Lemma 4 shows that to prove the theorem it suffices to show that every point of X has an open neighbourhood on which R induces a regular equivalence relation. This is what we do in the following lemmas.

Lemma 5 *Set $\dim X = n$ and $\dim R = n + p$. For all $a \in X$, there are charts ξ and η of X in the neighbourhood of a such that the restrictions of the $n + p$ functions*

$$
\begin{aligned}
\xi_i \circ \text{pr}_1 : (x, y) \mapsto \xi_i(x) \quad (1 \leq i \leq n), \\
\eta_j \circ \text{pr}_2 : (x, y) \mapsto \eta_j(y) \quad (1 \leq i \leq p)
\end{aligned}
\tag{4.9.4}
$$

to R form a chart of R in the neighbourhood of (a, a).

Let us start with a chart (U, ξ) of X such that $a \in U$. Since $\text{pr}_1 : R \to X$ is a submersion, in the neighbourhood of (a, a) in R it behaves as the projection from a Cartesian product onto one of its factors. Thus, the restrictions to R of the functions $\xi_i \circ \text{pr}_1$ defined in the neighbourhood of (a, a) are part of a local chart of R in the neighbourhood of (a, a). Hence, in the neighbourhood of (a, a) in $X \times X$, there are p other C^r functions $\zeta_1, \ldots, \zeta_p$ such that the restrictions of the $n + p$ functions $\xi_i \circ \text{pr}_1$ and ζ_j to R form a chart of R at (a, a).

To continue the proof of Lemma 5, we need a result whose proof is left as an exercise.

Exercise. Let E be an m-dimensional manifold, F a q-dimensional submanifold and $\psi_1, \ldots, \psi_q$ C^r functions defined in a neighbourhood (in E) of a point a of F. The restrictions of the functions ϕ_i to F form a local chart of F at a if and only if there are $m - q$ other C^r functions $\phi_{q+1}, \ldots, \phi_m$ in the neighbourhood of a, vanishing on F and such that the m functions $\phi_1, \ldots, \phi_m$ form a local chart of E at a.

The exercise can be solved by reducing the question—locally—to the case $E = \mathbb{R}^m$ and $F = \mathbb{R}^q$, and by writing down the Jacobian matrices of F at $a = 0$.

This exercise shows that, in the neighbourhood of (a, a) in $X \times X$, there are in fact n C^r functions ζ_i $(1 \leq i \leq n)$ such that:

(i) the $2n$ functions ζ_i and $\xi_i \circ \text{pr}_1$ form a chart of $X \times X$ at (a, a);
(ii) the restrictions to R of the $n + p$ functions ζ_i $(1 \leq i \leq p)$ and $\xi_i \circ \text{pr}_1$ form a chart of R at (a, a);

(iii) all $n - p$ functions ζ_j $(p + 1 \leq j \leq n)$ vanish at every point of R sufficiently
 near (a, a).

However, the $2n$ functions $\xi_i \circ \mathrm{pr}_1$ and $\xi_i \circ \mathrm{pr}_2$ also form a chart of $X \times X$ at (a, a).
So property (i) means that the Jacobian of the functions

$$(x, y) \mapsto \xi_i(x) \text{ and } (x, y) \mapsto \zeta_i(x, y) \qquad (4.9.5)$$

with respect to the functions

$$(x, y) \mapsto \xi_i(x) \text{ and } (x, y) \mapsto \xi_i(y) \qquad (4.9.6)$$

is non-zero at (a, a), or else that the partial derivatives at (a, a) of the functions
$(x, y) \mapsto \zeta_i(x, y)$ with respect to the coordinates $\xi_i(y)$ evidently form a matrix with
a non-zero determinant. But the values of these partial derivatives remain unchanged
at (a, a) if some of the functions $(x, y) \mapsto \zeta_i(x, y)$ are replaced by the corresponding
functions $(x, y) \mapsto \zeta_i(a, y)$. It follows that, regarded as functions of (x, y) defined
in the neighbourhood of (a, a), the $2n$ functions

$$\xi_i(x)\,(1 \leq i \leq n), \quad \zeta_j(a, y)\,(l \leq j \leq p), \quad \zeta_k(x, y)\,(p < k \leq n) \qquad (4.9.7)$$

again form a chart of $X \times X$ in the neighbourhood of (a, a). Taking into account the
preceding exercise once more and the above point (iii), in conclusion, the restrictions
to R of the $n + p$ functions

$$\xi_i(x)\,(1 \leq i \leq n) \quad \text{and} \quad \zeta_j(a, y)\,(1 \leq j \leq p) \qquad (4.9.8)$$

form a chart of R in the neighbourhood of (a, a). But we saw above that the Jacobian
of the n functions

$$\eta_j(y) = \zeta_j(a, y) \quad (1 \leq j \leq n) \qquad (4.9.9)$$

with respect to the n functions $\xi_i(y)$ is non-zero at $y = a$. Consequently, functions
(4.9.9) form a chart of X in the neighbourhood of a, proving the lemma.

   Note that with this construction of the functions η_j, chart (4.9.7) of $X \times X$ in the
neighbourhood of (a, a) now consists of the $2n$ functions

$$\xi_i(x)\,(1 \leq i \leq n), \quad \eta_j(x)\,(1 \leq j \leq p), \quad \zeta_k(x, y)\,(p + 1 \leq k \leq n). \qquad (4.9.10)$$

Let U and V be the domains of the charts ξ and η of X. These are open neighbourhoods of a in X isomorphically mapped onto open subsets of $\mathbb{R}^n$ under ξ and η. If
need be by making U and V smaller, it may be assumed, on the one hand, that the
functions (4.9.10) form a chart of $U \times V$ (i.e. that the ζ_k are defined on $U \times V$ and
that functions (4.9.10) isomorphically map $U \times V$ onto an open subset of $\mathbb{R}^{2n}$), and
on the other, that $R \cap (U \times V)$ is the subset of $U \times V$ defined by the equations
$\zeta_{p+1}(x, y) = \cdots = \zeta_n(x, y) = 0$. All ξ_i and η_j will be assumed to vanish at a.

Since the $n + p$ functions

$$\xi_i(x)\,(1 \le i \le n), \quad \eta_j(y)(1 \le j \le p) \tag{4.9.11}$$

form a chart of $R \cap (U \times V)$—indeed, $R \cap (U \times V)$ is the subset of $U \cap V$ on which the last $n - p$ coordinates (4.9.10) vanish, the image of $R \cap (U \times V)$ under this chart contains a cube centered at 0 in $\mathbb{R}^n \times \mathbb{R}^p$ and in particular contains $\xi(U') \times \{0\}$ if $U' \subset U$ is a sufficiently small neighbourhood of a in X. Thus, for all $x \in U'$, there *exists a unique* point y satisfying the following conditions: $(x, y) \in R \cap (U' \times V)$ and $\eta_j(y) = 0$ for $1 \le j \le p$. Besides, y can still be characterized as *the* point of V such that

$$\eta_1(y) = \cdots = \eta_p(y) = \zeta_{p+1}(x, y) = \cdots = \zeta_n(x, y) = 0; \tag{4.9.12}$$

indeed, the last $n - p$ conditions state that $(x, y) \in R \cap (U' \times V)$.

Setting $y = s(x)$, we then get a *morphism* s from U' to V. As $s(x) = \mathrm{pr}_2(x, s(x))$, this will follow once the map $x \mapsto (x, s(x))$ from U' to $R \cap (U' \times V)$ will have been shown to be of class C^r. But if $R \cap (U' \times V)$ is identified with its image in chart (4.9.11), the map considered becomes the map $x \mapsto (\xi(x), 0)$ from U' to $\mathbb{R}^n \times \mathbb{R}^p$ and the desired result is then obvious.

We next show that if x_1 and x_2 are two points of U', then

$$(x_1, x_2) \in R \iff s(x_1) = s(x_2). \tag{4.9.13}$$

As x_1 (resp. x_2) belongs to the equivalence class of $s(x_1)$ (resp. $s(x_2)$), the right-hand side trivially implies the left-hand side. Conversely, if $(x_1, x_2) \in R$, then $(x_1, s(x_2)) \in R \cap (U' \times V)$ and moreover, $\eta_j(s(x_2)) = 0$ for $1 \le j \le p$, and so $s(x_2) = s(x_1)$ since, as seen above while defining s, these two conditions characterize $s(x_1)$.

Finally, we show that s *is of constant rank in the neighbourhood of a*. For this, we need to compute the Jacobian matrix of the function $y = s(x)$, which we do by taking the chart ξ for x and the chart η for y. Hence it is a matter of computing the partial derivatives of all $\eta_j(y)$ with respect to the $\xi_i(x)$, and given the first p equalities (4.9.12), we need only attend to the η_j with index $j \ge p + 1$. For the computation, consider the $\xi_k(x, y)$ as a function of the coordinates $\xi_i(x)$ and $\eta_j(y)$ and differentiate the equality

$$\zeta_k(x, s(x)) = 0. \tag{4.9.14}$$

The chain rule shows that

$$\frac{\partial \zeta_k}{\partial \xi_i}(x, s(x)) + \sum_{j=p+1}^{n} \frac{\partial \zeta_k}{\partial \eta_j}(x, s(x)) \cdot \frac{\partial \eta_j(s(x))}{\partial \xi_i} = 0. \tag{4.9.15}$$

But as functions (4.9.10) form a chart of $X \times X$ in the neighbourhood of (a, a), the $n - p \times n - p$ square matrix $(\partial \zeta_k / \partial \eta_j)$ $(p + 1 \le j, k \le n)$ is clearly invertible in the neighbourhood of (a, a). Formula (4.9.15) then readily shows that for x sufficiently near a, the matrices

$$\left(\frac{\partial \eta_j(s(x))}{\partial \xi_i}\right)_{\substack{1 \leq i \leq n \\ p+1 \leq j \leq n}} \text{ and } \left(\frac{\partial \zeta_k}{\partial \xi_i}(x, s(x))\right)_{\substack{1 \leq i \leq n \\ p+1 \leq j \leq n}} \tag{4.9.16}$$

have the same rank. But as functions (4.9.10) form a chart of $X \times X$, the second matrix in (4.9.16) is clearly of maximal rank, i.e. $n - p$, in the neighbourhood of $x = a$. Hence so is the former one, and therefore, finally,

$$\mathrm{rk}_x(s) = n - p \tag{4.9.17}$$

in the neighbourhood of a, as claimed.

However, s is then a subimmersion in the neighbourhood of a and if U' is taken to be sufficiently small, then Theorem 1 shows that $s = j \circ p$, where p is a *submersive* map from U' onto a submanifold S of X and j is the canonical immersion from S to X. (4.9.13) can obviously be replaced by

$$(x_1, x_2) \in R \iff p(x_1) = p(x_2). \tag{4.9.18}$$

So in conclusion, if we confine ourselves to the open set U', then there is a *submersion* $p : U' \to S$ such that R is the equivalence relation $p(x_1) = p(x_2)$ in U'. This proves that the restriction of R to U' is a regular equivalence relation, and as every point of X has a neighbourhood U' satisfying this condition, Lemma 4 now implies Theorem 5, completing its proof.

Corollary *Let G be a Lie group and H a (closed) Lie subgroup of G. There is a unique manifold structure on G/H with the property that the canonical map p from G onto G/H is a submersion. The map $(g, x) \mapsto gx$ from $G \times G/H$ to G/H is then a morphism.*

Indeed, G/H is the quotient of G by the equivalence relation $x^{-1}y \in H$. Let R be its graph, and let us consider the maps u and v from $G \times G$ to $G \times G$ given by

$$u(x, y) = (x, x^{-1}y), \quad v(x, y) = (x, xy); \tag{4.9.19}$$

these are mutually inverse *automorphisms* of the manifold $G \times G$. Since $R = v(G \times H)$ plainly holds and as $G \times H$ is a closed subset of $G \times G$, R is also a closed submanifold of $G \times G$. It remains to check that pr_2 or, what amounts to the same by symmetry, pr_1 induces a submersion from R onto G. As v induces an isomorphism from $G \times H$ onto R, R can be replaced by $G \times H$ and pr_1 by $\mathrm{pr}_1 \circ v$. But $\mathrm{pr}_1 \circ v = \mathrm{pr}_1$ clearly holds. Hence it remains to check that $\mathrm{pr}_1 : G \times H \to G$ is a submersion, which is obvious.[3]

---

[3]More generally, let us consider the equivalence relation R defined by a Lie group H acting on the right on a manifold X; it also satisfies condition (RER 2). Indeed, let (a, b) be a point of R. Then

Finally, to show that $(g, x) \mapsto gx$ is of class C^r like G and H, it suffices to check that if ϕ is a C^r function on an open subset of G/H then the function $\phi(gx)$ is of class C^r on the open subset of $G \times G/H$ where it is defined. But as a manifold, $G \times G/H$ is clearly (exercise!) canonically identifiable with the quotient of $G \times G$ by the subgroup $\{e\} \times H$. Denoting by p the canonical map from G onto G/H, it therefore suffices to check that the composition of the function $(g, x) \mapsto \phi(gx)$ and the map $(g, g') \mapsto (g, p(g'))$—in other words the function $(g, g') \mapsto \phi(g.p(g')) = \phi \circ p(gg')$—is of class C^r on the open subset of $G \times G$ where it is defined. However, it is obtained by taking the composition of the function $\phi \circ p$, of class C^r on the open subset of G where it is defined since p is a morphism, with the map $(g, g') \mapsto gg'$ from $G \times G$ to G. Hence the corollary.

As an example, let us consider an n-dimensional real vector space E and choose integers $p_1, \ldots, p_r \geq 1$ such that $p_1 + \cdots + p_r = n$. We will call a *flag* of type $(p_1, \ldots, p_r)$ in E any family $(F_1, \ldots, F_{r-1})$ of vector subspaces of E satisfying

$$F_1 \subset F_2 \subset \cdots \subset F_{r-1}, \quad \dim F_i = p_1 + \cdots + p_i \text{ for } 1 \leq i \leq r - 1. \quad (4.9.20)$$

The group $G = GL(E)$ acts transitively on these flags, the set of which can thus be identified with $GL(E)/H$, where H is the stabilizer of a given flag. Taking $E = \mathbb{R}^n$ and $F_i = \mathbb{R}^{p_1 + \cdots + p_i}$, the stabilizer H of $(F_1, \ldots, F_{r-1})$ is the set of triangular block matrices with blocks corresponding to the partition $n = p_1 + \cdots + p_r$ of n. Consequently H is a Lie subgroup of G, and a manifold structure can be deduced on the set of flags of a given type (*Grassmannian manifolds*). For the partition $n = 1 + (n - 1)$, we recover the projective space $P(E)$.

In conclusion, note that the corollary of Theorem 5 can very well be proved directly without using the latter; this is for example what Dieudonné (Chapter XVI, §10) and Varadarajan (pp. 74–84, where some interesting additions on orbit spaces can be found) do following many others and mainly Chevalley, the ancestor of all these authors as regards the modern theory of Lie groups. Attention should be paid to the fact that for Varadarajan as well as for some other authors, "submanifolds" are not defined as in these Notes: for Varadarajan, a submanifold Y of a manifold X is the object consisting of a subset Y of X and of a manifold structure on Y satisfying the property that the canonical injection of Y in X is an immersion. With this definition, an everywhere dense geodesic in a torus, endowed with the manifold structure of $\mathbb{R}$, is again a submanifold of the torus (although it is not always a locally closed subspace contrary to the conventions adopted here). The introduction of this more general notion (better called the notion of a *submanifold embedded* in X to avoid confusing

---

(Footnote 3 continued)

there exists an $h \in H$ such that $a = bh$. The map $x \mapsto (xh, x)$ is then a morphism from X to $X \times X$ which maps X to R (so it will be a morphism from X to R if R is a closed submanifold of $X \times X$), and a to the point (a, b) considered, and which satisfies $\mathrm{pr}_2(xh, x)$. Theorem 4 then shows that $\mathrm{pr}_2 : R \to X$ is submersive at the point of R under consideration.

This argument shows that to define a natural manifold structure on the quotient X/H, it suffices to check that the set R of pairs (x, xh) is a closed submanifold of $X \times X$.

it with "good" locally closed submanifolds) is warranted from the perspective of systems of differential equations, as will be seen at the end of Chap. 6, in the context of subgroups of Lie groups. It is, however, pointless to be more concerned about this at this point.

Chapter 5
The Lie Algebra of a Lie Group

Every manifold considered in this chapter is assumed to be of class C^∞; the definitions and results hold for analytic manifolds, provided they are endowed with the corresponding C^∞ structure. The reader will certainly realize that many of the definitions and results presented for C^∞ manifolds can in fact be extended to C^r manifolds for finite r, if need be with suitable modifications.

5.1 Contacts of Order n, Punctual Distributions

Let X be a manifold, a a point of X and $\mathscr{F}_a$ a set of (real-valued) C^∞ functions whose domains of definition are open and contain a. If $f, g \in \mathscr{F}_a$, then $f + g$ and fg denote the functions $x \mapsto f(x) + g(x)$ and $x \mapsto f(x)g(x)$, defined in the intersection of the domains of definition of f and g; similarly the product of an element $f \in \mathscr{F}_a$ by a scalar is defined in the obvious way. This seemingly gives an associative algebra structure on $\mathscr{F}_a$, but in fact, this is not at all the case because the identity element for addition can only be the function which vanishes *everywhere*, so that equality $f + g = 0$ assumes that f and g are defined *everywhere*; $\mathscr{F}_a$ is therefore not even a group with respect to addition...

There are many useful equivalence relations in $\mathscr{F}_a$ which, after passing to the quotient, effectively lead to genuine associative algebras. The most straightforward among these equivalence relations is the one which consists in considering two functions f and g as equivalent if they coincide *in some* (unspecified) *neighbourhood* of a. The corresponding equivalence classes are then called the *germs* of the C^∞ functions at a.

To define the others, we first introduce a chart (U, ξ) at a, with $n = \dim X$ coordinate functions $\xi_1, \ldots, \xi_n$. Let $\mathrm{Def}(f)$ be the domain of definition of $f \in \mathscr{F}_a$. Then f is necessarily defined on the open set $U \cap \mathrm{Def}(f)$, and the following equality holds in it:

© Springer International Publishing AG 2017
R. Godement, *Introduction to the Theory of Lie Groups*,
Universitext, DOI 10.1007/978-3-319-54375-8_5

$$f(x) = \phi(\xi_1(x), \ldots, \xi_n(x)), \tag{5.1.1}$$

where ϕ is a well-defined C^∞ function on the open subset of $\mathbb{R}^n$ corresponding to $U \cap \mathrm{Def}(f)$ under ξ. If $\alpha = (\alpha_1, \ldots, \alpha_n)$ is a multi-index, i.e. a sequence of n natural integers, then for all $x \in U \cap \mathrm{Def}(f)$, it is possible to define the number

$$D_\xi^\alpha f(x) = \text{ value of the function } \partial^{|\alpha|}\phi / \partial\xi_1^{\alpha_1} \ldots \partial\xi_n^{\alpha_n} \text{ at } \xi(x) \in \mathbb{R}^n, \tag{5.1.2}$$

where

$$|\alpha| := \alpha_1 + \ldots + \alpha_n \tag{5.1.3}$$

for every multi-index α. The function $D_\xi^\alpha f$, which is also C^∞ on $U \cap \mathrm{Def}(f)$, evidently depends on the choice of the chart (U, ξ). But if it is replaced by another chart (V, η), the usual chain rules show that, in the open set $U \cap V \cap \mathrm{Def}(f)$, where the $D_\xi^\alpha f$ and D_η^α are simultaneously defined, equalities of the form

$$D_\xi^\alpha f(x) = \sum_{|\beta| \le |\alpha|} c_\beta^\alpha(x) \cdot D_\eta^\beta f(x) \tag{5.1.4}$$

hold, where the functions c_β^α are defined on $U \cap V$. These do not depend on f and can be written as algebraic formulas easily expressible using partial derivatives of the coordinates η with respect to the coordinates ξ. Conversely, there are formulas analogous to formulas (5.1.4) to go from the $D_\xi^\alpha f$ to the $D_\eta^\beta f$.

Exercise. Let η^β denote the function $\eta_1^{\beta_1} \ldots \eta_n^{\beta_n}$. Show that if $\eta(a) = 0$, then $c_\beta^\alpha(a) = D_\xi^\alpha(\eta^\beta)/\beta_1! \ldots !\beta_n!$, and that the same formula holds in the general case provided the functions η_i are replaced by the functions $\eta_i - \eta_i(a)$.

Two functions $f, g \in \mathscr{F}_a$ are said *to have at least a contact of order p at a* if

$$D_\xi^\alpha f(a) = D_\xi^\alpha g(a) \quad \text{for all } \alpha \text{ such that } |\alpha| \le p. \tag{5.1.5}$$

This is well-defined by formulas (5.1.4), i.e. is independent of the choice of the local chart (U, ξ). If $g = 0$, f is said to have a *zero of order at least $p + 1$ at a*. For each p, (5.1.5) is clearly an equivalence relation on the set $\mathscr{F}_a$; the corresponding quotient set will be denoted $\mathscr{F}_a^{(p)}$, and its elements will be called the *functional elements of order p at a*. Obviously, the composition laws defined above in $\mathscr{F}_a$ are transferred to the quotient determined by equivalence relation (5.1.5), and as in the case of germs of functions at a, enable us to endow $\mathscr{F}_a^{(p)}$ with a genuine associative algebra structure. It can be easily formulated:

Exercise. Let I^{p+1} be the ideal of the ring of formal series $\mathbb{R}[[X_1, \ldots, X_n]]$ generated by the monomials of degree $p + 1$ in X_i (so that I^{p+1} consists of the series all of whose terms of degree $\le p$ are zero). Choose a chart (U, ξ) of the given manifold at a, and to every $f \in \mathscr{F}_a$ associate the image in the quotient ring

$$\mathbb{R}[[X_1, \ldots, X_n]]/I^{p+1} \tag{5.1.6}$$

of the formal series

$$\sum_\alpha D_\xi^\alpha f(a) \cdot X^\alpha/\alpha!, \tag{5.1.7}$$

where

$$\alpha! = \alpha_1! \ldots \alpha_n! \tag{5.1.8}$$

and where the monomial X^α in the indeterminates X_i is defined in the obvious way. It is easy to understand why (5.1.7) is called the *Taylor series of the function f at a* with respect to the given chart (it does not always converge—here we obviously have a formal series). Thus, (5.1.5) states that the Taylor series of f and g at a are equal mod I^{p+1}, and the exercise then consists in showing that the map from $\mathscr{F}_a$ to the algebra (5.1.6) defined above, leads, by passage to the quotient, to an *isomorphism* from the algebra $\mathscr{F}_a^{(p)}$ onto the algebra (5.1.6).

Note that the algebra (5.1.6) is finite-dimensional—the images of the monomials X^α with $|\alpha| \le p$ form a basis—and that it could also be defined using the polynomials in X_i instead of formal series (but obviously the Taylor series of a C^∞ function rarely reduces to a polynomial, which is why we have opted to start with formal series rather than polynomials in the previous *Exercise*). In concrete terms, the elements of the algebra (5.1.6) are polynomials of degree $\le p$ in X_i with the usual operations, but ignoring the terms of degree $> p$. This is what is done in the theory of finite expansions, so that the image of a function $f \in \mathscr{F}_a$ in (5.1.6), i.e. the polynomial

$$\sum_{|\alpha| \le p} D_\xi^\alpha f(a) \cdot X^\alpha/\alpha!, \tag{5.1.9}$$

could be called the *limited expansion of order p of f at a* with respect to the given chart.

We are at last in a position to define an intermediate equivalence relation in $\mathscr{F}_a$ between (5.1.5) and the one defining germs of functions, by stating that two functions f and g have *an infinite order contact at a* if

$$D_\xi^\alpha f(a) = D_\xi^\alpha g(a) \quad \text{for all } \alpha. \tag{5.1.10}$$

The corresponding quotient set is denoted by $\mathscr{F}_a^{(\infty)}$, and it is also an associative algebra with the obvious composition laws. To say that f and g have an infinite order contact at a obviously amounts to saying that their Taylor series at a are identical. Associating its Taylor series (5.1.7) to each $f \in \mathscr{F}_a$ clearly gives, by passage to the quotient, a map from the algebra $\mathscr{F}_a^{(\infty)}$ to the algebra of formal series $\mathbb{R}[[X_1, \ldots, X_n]]$, which is clearly an injective homomorphism. In fact, it is even an *isomorphism*—in other words, the map in question is surjective. So (the problem being a local one) we need to prove that, in the neighbourhood of 0 in $\mathbb{R}^n$, it is always possible to construct a C^∞ function *all* of whose partial derivatives at the origin take specified values, a well-known result.[1]

Finally, observe that the notion of a pth or infinite order contact can be defined more generally for manifold morphisms. Let X and Y be two manifolds, a and b points of X and Y, U and V open neighbourhoods of a in X and finally ϕ and ψ morphisms from U and V to Y such that $\phi(a) = \psi(a) = b$. If local charts of X (resp. Y) at a (resp. b) defined by the coordinate functions ξ_i $(1 \leq i \leq m)$ (resp. η_j $(1 \leq i \leq n)$) are chosen, in the neighbourhood of a, the map ϕ (resp. ψ) will be defined by equations of type $\eta_j = \phi_j(\xi_1, \ldots, \xi_m)$ (resp. $\eta_j = \psi_j(\xi_1, \ldots, \xi_m)$); ϕ *and ψ are then said to have a contact of order at least p at a* if the partial derivatives of order $\leq p$ at a of all ϕ_j are equal to those of all ψ_j; a contact of infinite order being defined likewise. The definition is clearly independent of the local charts chosen.

Keeping the same assumptions, let us now consider the algebra $\mathscr{F}_a^{(p)}(X)$ of pth order functional elements on X at a and the analogous algebra $\mathscr{F}_b^{(p)}(Y)$ relative to Y at b. To every C^∞ function g defined in an open neighbourhood of b in Y, associate the C^∞ function $g \circ \phi$ defined in an open neighbourhood of a—namely the set of x for which ϕ is defined at x and such that g is defined at $\phi(x)$. In turn, replacing g by a function g' having at least a pth order contact with g at b clearly transforms the function $g' \circ \phi$ into a function $g \circ \phi$ which has at least a pth order contact at a with the preceding function. So, by passage to the quotient, the map $g \mapsto g \circ \phi$ defines a map from $\mathscr{F}_b^{(p)}(Y)$ to $\mathscr{F}_a^{(p)}(X)$, which is obviously an algebra homomorphism (since $g \mapsto g \circ \phi$ is compatible with the algebraic operations on functions). Thus, ϕ and ψ have at least a pth order contact at a if and only if the homomorphisms $\mathscr{F}_b^{(p)}(Y) \to \mathscr{F}_a^{(p)}(X)$ defined by ϕ and ψ are equal. Hence the class of ϕ up to pth order contacts at a can be defined by associating to ϕ the homomorphism $\mathscr{F}_b^{(p)}(Y) \to \mathscr{F}_a^{(p)}(X)$ it specifies.

Exercise. ϕ can always be chosen so that its corresponding homomorphism is a specified one.

Exercise. Endow the set of homomorphisms from $\mathscr{F}_b^{(p)}(Y)$ to $\mathscr{F}_a^{(p)}(X)$ with a C^∞ or even C^ω manifold structure.

Now that these preliminaries have been established, we are in a position to define punctual distributions. Let a be a point of a manifold X and $\mathscr{F}_a$ the set of C^∞ functions defined on an open set containing a. A map

$$\mu : \mathscr{F}_a \to \mathbb{R} \tag{5.1.11}$$

is called a *distribution at a on X* if it satisfies the following two properties: μ is *linear* in the obvious sense, and moreover there is an integer p such that $\mu(f) = 0$ if f has a zero of order $\geq p + 1$ at a. Clearly, $\mu(f)$ is then independent of the image of f in the algebra $\mathscr{F}_a^{(p)}$, and μ can be canonically identified with a linear form on the latter. It is also possible to avoid specifying p and to regard μ as *a linear form on the*

---

[1] See, for example, W.F. Donoghue, Jr., *Distributions and Fourier Transforms* (Academic Press, 1969), p. 50, or MA V, Sect. 29, for the case $n = 1$.

algebra $\mathscr{F}_a^{(\infty)}$ which is identically zero on a sufficiently large power of the ideal I of $\mathscr{F}_a^{(\infty)}$ consisting of formal series without constant term. Indeed, $\mathscr{F}_a^{(p)} = \mathscr{F}_a^{(\infty)}/I^{p+1}$ clearly holds for all finite p.

The distributions at a can all be easily formulated using a chart (U, ξ) of X at a such that $\xi(a) = 0$. For all $f \in \mathscr{F}_a$, Taylor's formula indeed shows that

$$f(x) = \sum_{|\alpha| \leq p} D_\xi^\alpha f(a) \cdot \xi^\alpha(x)/\alpha! + r(x), \qquad (5.1.12)$$

where the function $r \in \mathscr{F}_a$ has a zero of order at least $p + 1$ at a. If μ is at most of order p, then $\mu(r) = 0$, and so

$$\mu(f) = \sum_{|\alpha| \leq p} D_\xi^\alpha f(a) \cdot \mu(\xi^\alpha)/\alpha! \qquad (5.1.13)$$

or

$$\boxed{\sum_{|\alpha| \leq p} c_\alpha \cdot D_\xi^\alpha f(a)} \quad \text{for all } f \in \mathscr{F}_a, \qquad (5.1.14)$$

with coefficients

$$c_\alpha = \mu(\xi^\alpha)/\alpha! \qquad (5.1.15)$$

independent from f (but dependent on the chosen chart) and characterizing μ. It is well known that this is also the general form of punctual distributions on a space $\mathbb{R}^n$.

A *distribution of order 0* at a is a linear form of type $f \mapsto c \cdot f(a)$, where c is a constant. This is easily seen to be a "Dirac measure" consisting of the mass c assigned to the point a of X. *Distributions of order ≤ 1* are of the form

$$f \mapsto c \cdot f(a) + \sum c_i \cdot D_\xi^i f(a), \qquad (5.1.16)$$

where $D_\xi^i f$ denotes the partial derivative function of f with respect to coordinates ξ_i. Denoting distribution (5.1.16) by μ,

$$c = \mu(1), \quad c_i = \mu(\xi_i) \qquad (5.1.17)$$

clearly hold. By subtracting a Dirac measure form μ, we may assume that $\mu(1) = 0$; then

$$\mu(f) = \sum c_i \cdot D_\xi^i f(a). \qquad (5.1.18)$$

By definition, distributions of this type are just *tangent vectors to X at a*, a result which we will recover in the next section, but differently.

5.2   Tangent Vectors to a Manifold at a Point

Let E be a finite-dimensional real vector space and X a locally closed submanifold of E. The notion of a *tangent vector to X at a given point $a \in X$* is geometrically obvious: a vector $h \in E$ is tangent to X at a if there is a sequence of points $x_p \in X$ tending to a and such that the *direction* of the vector with initial point a and terminal point x_p converges to the direction of the vectors h. This means that there is a sequence of numbers $\varepsilon_p > 0$ tending to 0 and such that

$$h = \lim \frac{1}{\varepsilon_p}(x_p - a); \qquad (5.2.1)$$

as already pointed out in Chap. 3, the similarity with von Neumann's viewpoint is more than mere coincidence.

Instead of using a sequence tending to a, we could equally consider a path $t \mapsto \gamma(t)$ in X with origin a and whose derivative at the origin

$$h = \lim \frac{1}{t}(\gamma(t) - \gamma(0)) = \gamma'(0) \qquad (5.2.2)$$

exists. The existence of a path satisfying (5.2.2) obviously implies that of a sequence satisfying (5.2.1); we will shortly see that the converse also holds.

For the moment let us keep to definition (5.2.1) and consider a real-valued C^1 function f defined in a neighbourhood of a in X. By (5.2.1),

$$x_p = a + \varepsilon_p h + o(\varepsilon_p) \qquad (5.2.3)$$

as p increases indefinitely; but since X is a submanifold of E, in the neighbourhood of a, the function f is the restriction to X of a function ϕ of class C^1 in an open neighbourhood of a in E. The mean value formula for functions of several variables then shows that

$$\begin{aligned} f(x_p) - f(a) = \phi(x_p) - \phi(a) &= \phi(a + \varepsilon_p h + o(\varepsilon_p)) - \phi(a) \\ &= u(\varepsilon_p h + o(\varepsilon_p)) + o(\varepsilon_p h + o(\varepsilon_p)) \\ &= u(h)\varepsilon_p + o(\varepsilon_p), \end{aligned} \qquad (5.2.4)$$

where $u = \phi'(a) = d\phi(a) \in E^*$ is a tangent linear form to ϕ at a, given by

$$u(h) = \frac{d}{dt}\phi(a + th)_{t=0} \qquad (5.2.5)$$

for all $h \in E$.

Formula (5.2.4) is also equivalent to the definition

$$\lim \frac{1}{\varepsilon_p}[f(x_p) - f(a)] = u(h), \qquad (5.2.6)$$

which shows not only the existence of the limit appearing on the left-hand side but also that it only depends on the tangent vector h. In fact:

Lemma *Let h' and h'' be two tangent vectors to X at a, defined by equations*

$$h' = \lim \frac{1}{\varepsilon_p'}(x_p' - a), \quad h'' = \lim \frac{1}{\varepsilon_p''}(x_p'' - a), \qquad (5.2.7)$$

where $x_p', x_p'' \in X$ for all p. Then, $h' = h''$ if and only if

$$\lim \frac{1}{\varepsilon_p'}[f(x_p') - f(a)] = \lim \frac{1}{\varepsilon_p''}[f(x_p'') - f(a)] \qquad (5.2.8)$$

for all C^1 functions f defined in the neighbourhood of a in X.

Indeed, if (5.2.8) holds, then it can be applied to the case where f is the restriction to X of a linear form ϕ on E. Since in this case,

$$\frac{1}{\varepsilon_p}[\phi(x_p - \phi(a))] = \phi\left(\frac{x_p - a}{\varepsilon_p}\right), \qquad (5.2.9)$$

Equation (5.2.8) clearly means that $\phi(h') = \phi(h'')$; so $h' = h''$ since ϕ is an arbitrary element of the dual of E, qed.

Returning to the tangent vector (5.2.1), it thus follows that it is determined by the map which associates the number

$$\boxed{h(f) = \lim \frac{1}{\varepsilon_p}[f(x_p) - f(a)]} \qquad (5.2.10)$$

to each C^1 function f in the neighbourhood of a in X. It depends in a very simple way on f; indeed,

$$h(\lambda f) = \lambda \cdot h(f), \quad h(f + g) = h(f) + h(g), \qquad (5.2.11)$$

$$h(fg) = h(f)g(a) + f(a)h(g). \qquad (5.2.12)$$

We assume that after having so often heard about differentiation rules for sums and products of functions, the reader will have no difficulty in finding the proofs of (5.2.11) and (5.2.12) on his own.

These definitions and computations assume that X is a submanifold of a vector space, but they can be extended to any C^r manifold X with $r \geq 1$, without using any embedding of X into a vector space.[2] For a given point of X, a family (x_p, ε_p) where $x_p \in X$ tend to a and $\varepsilon_p > 0$ tend to zero, will be called a *deriving sequence* at a if the limit (5.2.10) exists for all C^1 functions f defined in the neighbourhood of a. The above lemma then prompts us to define the *tangent vectors* to X at a as the class of deriving sequences at a for the equivalence relation obtained by regarding as equivalent two deriving sequences (x'_p, ε'_p) and (x''_p, ε''_p) satisfying condition (5.2.8) for all f.

The class of a deriving sequence being by definition characterized by the limit (5.2.10), a tangent vector h to X at a could also be defined as a map $f \mapsto h(f)$, defined on the C^1 functions in the neighbourhood of a, and such that there exists a deriving sequence satisfying (5.2.10).

But let x_p be a sequence of points in X tending to a, ε_p a sequence of non-trivial real numbers tending to 0, and (U, ξ) a chart of X at a. If (x_p, ε_p) is a deriving sequence at a, and so defines a tangent vector h to X at a, then the limits

$$h(\xi_i) = \lim \frac{\xi_i(x_p) - \xi_i(a)}{\varepsilon_p} = h_i \qquad (5.2.13)$$

exist; similarly, in more geometric terms, so does the vector

$$h_{U,\xi} = \lim \frac{1}{\varepsilon_p}[\xi(x_p) - \xi(a)] \qquad (5.2.14)$$

of $\mathbb{R}^n$ if $n = \dim(X)$.

Conversely, suppose that limits (5.2.13) or (5.2.14) exist. Let $f(x) = \phi(\xi(X))$ be the composite of the chart $\xi : U \to \mathbb{R}^n$ and a C^1 function ϕ in the neighbourhood of $\xi(a)$ of $\mathbb{R}^n$; in particular, it is a C^1 function in the neighbourhood of a. Then

$$\begin{aligned}
f(x_p) &= \phi[\xi(x_p)] = \phi[\xi(a) + \varepsilon_p h_{U,\xi} + o(\varepsilon_p)] \\
&= \phi[\xi(a)] + \varepsilon_p \sum h_i \cdot D_i \phi[\xi(a)] + o(\varepsilon_p)
\end{aligned} \qquad (5.2.15)$$

by the mean value theorem. This implies the existence of the limit

$$\lim \frac{1}{\varepsilon_p}[f(x_p) - f(a)] = \sum h_i \cdot D_i \phi[\xi(a)] = \sum h_i \cdot D_\xi^i f(a), \qquad (5.2.16)$$

returning to the notation introduced at the end of the previous section. So finally, it follows that deriving sequences are characterized by the existence of limits (5.2.13)—it

---

[2] It is possible to show (H. Whitney) that every manifold countable at infinity is isomorphic to a submanifold of a finite-dimensional vector space. For the compact case, see M. Berger and B. Gostiaux, *Differential Geometry*, p. 104; for the general case J. Dieudonné, *Eléments d'Analyse*, XVI, §25, exercise 2.

suffices to check it in a particular chart—and thus that the map $f \mapsto h(f)$ corresponding to such a sequence is indeed a distribution of order 1 at a such that $h(1) = 0$, as claimed at the end of the previous section.

Conversely, given a distribution on X such as

$$h(f) = \sum h_i \cdot D_i \phi[\xi(a)] = \sum h_i \cdot D_\xi^i f(a), \qquad (5.2.17)$$

with fixed coefficients $h_i \in \mathbb{R}$, finding a deriving sequence (x_p, ε_p) defining (5.2.17) by formula (5.2.10) is not difficult: choose $\varepsilon_p = 1/p$ and for x_p the point of U whose coordinates are given by

$$\xi_i(x_p) = \xi_i(a) + h_i/p. \qquad (5.2.18)$$

Setting $\xi_i(a) = \alpha_i$, it then amounts to checking that

$$\lim_{p \to \infty} p[\phi(\alpha_1 + h_1/p, \ldots, \alpha_n + h_n/p) - \phi(\alpha_1, \ldots, \alpha_n)]$$
$$= \sum h_i \cdot D_i \phi(\alpha_1, \ldots, \alpha_n), \qquad (5.2.19)$$

which is clear...

Besides, we could consider the path $\gamma(t)$ with origin a, given by

$$\xi_i(\gamma(t)) = \alpha_i + t h_i \qquad (5.2.20)$$

for t sufficiently near zero, and which, in the chart (U, ξ), becomes the line through $\xi(a)$ with direction vector $(h_1, \ldots, h_n)$. Then,

$$h(f) = \frac{d}{dt} f(\gamma(t)) \Big|_{t=0} \qquad (5.2.21)$$

for all $f \in \mathcal{F}_a$, and h is called the *tangent vector to γ at $t = 0$*; it is written $\gamma'(0)$. We will omit stating

$$\gamma'(0) = \lim \frac{\gamma(t) - \gamma(0)}{t}$$

since this formula is not defined for an abstract manifold not embedded into a vector space.

Since the tangent vectors to X at a can be canonically identified with distributions of type (5.2.7), the set of these tangent vectors can be considered as an n-dimensional vector space; it is written $T_a(X)$ or $X'(a)$ depending on the context. Each chart (U, ξ) of X at a defines an isomorphism from the vector space $X'(a)$ onto $\mathbb{R}^n$: it associates to each $h \in X'(a)$ the vector $h_{U, \xi} \in \mathbb{R}^n$ given by formula (5.2.14), and which is obviously independent of the choice of the deriving sequence defining h. If (V, η) is another chart at a, so that in the neighbourhood of a, $\eta = \theta \circ \xi$ where θ is a diffeomorphism from a neighbourhood of $\xi(a) \in \mathbb{R}^n$ onto a neighbourhood of $\eta(a) \in \mathbb{R}^n$, it is somewhat obvious that the tangent linear map to θ at $\xi(a)$ enables us to go from $h_{U,\xi}$ to $h_{V,\eta}$. Conversely, given a vector $h_{U,\xi} \in \mathbb{R}^n$ for each chart, with the previous "covariance" condition, it is somewhat clear that there is a uniquely defined $h \in X'(a)$ which, in each chart (U, ξ), is represented by the vector $h_{U,\xi}$. These remarks could provide yet another possible definition of tangent vectors to a manifold.

These considerations show that, if the manifold X is given as an open subset of a vector space E, then, for all $a \in X$, the "abstract" vector space $T_a(X)$ can be *canonically* identified with E itself: it suffices to consider the map $h \mapsto h_{U,\xi}$ corresponding to the obvious chart, the one in which $U = X$ and $\xi(x) = x$ for all $x \in X$. Equivalently, as a distribution, every tangent vector to X at a is of the form

$$f \mapsto \frac{d}{dt} f(a + th)\bigg|_{t=0} \qquad (5.2.22)$$

for a unique vector $h \in E$. In particular, there is a canonical isomorphism

$$T_a(E) = E \qquad (5.2.23)$$

of the vector space E itself.

More generally, suppose that X is a locally closed submanifold of a vector space E. We show how $T_a(X)$ can be canonically identified with a vector subspace of E for all $a \in X$. This is obvious in interpretation (5.2.1) for tangent vectors. Another way of viewing the situation would be to observe that if $f \in E^*$ is a linear form on E, and if h is a tangent vector to X at a in the sense of distribution theory, then h can be applied to the restriction $f|X$ of f to X, a restriction which is obviously of class C^1, possibly even C^∞, on X. The map $f \mapsto h(f|X)$ is obviously a linear form on E^*, and so an element of E which must be the element of E associated to h by (5.2.10).

Besides, given an "abstract" manifold X, a point a of X and a C^1 map ϕ from a neighbourhood of a in X to a vector space E, a linear map from $T_a(X)$ to E can be immediately derived (which again gives the above identification when ϕ is the identity map). It suffices to associate the number $h(f \circ \phi)$ to each $f \in E^*$; this gives a linear form on E^*, which by biduality, provides the image of h under the desired map. In simpler terms, setting $\phi(x) = \sum \phi_i(x)a_i$ where (a_i) is a basis of E, the vector of E associated to h is just $\sum h(\phi_i)a_i$.

5.3 Tangent Linear Map to a Morphism

Even more generally, consider two manifolds X and Y, points a and b of X and Y, a C^1 map ϕ from a neighbourhood of a in X to Y such that $\phi(a) = b$. If g is a numerical function of class C^1 in the neighbourhood of b, the composite functions $g \circ \phi$ is of class C^1 in the neighbourhood of a, so that if $h \in T_a(X)$, the map $g \mapsto h(g \circ \phi)$ can be considered. If h is defined by a deriving sequence (x_p, ε_p), this clearly gives the tangent vector $k \in T_b(Y)$ defined by the deriving sequence $(\phi(x_p), \varepsilon_p)$—which is indeed such a sequence since the limit

$$k(g) = \lim \frac{1}{\varepsilon_p}[g(y_p) - g(b)], \quad \text{where } y_p = \phi(x_p), \tag{5.3.1}$$

exists for all g. Then, k will be said to be the direct image (or simply the image) of h under ϕ, and the resulting map from $T_a(X)$ to $T_b(Y)$, which is obviously linear, will be written either as $\phi'(a)$—in which case it is called the *tangent linear map* to ϕ at a—or as $d\phi(a)$—in which case it is called the *differential* of ϕ at a. It is usually written as

$$k = \phi'(a)h = d\phi(a; h) \tag{5.3.2}$$

to indicate that $k \in T_b(Y)$ is the image of $h \in T_a(X)$ under ϕ. Obviously, given three manifolds X, Y and Z and morphisms $\phi : X \to Y$ and $\psi : Y \to Z$, and hence a composite morphism $\theta : X \to Z$,

$$\theta'(a) = \psi'((a)) \circ \phi'(a), \tag{5.3.3}$$

an "abstract" version of the chain rule.

In the above, setting Y to be a vector space E, in which case $T_b(Y)$ can be canonically identified with E for all b, the mapping $\phi'(a)$ can be identified with a linear map from $T_a(X)$ to E. The reader will easily check that it is the one introduced at the end of the preceding section. Manifold theory is full of canonical identifications and functorial constructions which, at each step, seemingly raise compatibility questions meant to reassure the reader on the non-contradictory nature of the notions and abuse of language introduced. Having exhausted these innocent delights several decades ago, the author gladly refers the sceptical reader to Dieudonné, not to mention Bourbaki. We are not on Earth to have fun.

[It is nonetheless useful to note that, a significant generalization of the previous constructions leads to the notion of a *direct image*, or simply of an image, of a *punctual distribution* under a morphism. Given a morphism ϕ from X to Y such that $\phi(a) = b$, and a distribution μ at a on X, its direct image, often written $\phi_*(\mu)$, is defined to be the distribution $g \mapsto \mu(g \circ \phi)$ at point b on Y. Since $\mu(g \circ \phi)$ is a linear combination of derivatives of order $\leq s$ of the function $g \circ \phi$ with respect to a chart at a, it is clearly also a linear combination of the derivatives of order $\leq s$ of g with respect to a chart of Y at b, justifying the definition.]

Tangent linear maps are particularly useful for stating that a morphism ϕ is an immersion or a submersion. Indeed, keeping the above assumptions and notation,

$$\boxed{\text{rk}_a(\phi) = \text{rk}(\phi'(a)) = \dim(\text{Im}\,\phi'(a)),} \tag{5.3.4}$$

where the rank of the linear map $\phi'(a)$ from $T_a(X)$ to $T_b(Y)$ appears on the right-hand side. Indeed, let us consider local charts (U, ξ) and (V, η) of X and Y at a and b.

Then there is a basis of $T_a(X)$ consisting of the vectors

$$u_i : f \mapsto D_\xi^i f(a), \tag{5.3.5}$$

as well as a basis of $T_b(Y)$ consisting of the vectors

$$v_j : g \mapsto D_\eta^j g(a). \tag{5.3.6}$$

To compute the rank of $\phi'(a)$, it suffices to compute its matrix with respect to these bases of $T_a(X)$ and $T_b(Y)$. For this, it is necessary to write

$$\phi'(a)u_i = \sum c_{ij} v_j \tag{5.3.7}$$

and to find the c_{ij}. But if $v \in T_b(Y)$, then the coordinates of v with respect to basis (5.3.6) are, as we know, the numbers $v(\eta_j)$. Hence the value of the left-hand side of (5.3.6) at the function η_j needs to be calculated, i.e.—use definition (5.3.2) of the image of a distribution under a morphism—the value of the distribution u_i at the function $\eta_j \circ \phi$, where

$$c_{ij} = u_i(\eta_j \circ \phi). \tag{5.3.8}$$

As the coordinates of $y = \phi(x)$ can be computed using formulas

$$\eta_j[\phi(x)] = \phi_j[\xi_1(x), \dots, \xi_p(x)], \tag{5.3.9}$$

the desired number c_{ij} is obtained by applying (5.3.5) to function (5.3.9), and so finally

$$c_{ij} = \frac{\partial \phi_j}{\partial \xi_i}(a). \tag{5.3.10}$$

In other words, *the matrix of $\phi'(a)$ with respect to bases (5.3.5) and (5.3.6) is the Jacobian matrix of ϕ at a in the charts considered* (Chap. 4, Sect. 4.2), and now, equalities (5.3.4) become obvious.

Relation (5.3.4) shows that in particular *a morphism $\phi : X \to Y$ is an immersion (resp. submersion) at a point $a \in X$ if and only if the tangent mapping $\phi'(a) : T_a(X) \to T_b(Y)$, where $b = \phi(a)$, is injective (resp. surjective).* For example, if X is a locally closed submanifold of Y and ϕ the identity map, which is therefore an immersion, the derived map $\phi'(a)$ enables us to *identify $T_a(x)$ with a vector subspace* of $T_a(Y)$, implying that a tangent vector $h \in T_a(Y)$ belongs to the subspace $T_a(X)$ if and only if

$$df(a; h) = 0 \text{ for all } f \in \mathscr{F}_a(Y) \text{ whose restriction to } X \tag{5.3.11}$$
$$\text{vanishes in the neighbourhood of } a.$$

More precisely, choose a local chart $(U, \xi_1, \ldots, \xi_n)$ of Y at a such that $U \cap X$ is defined by the equations $\xi_{p+1} = \ldots = \xi_n = 0$; then the subspace $T_a(X)$ of $T_a(Y)$ consists of h such that

$$d\xi_{p+1}(a; h) = \ldots = d\xi_n(a; h) = 0. \qquad (5.3.12)$$

It is a matter of showing that, if condition (5.3.12) holds, for $f \in \mathscr{F}_a(Y)$, $h(f)$ can be uniquely expressed using the restriction of f to X, which is obvious since $h(f)$ is then a linear combination of the first p partial derivatives of f at a.

Exercise. Let $f : X \to Y$ be a subimmersion and let $Z = f^{-1}(b)$ for a given $b \in Y$. Show that for all $a \in Z$, $T_a(Z) = \mathrm{Ker}(f'(a))$.

5.4   Tangent Vectors to a Cartesian Product

We next consider two manifolds X and Y and the product manifold $X \times Y$. Our aim is to construct a canonical isomorphism

$$\boxed{T_{a,b}(X \times Y) = T_a(X) \times T_b(Y)} \qquad (5.4.1)$$

for each point $(a, b) \in X \times Y$. First of all, the morphisms pr_1 and pr_2 from $X \times Y$ to X and Y respectively define canonical linear maps

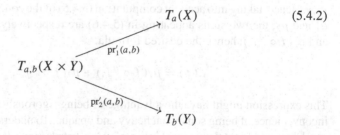

$$\qquad\qquad T_a(X) \qquad\qquad\qquad (5.4.2)$$

from which the canonical homomorphism

$$T_{a,b}(X \times Y) \to T_a(X) \times T_b(Y) \qquad (5.4.3)$$

can be derived. Its bijectivity remains to be shown. As we will anyhow need to compute it explicitly, it is necessary to choose charts (U, ξ) and (V, η) of X and Y at a and b. This gives a chart of $X \times Y$ whose domain is $U \times V$ and whose coordinate functions are $\xi_i(x)$ and $\eta_j(y)$ considered as functions on $U \times V$, in other words the functions $\xi_i \circ \mathrm{pr}_1$ and $\eta_j \circ \mathrm{pr}_2$. The components of an element $\mu \in T_{a,b}(X \times Y)$ with respect to the basis of $T_{a,b}(X \times Y)$ defined by the chart considered, see (5.1.17) and (5.1.18), are the numbers

$$c_i = \mu(\xi_i \circ \mathrm{pr}_1) = \mu_1(\xi_i), \quad d_j = \mu(\eta_j \circ \mathrm{pr}_2) = \mu_2(\eta_j) \qquad (5.4.4)$$

where $\mu_1 \in T_a(X)$ and $\mu_2 \in T_b(Y)$ are the images of μ under the maps (5.4.2). As all c_i and d_j can be chosen arbitrarily, surjectivity of (5.4.3) follows.

The notation remaining unchanged, in terms of the coordinate functions $\xi_i(x)$ and $\eta_j(y)$ in the neighbourhood of $(a, b) \in X \times Y$, the distribution μ is clearly given by

$$\mu(f) = \sum c_i \cdot D^i_\xi f(a, b) + \sum d_j \cdot D^j_\eta f(a, b), \qquad (5.4.5)$$

where the functions $D^i_\xi f$ (resp. $D^j_{\eta_j} f$) are obtained by considering $f(x, y)$ as a function of the $\xi_i(x)$ and $\eta_j(y)$ and differentiating with respect to $\xi_i(x)$ (resp. $\eta_j(y)$). This equality can also be interpreted using projections $\mu_1 \in T_a(X)$ and $\mu_2 \in T_b(Y)$ of μ. Indeed, consider the immersions

$$'j_b : x \mapsto (x, b), \quad ''j_a : x \mapsto (a, y), \qquad (5.4.6)$$

from X and Y to $X \times Y$ (the indices $'$ and $''$ have been placed in front of the letter j so as to avoid confusion with derivatives). The following functions can be respectively derived in the neighbourhood of a in X and of b in Y from a C^r function f in the neighbourhood of (a, b):

$$f \circ 'j_b : x \mapsto f(x, b), \quad f \circ ''j_a : y \mapsto f(a, y). \qquad (5.4.7)$$

To differentiate $f(x, y)$ with respect to X at (a, b) it is not really necessary to make y vary. Hence, taking into account computation (5.4.5) of the components of the vectors μ_1 and μ_2, the two sums appearing in (5.4.6) are respectively equal to $\mu_1(f \circ 'j_b)$ and $\mu_2(f \circ ''j_a)$; hence the desired formula:

$$\mu(f) = \mu_1(f \circ ''j_b) + \mu_2(f \circ ''j_a). \qquad (5.4.8)$$

This expression might have the advantage of being rigorously correct, but it has the inconvenience of being somewhat heavy and opaque. To understand the formula, it is much better to adopt the *integral expression for distributions*. It consists in pretending to believe that a distribution is a measure: given a distribution μ on a manifold, write

$$\mu(f) = \int f(x)d\mu(x), \qquad (5.4.9)$$

and in the case of a product manifold, as in integration theory, use the expression

$$\int \int f(x, y)d\mu(x, y). \qquad (5.4.10)$$

Then, for a tangent vector μ to $X \times Y$ at (a, b), (5.4.8) becomes

$$\boxed{\int \int f(x, y)d\mu(x, y) = \int f(x, b)d\mu_1(x, y) + \int f(a, y)d\mu_2(x, y).} \quad (5.4.11)$$

Exercise. Let $t \mapsto (\gamma_1(t), \gamma_2(t))$ be a path with origin (a, b) in $X \times Y$ defining μ. Show that μ_1 and μ_2 are defined by the paths $t \mapsto \gamma_1(t)$ and $t \mapsto \gamma_2(t)$ and that (5.4.8) is equivalent to

$$\frac{d}{dt} f(\gamma_1(t), \gamma_2(t))\Big|_{t=0} = \frac{d}{dt} f(\gamma_1(t), b)\Big|_{t=0} + \frac{d}{dt} f(a, \gamma_2(t))\Big|_{t=0}. \quad (5.4.12)$$

The likeness with the chain rule is evident; is it a "resemblance" rather than the same result?

Exercise. More generally, let $z \mapsto (\phi_1(z), \phi_2(z))$ be a morphism from a manifold Z to $X \times Y$, which maps $c \in Z$ onto $(a, b) \in X \times Y$. Show that the tangent map to this morphism at c is the sum of the tangent maps to the partial maps $z \mapsto (\phi_1(z), b)$ and $z \mapsto (a, \phi_2(z))$, in other words that if μ is a tangent vector to Z at c, then

$$\int f(\phi_1(z), \phi_2(z))d\mu(z) = \int f(\phi_1(z), b)d\mu(z) + \int f(a, \phi_2(z))d\mu(z) \quad (5.4.13)$$

for every C^r function in the neighbourhood of (a, b). Suppose that $Z = \mathbb{R}$ and that $c = 0$; how should μ be chosen so as to recover (5.4.12)?

Does formula (5.4.13) apply to distributions of order > 1?

Exercise. Let X and Y be manifolds, a and b be points of X and Y, and μ_1 and μ_2 be distributions at a and b on X and Y. Define the distribution $\mu = \mu_1 \otimes \mu_2$ at (a, b) on $X \times Y$ by the formula

$$\mu(f) = \int d\mu_1(x) \int f(x, y)d\mu_2(y). \quad (5.4.14)$$

Show that the order of "integration" can be permuted, and compute μ in terms of the expression of μ_1 and μ_2 in the local charts (U, ξ) and (V, η). Show that every distribution at (a, b) on $X \times Y$ is a finite sum of distributions of the previous type.

Show that (5.4.8) can also be written as

$$\boxed{\mu = \mu_1 \otimes \varepsilon_b + \varepsilon_a \otimes \mu_2} \quad (5.4.15)$$

where ε_a and ε_b are the Dirac measures at a and b on X and Y respectively.

5.5   Tangent Manifold

Let X be an n-dimensional manifold and s a natural number. Set $P^{(s)}(X)$ to be the set of punctual distributions of order $\leq s$ on X, and for all $a \in X$, let $P_a^{(s)}(X)$ be the subset of distributions of order $\leq s$ at a, so that $P^{(s)}(X)$ is the disjoint union of the "fibers" $P_a^{(s)}(X)$. Write π for the map from $P^{(s)}(X)$ onto X which associates to each punctual distribution the corresponding point of X; hence $P_a^{(s)}(X) = \pi^{-1}(a)$ for all $a \in X$.

If X is of class C^r, where $r = \infty$ or ω, then endowing $P^{(s)}(X)$ with a C^r manifold structure is easy; (for finite r, we would need to take $s \leq r$, and we would then get a C^{r-s} structure on $P^{(s)}(X)$). For this, choose a local chart (U, ξ) of X and first consider the set $\pi^{-1}(U)$ of distributions having a point of U as support. As seen earlier, such a distribution is characterized by its support $\pi(\mu)$, i.e. by the corresponding point $\xi(\pi(\mu))$ of $\mathbb{R}^n$ and by the values $\mu(\xi^\alpha)$ it takes on the monomials of total degree $|\alpha| \leq s$ in the coordinate functions ξ_i, values that can moreover be chosen arbitrarily. Hence, denoting by N the number of multi-exponents α such that $|\alpha| \leq s$, we get a *bijective* map

$$\pi^{-1}(U) \to \xi(U) \times \mathbb{R}^N \subset \mathbb{R}^{n+N} \tag{5.5.1}$$

by associating to each $\mu \in \pi^{-1}(U)$ the element of $\xi(U) \times \mathbb{R}^N$ whose first component is the vector $\xi(\pi(\mu)) \in \xi(U)$, and the second one is the vector whose coordinates are the N numbers $\mu(\xi^\alpha)$ for α such that $|\alpha| \leq s$.

Denoting the map (5.5.1) by $\xi^{(s)}$, we next define the desired manifold structure on $P^{(s)}(X)$ in such a way that, for any chart (U, ξ) of X, the pair $(\pi^{-1}(U), \xi^{(s)})$ is a *chart* of $P^{(s)}(X)$. To show that we thereby obtain a manifold structure on $P^{(s)}(X)$, as in Chap. 4, Sect. 4, we need to show that the "charts" of $P^{(s)}(X)$ corresponding under this construction to two arbitrary charts of X are "compatible".

This is a matter of checking that, if two charts (U, ξ) and (V, η) of X are chosen, thereby defining C^r structures on $\pi^{-1}(U)$ and $\pi^{-1}(V)$, then (i) the intersection $\pi^{-1}(U) \cap \pi^{-1}(V)$ is open in $\pi^{-1}(U)$ and in $\pi^{-1}(V)$, and (ii) the two C^r structures induced on this intersection by those of $\pi^{-1}(U)$ and $\pi^{-1}(V)$ are identical. (i) is obvious because $\pi^{-1}(U) \cap \pi^{-1}(V) = \pi^{-1}(U \cap V)$; and it is clear that the restrictions of π to $\pi^{-1}(U)$ and $\pi^{-1}(V)$ are continuous maps since, up to (5.5.1), the restriction of π to $\pi^{-1}(U)$ is the projection $\xi(U) \times \mathbb{R}^N \to \xi(U)$. As for condition (ii), it is a matter of showing that if we consider the bijections

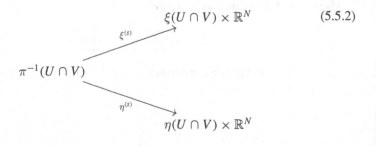

$$\xi(U \cap V) \times \mathbb{R}^N \tag{5.5.2}$$

defined by the two given charts, then there is a C^r map from $\xi(U \cap V) \times \mathbb{R}^N$ to $\eta(U \cap V) \times \mathbb{R}^N$ enabling us to go from the former to the latter. However, this map is the Cartesian product of the map from $\xi(U \cap V)$ to $\eta(U \cap V)$ describing the change of charts, and which is therefore of class C^r, and of the map from $\mathbb{R}^N$ to $\mathbb{R}^N$, which indicates how, for each distribution $\mu \in \pi^{-1}(U \cap V)$, one goes from the $\mu(\xi^\alpha)$ to the $\mu(\eta^\beta)$. So we only need to show that it is also of class C^r. But if $x = \pi(\mu)$, supposing first that $\xi(x) = 0$, equality (5.2.14) shows that

$$\mu(\eta^\beta) = \sum_{|\alpha| \leq s} D_\xi^\alpha \eta^\beta(x) \mu(\xi^\alpha)/\alpha! \tag{5.5.3}$$

If $\xi(x) \neq 0$, replace the coordinates ξ_i by the functions $\xi_i - \xi_i(x)$ vanishing at x, and define another chart $(U, \xi - \xi(x))$. The partial derivatives D_ξ^α appearing in (5.5.3) clearly remain unchanged for $|\alpha| > 0$ and the monomials ξ^α are replaced by the expressions

$$(\xi - \xi(x))^\alpha = \prod (\xi_i - \xi_i(x))^{\alpha_i}. \tag{5.5.4}$$

Expanding these expressions by applying the binomial formula and by inserting the result into

$$\mu(\eta^\beta) = \sum_{|\alpha| \leq s} D_\xi^\alpha \eta^\beta(x) \cdot \mu[(\xi - \xi(x))^\alpha]/\alpha!, \tag{5.5.5}$$

we obviously get an equality of the form

$$\mu(\eta^\beta) = \sum_{|\alpha|,|\gamma|,|\delta| \leq s} c_{\alpha\gamma\delta} D_\xi^\alpha \eta^\beta(x) \cdot \xi^\gamma(x) \cdot \mu(\xi^\delta), \tag{5.5.6}$$

the η^β being expressible using the $\mu(\xi^\alpha)$ and $\xi(\pi(\mu))$ with constants $c_{\alpha\gamma\delta}$ independent of both μ and $x = \pi(\mu)$. However, the numbers $\xi^\gamma(x) = \xi^\gamma(\pi(\mu))$ are monomials in $\xi_i(x) = \xi_i(\pi(\mu))$ and hence in the coordinates of μ relative to the first chart of $\pi^{-1}(U \cap V)$. Similarly, the partial derivatives $D_\xi^\alpha \eta^\beta(x)$ are C^r functions in $\xi_i(x) = \xi_i(\pi(\mu))$, hence in the coordinates of μ relative to the first chart. Finally, the numbers $\mu(\xi^\delta)$ are, by definition of the map (5.5.1), some of the coordinates of μ in the first chart. Hence, the coordinates $\mu(\eta^\beta)$ of μ in the second chart are indeed C^r functions in the coordinates of μ relative to the first one, and are independent from μ.

These arguments apply notably to *tangent vectors*, the computations being slightly simpler in this case. Let $T(X) = \bigcup T_a(X)$ be the set of tangent vectors at the various points of X and let π be once again the map which associates to each $h \in T(X)$ its origin in X. If (U, ξ) is a chart of X, then all $h \in \pi^{-1}(U)$ are characterized by their origins, i.e. by $\xi(\pi(h)) \in \xi(U)$, and by the values $h(\xi_i)$ they take at the coordinate functions. Here, bijection (5.5.1) reduces to the map

$$\xi' : \pi^{-1}(U) \to \xi(U) \times \mathbb{R}^n \subset \mathbb{R}^{2n} \tag{5.5.7}$$

which associates to $h \in \pi^{-1}(U)$ the point whose coordinates are the $2n$ numbers

$$\xi_i(\pi(h)), \quad h(\xi_i). \tag{5.5.8}$$

The C^r structure of $T(X)$ is obtained by regarding (5.5.7) as a chart of $\pi^{-1}(U)$. When the chart (U, ξ) is replaced by another chart (V, η), the coordinate transformation formulas (5.5.8) become particularly simple, and are reduced to

$$h(\eta_j) = \sum_i \frac{\partial \eta_j}{\partial \xi_i}(x) \cdot h(\xi_i), \tag{5.5.9}$$

as can be seen by applying formula (5.2.7), which, contrary to (5.2.3), *does not* assume that $\xi(x) = 0$, since $h(1) = 0$ for all tangent vectors. The set $T(X)$, endowed with this structure, is called the *tangent manifold* to X.

These constructions are evidently "natural" or "functorial", i.e. whenever there is a morphism $\phi : X \to Y$ associating to each punctual distribution on X its image under ϕ, there is a morphism $P^{(s)}(X) \to P^{(s)}(Y)$ or $T(X) \xrightarrow{\phi'} T(Y)$ such that the diagram

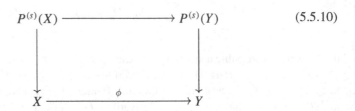

$$\tag{5.5.10}$$

is commutative, etc.

Exercise. If ϕ is an immersion (resp. submersion) so is the corresponding map $P^{(s)}(X) \to P^{(s)}(Y)$. (Use the fact that ϕ admits a local left (resp. right) inverse.) What about the subimmersion case?

Exercise. If Y is a submanifold of X, then $T(Y)$ can be canonically identified with a submanifold of $T(X)$.

Exercise. Let X and Y be two manifolds. Construct a canonical isomorphism from $T(X \times Y)$ onto $T(X) \times T(Y)$.

The reader seeking more substantial questions to think about can for example try to understand the construction of $T(T(X))$.

5.6 The Adjoint Representation of a Lie Group

Let G be a Lie group,[3] and consider the map

$$m : G \times G \to G \qquad (5.6.1)$$

given by $m(x, y) = xy$. A map similar to m can be canonically derived:

$$\boxed{m' : T(G \times G) = T(G) \times T(G) \to T(G).} \qquad (5.6.2)$$

It enables us to endow $T(G)$ with a composition law. We show that $T(G)$, together with this law, is again a *Lie group* whose structure can be explicitly written in terms of that of G.

For this, let us consider vectors $h \in T_a(G)$ and $k \in T_b(G)$, where a and b are arbitrary points of G. Then $(h, k) = (h, 0) + (0, k)$ in $T_a(G) \times T_b(G) = T_{a,b}(G \times G)$, and so

$$m'(h, k) = m'(h, 0) + m'(0, k). \qquad (5.6.3)$$

If, as in Sect. 4, we consider the immersions $'j_b : x \to (x, b)$ and $''j_a : y \to (a, y)$ from G to $G \times G$, then it is readily seen that, up to identification of $T_{a,b}(G \times G)$ with $T_a(G) \times T_b(G)$, according to (5.4.12),

$$(h, 0) = \text{ image of } h \in T_a(G) \text{ under the tangent map to } 'j_b \text{ at } a, \\ (0, k) = \text{ image of } k \in T_b(G) \text{ under the tangent map to } ''j_a \text{ at } b. \qquad (5.6.4)$$

Likewise, $m'(h, 0)$ and $m'(0, k)$ are the images of $(h, 0)$ and $(0, k)$ under the tangent map

$$T_{a,b}(G \times G) = T_a(G) \times T_b(G) \to T_{ab}(G) \qquad (5.6.5)$$

to m at $(a, b) \in G \times G$. Hence, since the composition of tangent maps is similar to that of the maps themselves (chain rule formula (5.3.3)),

$$m'(h, 0) = \text{ image of } h \text{ under the tangent map to } m \circ 'j_b \text{ at } a, \\ \text{ i.e. to } x \mapsto xb, \\ m'(0, k) = \text{ image of } k \text{ under the tangent map to } m \circ ''j_a \text{ at } b, \\ \text{ i.e. to } y \mapsto ay. \qquad (5.6.6)$$

For all $b \in G$, passing to the derivative at a, the automorphism $x \mapsto xb$ of the manifold G defines an isomorphism from $T_a(G)$ onto $T_{ab}(G)$. It is natural to set

---

[3] Of class C^r with $r = \infty$ or ω to simplify the presentation.

$$\boxed{\begin{aligned} hb &= \text{image of } h \in T_a(G) \text{ under the tangent map to } x \mapsto xb \text{ at } a, \\ ak &= \text{image of } k \in T_b(G) \text{ under the tangent map to } y \mapsto ay \text{ at } b, \end{aligned}} \tag{5.6.7}$$

and hence, equality (5.6.3) immediately becomes

$$\boxed{m'(h, k) = hb + ak} \quad \text{if } h \in T_a(G) \quad \text{and} \quad k \in T_b(G). \tag{5.6.8}$$

To go further, set

$$\mathfrak{g} = T_e(G) \tag{5.6.9}$$

to be the tangent space at the identity. For all $c \in G$, the translation $x \mapsto cx$ mapping e to c enables us to define an isomorphism from $\mathfrak{g}$ onto $T_c(G)$, namely $X \mapsto cX$ (the elements of $\mathfrak{g}$ are traditionally written X, Y, etc., but this is not obligatory...). The map

$$(c, X) \mapsto cX \tag{5.6.10}$$

is obviously a bijection from $G \times \mathfrak{g}$ onto $T(G)$, and it is a manifold isomorphism (exercise!). Setting

$$m'(h, k) = hk \tag{5.6.11}$$

in order to highlight that the composition law m' on $T(G)$ extends the composition law m on G, and writing $h = aX$ and $k = bX$ with $X, Y \in \mathfrak{g}$, (5.6.8) becomes

$$(aX)(bY) = (aX)b + a(bY). \tag{5.6.12}$$

Differentiating equality $a(by) = (ab)y$ or equality $a(yb) = (ay)b$ or equality $(ya)b = y(ab)$ at the identity, we get

$$a(bY) = (ab)Y, \quad a(Yb) = (aY)b, \quad (Ya)b = Y(ab) \tag{5.6.13}$$

for all $a, b \in G$ and $Y \in \mathfrak{g}$. Hence, the brackets are pointless and can be removed since (5.6.13) justifies the obvious formal calculations. This gives

$$(aX)(bY) = aXb + abY = abb^{-1}Xb + abY = ab(b^{-1}Xb + Y) \tag{5.6.14}$$

where

$$\boxed{b^{-1}Xb = \text{image of X under the tangent map to } x \mapsto b^{-1}xb \text{ at } e.} \tag{5.6.15}$$

The usual notation is

$$\boxed{\text{ad}\,(g)X = gXg^{-1}} \text{ for } g \in G \text{ and } X \in \mathfrak{g}. \tag{5.6.16}$$

As $c(bxb^{-1})c^{-1} = cbx(cb)^{-1}$, the chain rule shows that

$$\mathrm{ad}\,(xy) = \mathrm{ad}\,(x)\,\mathrm{ad}\,(y), \qquad (5.6.17)$$

with obviously ad $(e) = 1$. Consequently, $x \mapsto \mathrm{ad}\,(x)$ is a homomorphism from the group G to the linear group $GL(\mathfrak{g})$ of the vector space $\mathfrak{g}$; it is called the *adjoint representation* of G (it being understood that, in general, a *representation* of G in a finite-dimensional vector space E over $\mathbb{R}$ is a C^r homomorphism from G to the linear group of E). The fact that ad is of class C^{r-1} like G can be seen by using a chart (U, ξ) at e such that $\xi(e) = 0$. If (X_i) is the corresponding basis of $\mathfrak{g}$ consisting of the distributions X_i given by

$$X_i(f) = D_\xi^i f(e), \qquad (5.6.18)$$

we get ad $(g)X_i$ for fixed $g \in G$, by applying X_i to the composite map of f and the automorphism $x \mapsto gxg^{-1}$, i.e. by differentiating the function $x \mapsto f(gxg^{-1})$ with respect to the ξ_i at the origin. Therefore,

$$\mathrm{ad}\,(g)X_i = \sum c_{ij}(g)X_j \qquad (5.6.19)$$

where

$$c_{ij}(g) = \left.\frac{\partial \xi_j(gxg^{-1})}{\partial \xi_i(x)}\right|_{x=e}; \qquad (5.6.20)$$

as the expressions $\xi_i(gxg^{-1})$ are C^{r-1} functions of the pair (x, g), the coefficients (5.6.20) are themselves C^{r-1} functions of g; hence the desired result.

We are now in a position to formulate the composition law on $T(G)$ by identifying $T(G)$ with the product manifold $G \times \mathfrak{g}$ under the map (5.6.10), and by using (5.6.14) and (5.6.16); on $G \times \mathfrak{g}$, the composition law is evidently given by

$$\boxed{(a, X)(b, Y) = (ab, \mathrm{ad}\,(b)^{-1}X + Y).} \qquad (5.6.21)$$

This composition law is immediately seen to turn $T(G)$ into a new *Lie group* whose identity element is $(e, 0)$. The maps $g \mapsto (g, 0)$ and $X \mapsto (e, X)$ enable us to canonically identify G and the additive group $\mathfrak{g}$ with subgroups of $T(G)$, and formula

$$(a, 0)(e, X) = (a, X) = aX \qquad (5.6.22)$$

shows that every element of $T(G)$ is expressible in a unique way as the product of an element of G and an element of $\mathfrak{g}$. Note that in general the elements of G do *not* commute with those of $\mathfrak{g}$, so that, as a group, $T(G)$ is not the direct product of G and $\mathfrak{g}$. The subgroup $\mathfrak{g}$ is normal in $T(G)$—besides it is the kernel of the homomorphism $(a, X) \mapsto a$ from $T(G)$ onto G—but the same is not true for G.

Exercise. Let G and H be two (abstract) groups and σ a homomorphism from G to the group of automorphisms of H. Endow $G \times H$ with the composition law given by

$$(g, h)(g', h') = (gg', (\sigma(g')^{-1}h)h'). \tag{5.6.23}$$

Show that $G \times H$ is a group (the "semi-direct product" of G and H relative to σ).

These constructions can be easily formulated when $G = GL(E)$, where E is a finite-dimensional real vector space, so that $\mathfrak{g} = \mathscr{L}(E)$, the algebra of endomorphisms of E, the distribution on G defined by $X \in \mathscr{L}(E)$ being given by

$$X(f) = \frac{d}{dt} f(1 + tX)_{t=0}. \tag{5.6.24}$$

If $Y = \mathrm{ad}\,(g)X$, then, as seen above, using the integral notation which is particularly convenient here,

$$\begin{aligned}
Y(f) &= \int f(gxg^{-1})dX(x) = \frac{d}{dt} f[g(1 + tX)g^{-1}]_{t=0} \\
&= \frac{d}{dt} f(1 + t \cdot gXg^{-1})_{t=0}
\end{aligned} \tag{5.6.25}$$

and so identifying $\mathfrak{g}$ with $\mathscr{L}(E)$ leads to the formula

$$\mathrm{ad}\,(g)X = g \circ X \circ g^{-1}, \tag{5.6.26}$$

where these are products of endomorphisms of E [i.e. ordinary products of matrices when $G = GL_n(\mathbb{R})$]. For example, if $G = GL_n(\mathbb{R})$, then

$$T(G) = GL_n(\mathbb{R}) \times M_n(\mathbb{R}) \quad \text{with} \quad (a, X)(b, Y) = (ab, b^{-1}Xb + Y). \tag{5.6.27}$$

This result holds more generally for any closed (hence Lie) subgroup H of $G = GL_n(\mathbb{R})$ or $GL(E)$. First, note that $T_e(H)$ is identified with a vector subspace of $T_e(G) = \mathscr{L}(E)$ under the immersion of H in G; this is consistent with the general theory of manifolds (end of Sect. 3). If, as in Sect. 1 of this chapter, $X \in T_e(H)$ is defined by a deriving sequence (h_p, ε_p)—so that the $h_p \in H$ tend to e and the $\varepsilon_p > 0$ to 0. Clearly, (exercise and generalization to arbitrary manifold morphisms) the image of X in $T_e(G)$ will be defined by the same sequence. The identification of $T_e(G)$ with $\mathscr{L}(E)$—which follows from the fact that $GL(E)$ is an open submanifold of the vector space $\mathscr{L}(E)$—therefore transforms X into the matrix or the endomorphism

$$X = \lim \frac{1}{\varepsilon_p}(h_p - 1). \tag{5.6.28}$$

Here too, the similarity of this equality with formula (3.4.1) of Chap. 3 is not a mere coincidence... We first deduce that *the image of $T_e(H)$ in $\mathscr{L}(E)$ is in fact the* *subalgebra* $\mathfrak{h}$ *of* $\mathscr{L}(E)$ *associated to* H *by von Neumann's theory, i.e. the set of* $X \in \mathscr{L}(E)$ *such that* $\exp(tX) \in H$ *for all* t. It is now clear that the diagram

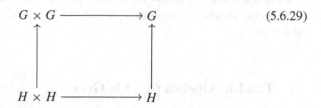

$$(5.6.29)$$

whose horizontal lines denote the composition laws given in G and H, is commutative since the canonical injection of H in G is a Lie group morphism. So the canonical injection from $T(H)$ to $T(G)$ it gives rise to by differentiation is again a Lie group morphism. Formula (5.6.27), which expresses explicitly the composition law in $T(G)$, therefore still holds in $T(H) = H \times \mathfrak{h}$.

For example, let H be the *orthogonal group of a non-degenerate symmetric matrix* $S = S' \in M_n(\mathbb{R})$. Making $G = GL_n(\mathbb{R})$ act on $M_n(\mathbb{R})$ by $g(U) = gUg'$ shows that H is the stabilizer of S, and as the map $g \mapsto g(S)$ is a subimmersion (see p. 117), it follows that

$$T_e(H) = \text{kernel of the tangent map of } g \mapsto g(S) \text{ at } e. \qquad (5.6.30)$$

The latter can be seen using one of the exercises of Sect. 3. But the explicit formulation of this tangent map is easy when $M_n(\mathbb{R})$ is identified with tangent spaces to $GL_n(\mathbb{R})$ and to $M_n(\mathbb{R})$ at 1 and at S: to find the image of an element $X \in M_n(\mathbb{R}) = T_e(G)$ under this map, choose a tangent curve to X at $e = 1$ in G, for example $t \mapsto 1 + tX$; take its composite with the map $g \mapsto g(S)$ under consideration. The outcome is the curve $t \mapsto (1 + tX)S(1 + tX)'$ in $M_n(\mathbb{R})$, and its tangent vector at $t = 0$ remains to be found, i.e.

$$\frac{d}{dt}(1 + tX)S(1 + tX')\bigg|_{t=0} = XS + SX' \qquad (5.6.31)$$

according to Chap. 3. It then follows that the Lie group $T(H)$ consists of the pairs (g, X) with $gSg' = S$ and $XS + SX' = 0$, the composition law in $T(H)$ being given by

$$(g, X)(h, Y) = (gh, h^{-1}Xh + Y), \qquad (5.6.32)$$

where $h^{-1}Yh$ is now the ordinary product of matrices.

It is sufficiently obvious that these considerations apply more generally to the locally linear groups of Chap. 3 because if H is such a group and j a local embedding of H into a group $GL_n(\mathbb{R})$, then j is notably a morphism from $GL_n(\mathbb{R})$ to an open submanifold of H, which suffices to define a tangent linear map

$$J'(e) : T_e(H) \to T_e(GL_n(\mathbb{R})) = M_n(\mathbb{R}); \qquad (5.6.33)$$

it is clearly injective since j in an immersion. Obviously the image of $T_e(H)$ under $j'(e)$ is again the Lie subalgebra $\mathfrak{h}$ of $M_n(\mathbb{R})$ defined in Chap. 3, Theorem 6.

We next show how to endow the tangent space $T_e(G)$ of an arbitrary Lie group with a Lie algebra structure coinciding with that of Chap. 3, without using one-parameter subgroups.

5.7   The Lie Algebra of a Lie Group

Let G be a Lie group and $\mathfrak{g} = T_e(G)$ the tangent space to G at the origin. We now intend to endow $\mathfrak{g}$ with an internal composition law $(X, Y) \mapsto [X, Y]$ which, in the case of the group $GL_n(\mathbb{R})$, will be given by $[X, Y] = XY - YX$, where these are the usual operations on matrices X and Y. Together with this composition law, $\mathfrak{g}$ will be called the *Lie algebra* of G.

There are (at least) half a dozen known methods to do so; here we choose one consisting in *computing the derivative of the adjoint representation* $g \mapsto \operatorname{ad}(g)$ *at* e. Indeed, it is a morphism from the manifold G to the manifold $GL(\mathfrak{g})$, mapping e onto 1. The tangent map at e will therefore induce a linear map from $T_e(G) = \mathfrak{g}$ to $T_1(GL(\mathfrak{g})) = \mathscr{L}(\mathfrak{g})$, the space of endomorphisms of $\mathfrak{g}$. We will also denote it by ad —instead of $\operatorname{ad}'(e)$, which would be more correct but unusable. An endomorphism of the vector space $\mathfrak{g}$, namely $\operatorname{ad}(X) \in \mathscr{L}(\mathfrak{g})$, will then correspond to each $X \in \mathfrak{g}$. The desired composition law in $\mathfrak{g}$ will then be *defined* by setting

$$\boxed{[X, Y] = \operatorname{ad}(X)Y,} \tag{5.7.1}$$

the image of Y under the operator $\operatorname{ad}(X)$.

Before studying the general case, let us consider the case where $G = GL_n(\mathbb{R})$, and compute $\operatorname{ad}(X)$ for $X \in \mathfrak{g} = M_n(\mathbb{R})$. To do so, choose a tangent curve to X passing through e in G, for example $t \mapsto \exp(tX)$, and then apply ad to it. This gives the curve $t \mapsto \operatorname{ad}(\exp(tX)) = \gamma(t)$ in $GL(\mathfrak{g})$. We need to find its tangent vector at $t = 0$, i.e. the operator

$$\operatorname{ad}(X) = \frac{d}{dt}\gamma(t)_{t=0} = \frac{d}{dt}\operatorname{ad}(\exp(tX))_{t=0} \in \mathscr{L}(\mathfrak{g}). \tag{5.7.2}$$

Computation (5.6.26) of the adjoint representation in this case evidently gives

$$\operatorname{ad}(X)Y = \frac{d}{dt}\gamma(t)Y\big|_{t=0} = \frac{d}{dt}\exp(tX) \cdot Y \cdot \exp(tX)^{-1}\big|_{t=0}. \tag{5.7.3}$$

It is a question of differentiating an ordinary product of matrix-valued functions. Since obviously

$$\frac{d}{dt}\exp(tX) = X \cdot \exp(tX) = \exp(tX) \cdot X, \tag{5.7.4}$$

applying the usual formula and keeping track of the order of the factors gives

$$\text{ad}\,(X)Y = \exp(tX) \cdot X \cdot Y \cdot \exp(-tX)\big|_{t=0} - \exp(tX) \cdot Y \cdot X \cdot \exp(-tX)\big|_{t=0}$$
$$= XY - YX,$$

(5.7.5)

and hence the claimed formula

$$[X, Y] = XY - YX \tag{5.7.6}$$

in this case.

In the general case, choose a curve $t \mapsto \gamma(t)$ in G tangent to X at $t = 0$, and apply ad to it. This gives a curve $t \mapsto \text{ad}\,(\gamma(t))$ in $GL(\mathfrak{g})$. Its tangent at $t = 0$ remains to be found, whence the formula

$$\boxed{\text{ad}\,(X) = \frac{d}{dt}\text{ad}\,(\gamma(t))\bigg|_{t=0}}, \tag{5.7.7}$$

namely the ordinary derivative of a function with values in the vector space $\mathfrak{g}$. So for $Y \in \mathfrak{g}$,

$$\text{ad}\,(X)Y = \frac{d}{dt}\text{ad}\,(\gamma(t))Y\bigg|_{t=0}. \tag{5.7.8}$$

To show this, let us choose a tangent curve $s \mapsto \delta(s)$ to Y at $s = 0$, and to simplify the notation, set $Z = \text{ad}\,(X)Y$ and $Y_t = \text{ad}\,(\gamma(t))Y$. Considering these objects as distributions at e, for every C^r function f in the neighbourhood of e,

$$Y(f) = \int f(y)dY(y) = \frac{d}{ds}f(\delta(s))\bigg|_{s=0}, \tag{5.7.9}$$

and so, using the general formula (5.3.10),

$$Y_t(f) = \int f[\gamma(t)y\gamma(t)^{-1}]dY(y) = \frac{d}{ds}f[\gamma(t)\delta(s)\gamma(t)^{-1}]\bigg|_{s=0}. \tag{5.7.10}$$

For fixed f, the map $Y \mapsto Y(f)$ is a linear form on $T_e(G)$; the vector $Z = \frac{d}{dt}Y_t\big|_{t=0}$ is therefore given by

$$Z(f) = \frac{d}{dt}Y_t(f)\bigg|_{t=0} = \frac{d}{dt}\left\{\frac{d}{ds}f[\gamma(t)\delta(s)\gamma(t)^{-1}]_{s=0}\right\}_{t=0}$$
$$= \frac{\partial^2}{\partial t\partial s}f[\gamma(t)\delta(s)\gamma(t)^{-1}]_{s=t=0} \tag{5.7.11}$$
$$= \frac{\partial^2}{\partial s\partial t}f[\ldots].$$

Lemma *Let ϕ and ψ be C^r maps from $\mathbb{R}$ to G; for every C^r function f in the neighbourhood of the point $\phi(0)\psi(0)$,*

$$\left.\frac{d}{dt}f[\phi(t)\psi(t)]\right|_{t=0} = \left.\frac{d}{dt}f[\phi(t)\psi(0)]\right|_{t=0} + \left.\frac{d}{dt}f[\phi(0)\psi(t)]\right|_{t=0}. \tag{5.7.12}$$

The map $t \mapsto \phi(t)\psi(t)$ is indeed the composite of the map

$$t \mapsto (\phi(t), \psi(t)) \tag{5.7.13}$$

from $\mathbb{R}$ to $G \times G$ and the map $m : G \times G \to G$. The chain rule therefore shows that the tangent vector to $t \mapsto \phi(t)\psi(t)$ at $t = 0$ is the image under $m' : T(G) \times T(G) \mapsto T(G)$ of the tangent vector to (5.7.13) at $t = 0$. But this is (h, k) where h and k denote the tangent vectors to $t \mapsto \phi(t)$ and $t \mapsto \psi(t)$ at $t = 0$. The desired vector is therefore

$$m'(h, k) = h \cdot \psi(0) + \phi(0) \cdot k, \tag{5.7.14}$$

as shown in (5.6.8). Hence, to compute the left-hand side of (5.7.12), apply distribution $h \cdot \psi(0)$ to f. By definition of the product of a tangent vector and an element of G, this gives the first term of the right-hand side of (5.7.12). Apply distribution $\phi(0) \cdot k$, which gives the second term, then add the results obtained, qed. Besides, have a look at (5.4.13).

Replacing the curve $t \mapsto \psi(t)$ by $t \mapsto c\psi(t)$, where c is a fixed element of G, would give the more general (sic) formula

$$\frac{d}{dt}f[\phi(t)c\psi(t)]_{t=0} = \frac{d}{dt}f[\phi(t)c\psi(0)]_{t=0} + \frac{d}{dt}f[\phi(0)c\psi(t)]_{t=0}. \tag{5.7.15}$$

Returning to (5.7.11), it then follows that differentiation with respect to t gives

$$\begin{aligned}\frac{d}{dt}f[\gamma(t)\delta(s)\gamma(t)^{-1}]_{t=0} &= \frac{d}{dt}f[\gamma(t)\delta(s)]_{t=0} + \frac{d}{dt}f[\delta(s)\gamma(t)^{-1}]_{t=0}\\ &= \frac{d}{dt}f[\gamma(t)\delta(s)]_{t=0} - \frac{d}{dt}f[\delta(s)\gamma(t)]_{t=0}\end{aligned} \tag{5.7.16}$$

since in general,

$$\frac{d}{dt}f[\gamma(t)^{-1}]_{t=0} = -\frac{d}{dt}f[\gamma(t)]_{t=0}, \tag{5.7.17}$$

as can, for example, be seen by taking $\psi(t) = \phi(t)^{-1}$ in (5.7.12). As an aside, note that (5.7.17) means *the tangent map* to $x \mapsto x^{-1}$ at $x = e$ is $X \mapsto -X$.

The computation of $Z(f)$ can now be completed, and evidently leads to formula

$$Z(f) = \frac{d^2}{dsdt} f[\gamma(t)\delta(s)]_{s=t=0} - \frac{d^2}{dsdt} f[\delta(t)\gamma(s)]_{s=t=0}, \qquad (5.7.18)$$

which highlights the importance of non-commutativity in the computation of

$$Z = \text{ad}(X)Y = [X, Y]. \qquad (5.7.19)$$

Since, in integral notation,

$$\frac{d}{dt} f(\gamma(t))_{t=0} = \int f(x)dX(x), \quad \frac{d}{ds} f(\delta(t))_{s=0} = \int f(y)dY(y), \qquad (5.7.20)$$

Equation (5.7.18) can also be formulated as

$$Z(f) = \int \int f(xy)dX(x)dY(y) - \int \int f(xy)dY(x)dX(y). \qquad (5.7.21)$$

To interpret the latter, we need to define the *convolution* $\lambda * \mu = \nu$ of two punctual distributions λ and μ at points a and b: it is the distribution at ab given by

$$\lambda * \mu(f) = \int \int f(xy)d\lambda(x)d\mu(y). \qquad (5.7.22)$$

If, for example, $\lambda = \varepsilon_a$ and $\mu = \varepsilon_b$ are the Dirac measures at a and b, then clearly

$$\varepsilon_a * \varepsilon_b = \varepsilon_{ab}, \qquad (5.7.23)$$

so that the convolution of distributions extends the group multiplication (it is also one of the possible generalizations of the notion of group algebra in the theory of finite groups). Formula (5.7.21) then shows that

$$[X, Y] = X * Y - Y * X. \qquad (5.7.24)$$

Exercise. The existing similarity between (5.7.24) and (5.7.6) suggests that for the group $G = GL_n(\mathbb{R})$ the distribution $X * Y$ appearing in (5.7.24) is just the tangent vector $XY \in M_n(\mathbb{R}) = T_e(G)$ appearing in (5.7.6). True? False? (It might prove useful to first investigate the case $n = 1$.)

Exercise. Let $G = GL_n(\mathbb{R})$. Writing (e_i) for the canonical basis of $\mathbb{R}^n$, the matrix (g_i^j) of $g \in G$ with respect to this basis is therefore given by the formulas

$$g(e_i) = g_i^j e_j, \qquad (5.7.25)$$

where we have used Einstein's summation convention. This involves summing with respect to every index appearing twice in a product (here the index j) without mentioning it explicitly, yet mentioning it as we are doing at this very moment; see MA IX, Chap. 1. Since the g_i^j form a chart of G, if f is a function on G, in what follows, we set

$$D_j^i f = \partial f / \partial g_i^j \qquad (5.7.26)$$

and so there is an operator D_j^i acting, for example, on the C^∞ functions on an open subset of G. Let $X = (x_i^j) \in M_n(\mathbb{R})$ be a tangent vector to G at e; show that the distribution X at e on G is given by

$$X(f) = x_i^j \cdot D_j^i f(1), \qquad (5.7.27)$$

where we use Einstein's summation convention again. If $Y = (y_i^j)$ is another matrix, the distribution $X * Y$ at the origin is a linear form

$$\begin{aligned} f \to \frac{d^2}{ds dt} f[(1 + sX)(1 + ty)]_{s=t=0} \\ = x_k^j y_i^k D_j^i f(1) + x_i^j y_k^h D_h^k D_j^i f(1) \end{aligned} \qquad (5.7.28)$$

i.e. a second-order distribution; check that $X * Y - Y * X$ is the tangent vector defined by the matrix $XY - YX$ using direct arguments.

Exercise. generalizing (5.7.17). Let G be a Lie group; set $j(x) = x^{-1}$. Show that the tangent linear map to j at $a \in G$ is given by

$$j'(a)h = -a^{-1}ha^{-1}, \qquad (5.7.29)$$

where the right-hand side is computed in the Lie group $T(G)$, and that the automorphism j' of the manifold $T(G)$ which extends the automorphism j of the manifold G (see end of Sect. 5) is the map $x \mapsto x^{-1}$ in the group $T(G)$.

Exercise. Let U and V be two manifolds, X and Y tangent vectors to U and V at points a and b of U and V. This gives—see (5.4.15)—a distribution $X \otimes Y = T$ at (a, b) on the product manifold $U \times V$, which, by definition, is

$$T(f) = \int \int f(x, y) dX(x) dY(y) \qquad (5.7.30)$$

for any function f of class C^2 at least in the neighbourhood of (a, b). Suppose that X and Y are defined by deriving sequences (x_p, ε_p) and (y_q, η_q) in the sense of Sect. 1. Show that

$$T(f) = \lim_{p,q \to \infty} \frac{1}{\varepsilon_p \eta_q} [f(x_p, y_q) - f(x_p, b) - f(a, y_q) + f(a, b)]. \qquad (5.7.31)$$

[The question being a local one, it reduces to the case where U and V are open neighbourhoods of 0 in Cartesian products and where $a = 0, b = 0$; one can then use formula

$$f(x, y) - f(x, 0) - f(0, y) + f(0, 0)$$

$$= \sum x_i x_j \int_0^1 \int_0^1 D_1^i D_2^j f(sx, ty) ds dt, \qquad (5.7.32)$$

where the D_1^i (resp. D_2^j) are the partial differentiations with respect to the coordinates of the variable x (resp. y). Formula (5.7.31), where p and q independently tend to infinity then follows.]

Let G be a Lie group, X and Y two elements of its Lie algebra defined by deriving sequences (x_p, ε_p) and (y_q, η_q); show that the tangent vector $[X, Y]$ is the distribution

$$f \to \lim_{p,q \to \infty} \frac{1}{\varepsilon_p \eta_q} [f(x_p y_q) - f(y_q x_p)] \qquad (5.7.33)$$

at the origin.

5.8 Effect of a Homomorphism on a Lie Algebra

Let G and H be two (C^∞ or analytic) Lie groups and π a (C^∞ or analytic) homomorphism from G to H. As seen in a more general context at the end of Sect. 5, π canonically defines a map

$$\pi' : T(G) \to T(H) \qquad (5.8.1)$$

which induces, for each $a \in G$, a tangent linear map $\pi'(a) : T_a(G) \to T_b(H)$, where $b = \pi(a)$, and which is also, as will be seen, a Lie group homomorphism. The fact that π' is of class C^∞ or analytic is obvious from the construction of the manifolds $T(G)$ and $T(H)$ in Sect. 5. The compatibility of π' with the composition laws of $T(G)$ and $T(H)$ follows from the commutativity of the diagram

$$
\begin{array}{ccc}
G \times G & \xrightarrow{\pi \times \pi} & H \times H \\
\downarrow & & \downarrow \\
G & \xrightarrow{\pi} & H
\end{array}
\qquad (5.8.2)
$$

whose vertical arrows denote the composition laws of G and H. The "functorial" character of the map $X \mapsto T(X)$ then shows the commutativity of the diagram obtained from this one by replacing G and H by the manifolds $T(G)$ and $T(H)$, and π by π'. This result is anyhow not very deep. According to (5.6.20) which defines the composition laws in $T(G) = G \times \mathfrak{g}$ and $T(H) = H \times \mathfrak{h}$, where $\mathfrak{g}$ and $\mathfrak{h}$ denote the Lie algebras of G and H, it merely means that for $g \in G$, $X \in \mathfrak{g}$,

$$\boxed{\pi'(\text{ad}\,(g)X) = \text{ad}\,(h)Y \quad \text{where } h = \pi(g), \quad Y = \pi'(X);} \tag{5.8.3}$$

note that here the map π' acting on $\mathfrak{g} = T_e(G)$ is just the tangent linear map $\pi'(e)$.

Compatibility of π', or more exactly of $\pi'(e)$, with the Lie algebra structures of $\mathfrak{g}$ and $\mathfrak{h}$ is a more interesting and fundamental property—in other words,

$$\boxed{\pi'([X_1, X_2]) = [\pi'(X_1), \pi'(X_2)]} \quad \text{for all } X_1, X_2 \in \mathfrak{g}. \tag{5.8.4}$$

We are in a position to give several proofs of this equality.

The first proof consists in observing that (5.8.3) is also expressible as

$$u \circ \text{ad}\,(g) = ad(\pi(g)) \circ u \quad \text{where } u = \pi'(e), \tag{5.8.5}$$

and in finding the tangent linear maps to both sides at $g = e$. As the factor u is a linear map from $\mathfrak{g}$ to $\mathfrak{h}$ independent from g, it plays the role of a constant and the question is reduced to finding the tangent linear maps to $g \mapsto \text{ad}\,(g)$ and to $g \mapsto \text{ad}\,(\pi(g)) = \text{ad}\,\circ \pi(g)$ at $g = e$. By definition, the former is the map $U \mapsto \text{ad}\,(U)$ as agreed at the beginning of Sect. 7. The latter can be found by applying the chain rule, i.e. formula (5.3.3), by letting ϕ be the homomorphism π and ψ be the adjoint representation of H, with $a = \phi(a) = e$. As $\phi'(a) = \pi'(e) = u$ and as, by definition, $\psi'(b)$ is now also the map $V \mapsto \text{ad}\,(V)$ from $\mathfrak{h}$ to $\mathscr{L}(h)$, differentiating (5.8.5) at the origin we find

$$u \circ \text{ad}\,(U) = \text{ad}\,(u(U)) \circ u \tag{5.8.6}$$

for all $U \in \mathfrak{g}$. Since $u = \pi'$ on $\mathfrak{g}$, applying both sides to $X \in \mathfrak{g}$ gives $\pi'(\text{ad}\,(U)X) = \text{ad}\,(\pi'(U))\pi'(X)$. Its explicit formulation provides the desired equality $\pi'([U, X]) = [\pi'(U), \pi'(X)]$.

A second proof consists in starting from formula $[X, Y] = X * Y - Y * X$ which reduces the Jacobi bracket to the convolution (5.7.12) of punctual distributions and in applying π' to it. As seen in Sect. 3 in a much more general context, the image $\pi'(U)$ of $U \in \mathfrak{g} = T_e(G)$ under π'—as a punctual distribution on H—is just the direct image $\pi_*(U)$ under the morphism $\pi : G \to H$ of the distribution U on G. Then the formula in need of proof can be written

$$\pi_*(X * Y - Y * X) = \pi_*(X) * \pi_*(Y) - \pi_*(Y) * \pi_*(X). \tag{5.8.7}$$

As the map $\mu \mapsto \pi_*(\mu)$ is clearly linear for any manifold morphism ϕ, it suffices to show that more generally, if π is a (C^∞ or analytic) homomorphism from a Lie group G to a Lie group H, then

$$\boxed{\pi_*(\lambda * \mu) = \pi_*(\lambda) * \pi_*(\mu)} \tag{5.8.8}$$

for all punctual *distributions* λ and μ on G. This is obvious.

Indeed, set $\nu = \lambda * \mu$, and let λ', μ' and ν' be the images of λ, μ and ν under π. If g is a C^∞ function on H, then by definition of the direct image, $\nu'(g) = \nu(g \circ \pi)$ or, in integral notation,

$$\int_H g(z')d\nu'(z') = \int_G g \circ \pi(z)d\nu(z); \tag{5.8.9}$$

but integration with respect to a function ν of $z \in G$ reduces to integration with respect to the Cartesian product distribution $d\lambda(x)d\mu(y)$, namely the function obtained by replacing z by xy—see (5.7.22). Equality (5.8.9) then becomes

$$\begin{aligned}\int_H g(z')d\nu'(z') &= \int\int_{G \times G} g \circ \pi(xy)d\lambda(x)d\mu(y) \\ &= \int\int_{G \times G} g[\pi(x)\pi(y)]d\lambda(x)d\mu(y) \\ &= \int\int_{H \times H} g(x'y')d\lambda'(x')d\mu'(y')\end{aligned} \tag{5.8.10}$$

by definition of direct images λ' and μ'; we thereby recover definition (5.7.22) for H, whence (5.8.8). We will return to this later in the context of invariant differential operators.

Exercise. Find a third proof of (5.8.4) from the exercise at the end of Sect. 7, and in particular from definition (5.7.33) of the Jacobi bracket.

The result obtained applies notably to the canonical injection of a subgroup in a group: *if M is a closed subgroup[4] of a Lie group G, then the tangent map at e to the immersion j of M in G enables us to identify the Lie algebra $\mathfrak{m}$ of M with a Lie subalgebra of the Lie algebra $\mathfrak{g}$ of G.* We will see in the next chapter that conversely every subalgebra $\mathfrak{m}$ of $\mathfrak{g}$ corresponds to a subgroup M, which is not necessarily closed, but is endowed with a Lie group structure such that the canonical injection of M in G is an immersion and defines a Lie algebra isomorphism from the Lie algebra of M onto $\mathfrak{m}$.

In the above, let M be the kernel of a homomorphism π from G to another Lie group H. Then $M = \pi^{-1}(e)$, and we know that [Chap. 4, Sect. 4.8, make G act on H by $(g, h) \mapsto \pi(g)h$] π is a subimmersion. Consequently, (Chap. 4, Sect. 4.7), M is a submanifold—hence a Lie subgroup—of G, and the construction of the tangent spaces to the "fibers" of a subimmersion—a construction obvious in its standard form—implies that its Lie algebra is given by

$$\mathfrak{m} = \text{Ker}(\mathfrak{g} \xrightarrow{\pi'} \mathfrak{h}). \tag{5.8.11}$$

---

[4]Here we anticipate Chap. 6, where we will show that every closed subgroup of a Lie group is a manifold (hence a Lie group).

Note that, in this case, the subgroup M of G is normal, so that the subalgebra $\mathfrak{m}$ of $\mathfrak{g}$ is invariant under the adjoint representation of G. This can, for example, be checked by observing that more generally, if M is normal in G, then the tangent manifold $T(M)$ can be canonically identified with a normal subgroup of the tangent manifold $T(G)$, so that for $g \in G$, the automorphism $u \mapsto gug^{-1}$ of $T(G)$ leaves

$$\mathfrak{g} \cap T(M) = \mathfrak{m} \tag{5.8.12}$$

invariant. It is simpler to write down the commutative diagram

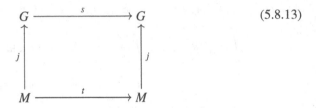

$$\tag{5.8.13}$$

where s is the automorphism $x \mapsto gxg^{-1}$ of G, t its restriction to M and j the canonical injection. Differentiating at the origin, we get

$$j'(e) \circ t'(e) = s'(e) \circ j'(e), \tag{5.8.14}$$

and hence that Im $(j'(e))$, i.e. the subalgebra $\mathfrak{m}$ of $\mathfrak{g}$, is invariant under $s'(e) = \operatorname{ad}(g)$, as claimed.

As a result, since $\operatorname{ad}(g)Y \in \mathfrak{m}$ for all $Y \in \mathfrak{m}$, the map $g \mapsto \operatorname{ad}(g)Y$ from G to $\mathfrak{g}$ actually takes its values in the subspace $\mathfrak{m}$ for all $Y \in \mathfrak{m}$. Therefore its differential at the origin,[5] which, according to the beginning of Sect. 7, is the map $X \mapsto \operatorname{ad}(X)Y = [X, Y]$, is also $\mathfrak{m}$-valued. Namely,

$$[X, Y] \in \mathfrak{m} \quad \text{for all } X \in \mathfrak{g} \text{ and } Y \in \mathfrak{m}. \tag{5.8.15}$$

In other words, the subalgebra of $\mathfrak{m}$ corresponding to a normal subgroup of G is said to be an *ideal* of $\mathfrak{g}$.

As an exercise, let us compute a subalgebra. Consider an element a of a Lie group G and the subgroup M of $g \in G$ commuting with a. It is the stabilizer of a in G when G is made to act on itself via the map $(g, x) \mapsto gxg^{-1}$, and so M is a closed Lie subgroup of G. As $g \mapsto gag^{-1}$ is a subimmersion, the subspace $\mathfrak{m}$ of $\mathfrak{g}$ is the kernel of the tangent linear map from $\mathfrak{g} = T_e(G)$ to $T_a(G)$ to the map $g \mapsto gag^{-1}$ from G to G at $g = e$. Hence, we need to find its differential at the origin.

Lemma *Let X be a manifold, G a Lie group, ϕ and ψ morphisms from X to G. Then the tangent linear map to the map $x \mapsto \phi(x)\psi(x)$ from X to G at $c \in X$ is*

---

[5]In this context, for us the term "differential" is synonymous with "tangent linear map". "Differentiating" a map is to compute its differential. See (5.3.2).

$$h \mapsto u(h)b + av(h), \tag{5.8.16}$$

where $a := \phi(c)$, $b := \psi(c)$, $u := \phi'(c)$, $v := \psi'(c)$. *["product derivative rule", which can also be written as* $d\mu(c; h) = d\phi(c; h)\psi(c) + \phi(c)d\psi(c; h)$, *where* $\mu(x) := \phi(x)\psi(x).$ *]*

In the above, (5.8.16) is well-defined since $u(h) \in T_a(G)$, so that the product $u(h)b$, defined by (6.7), is in $T_{ab}(G)$, and so is $av(h)$.

The proof of (5.8.16) is easy. Consider the map from X to $G \times G$ given by $\theta(x) = (\phi(x), \psi(x))$, and the map m from $G \times G$ to G given by $m(g, h) = gh$, so that the map we want is just $m \circ \theta$; its differential at c is therefore the composite of $\theta'(c)$ and $m'(a, b)$. Identifying the tangent space to $G \times G$ at (a, b) with $T_a(G) \times T_b(G)$ gives

$$\theta'(c)h = (u(h), v(h)). \tag{5.8.17}$$

On the other hand, the map $m'(a, b) : T_a(G) \times T_b(G) \to T_{ab}(G)$ is precisely the one that enabled us to define the composition law m' on the Lie group $T(G)$ in Sect. 6, and which is explicitly given by (5.6.8). It is necessary to observe that in the notation of (5.6.8) $m'(h, k)$ denotes the image of the pair $(h, k) \in T_a(G) \times T_b(G)$ under the map m' which reduces to the tangent map $m'(a, b)$ on $T_a(G) \times T_b(G)$. It should not be confused with the image under m', by the way equal to ab, of the pair (a, b), despite unintended notational similarities. Manifold and Lie group theory has always raised nearly intractable notational problems, and 'unimportant' contradictions are the best that can be hoped for. Having said this, to recover the right-hand side of (5.8.16), (5.6.8) still needs to be applied to the right-hand side of (5.8.17) by replacing h by $u(h)$ and k by $v(h)$ in the latter; hence the lemma; see also (5.7.12).

Returning to the computation of the differential of the map $g \mapsto gag^{-1}$ at the origin, the lemma can be applied taking $X = G$, $\phi(x) = xa$, $\psi(x) = x^{-1}$ and $c = e$. Then u is the map $X \mapsto Xa$ from $T_e(G) = \mathfrak{g}$ onto $T_a(G)$ and v the map $Y \mapsto -Y$ from $\mathfrak{g}$ onto $\mathfrak{g}$ as seen above. The differential we are looking for is therefore the map

$$X \mapsto Xa - aX = (X - \operatorname{ad}(a)X)a \tag{5.8.18}$$

from $\mathfrak{g}$ to $T_a(G)$. As a result, *if M is the centralizer of $a \in G$ in G, the Lie algebra $\mathfrak{m}$ of M is the set of elements $X \in \mathfrak{g}$ such that*

$$\operatorname{ad}(a)X = X. \tag{5.8.19}$$

Exercise. Let $\pi : G \to H$ be a Lie group homomorphism and H' a closed (hence Lie) subgroup of H. Show that $\pi^{-1}(H') = G'$ is a Lie subgroup of G whose Lie algebra is the inverse image under $\pi'(e)$ of the subalgebra of the Lie algebra of H associated to H' (make G act on the manifold H/H' using π and consider the stabilizer in G of the point of origin of H/H').

Exercise. If H_1 and H_2 are Lie subgroups of a Lie group G, so is $H_1 \cap H_2$ and, using obvious conventions,

$$\mathrm{Lie}(H_1 \cap H_2) = \mathrm{Lie}(H_1) \cap \mathrm{Lie}(H_2). \tag{5.8.20}$$

(Consider the map $g \mapsto (g, g)$ from G to $G \times G$.)

Exercise. (a) Let $j : X \to Y$ be an immersion; suppose that the image $j(X)$ is locally closed in Y and that j induces a homeomorphism from X onto $j(X)$. Show that $j(X)$ is a submanifold isomorphic to X under j. (b) Let G be a Lie group countable at infinity acting on a manifold X. Suppose that the orbit Ga of a point a of X is locally closed. Show it is a submanifold isomorphic (under the obvious map) to G/H, where H is the stabilizer of a in G. Describe the subspace of $T_a(X)$ consisting of the tangent vectors to Ga at a. (c) Let H and H' be two Lie subgroups of a Lie group G countable at infinity. Suppose that the set HH' of products xy ($x \in H$, $y \in H'$) is locally closed in G. Show it is a submanifold of G and that the subspace of the tangent Lie algebra of G to HH' at e is $\mathrm{Lie}(H) + \mathrm{Lie}(H')$. What about the case when H and H' commute?

Despite the few elementary results obtained in this section, the connections between Lie subalgebras and Lie subgroups cannot really be studied without using the exponential map of Chap. 6. To conclude this chapter, which is essentially devoted to "formal" aspects of the theory, we study invariant differential operators on Lie groups and beforehand differential operators on manifolds.

5.9   Differential Operators on Manifolds

It is well known that in the space $\mathbb{R}^n$, a (linear) differential operator L of class C^r associates to every sufficiently differentiable function f defined on an arbitrary open subset U of $\mathbb{R}^n$ a new function Lf defined on U and given by

$$Lf(X) = \sum_{|\alpha| \le s} c_\alpha(x) \cdot D^\alpha f(x),$$

where the $D^\alpha f$ are the usual partial derivatives of f and the c_α functions of class C^r on $\mathbb{R}^n$ depending only on L and called the coefficients of L. For each $x \in \mathbb{R}^n$, the equality

$$L_x(f) = \sum_{|\alpha| \le s} c_\alpha(x) \cdot D^\alpha f(x)$$

clearly defines a distribution of order $\le s$ at x, and if for each x, a distribution L_x is given, then this fully determines L. The notion of a differential operator can then obviously be extended to manifolds.

Let X be a C^r manifold (with $r = \infty$ or ω), s a natural number, and consider the manifold $P^{(s)}(X)$ of punctual distributions of order $\le s$ on X, defined in Sect. 5. A *differential operator of order $\le s$ and class C^r on X is a map

$$L : X \to P^{(s)}(X) \tag{5.9.1}$$

of class C^r such that, for all $x \in X$, the support of the distribution $L(x)$ is x. Denoting by π the canonical projection from $P^{(s)}(X)$ onto X, this obviously means that $\pi \circ L$ is the identity map from X to X. If these two conditions hold, (and if we know that it is a fiber space...) L is said to be a C^r *section* of the fibered manifold $P^{(s)}(X)$.

Exercise. Punctual distributions of order $\leq s - 1$ also being punctual distributions of order $\leq s$, $P^{(s-1)}(X) \subset P^{(s)}(X)$. Show that $P^{(s-1)}(X)$ together with its C^r structure is a closed submanifold of $P^{(s)}(X)$. Deduce that a differential operator of order $\leq s - 1$ is also a differential operator of order $\leq s$.

The previous exercise enables us (i) to define the notion of a *differential operator* with reference to any integer s, (ii) to define the exact *order* of a differential operator L: it is the smallest integer s such that L is of order $\leq s$.

Let L be a differential operator and f a C^r (or simply C^s if L is of order s) function on an open subset U of X. At each point x of U, apply the distribution $L(x) = L_x$ to the function f; this gives a number $L_x(f)$, and hence a function

$$Lf : x \mapsto L_x(f) \tag{5.9.2}$$

on U. Obviously Lf is of class C^{t-s} if f is of class C^t with $t \geq s$. Besides, let (U, ξ) be a chart of X and suppose that f is defined on U. As seen in Sect. 5, we get a chart of $P^{(s)}(X)$ in the open subset $\pi^{-1}(U)$ by taking for the coordinates of an element $\mu \in \pi^{-1}(U)$ the numbers

$$\xi_i(\pi(\mu)), \quad \mu(\xi^\alpha) \quad \text{for } |\alpha| \leq s. \tag{5.9.3}$$

Taking $\mu = L_x$ for $x \in U$, and taking into account that $\pi(L_x) = x$, the coordinates of L_x are seen to be the numbers

$$\xi_i(x), \quad L_x(\xi^\alpha) = L\xi^\alpha(x), \tag{5.9.4}$$

namely the values at x of the coordinate functions ξ_i and of the functions $L\xi^\alpha$ obtained by applying L to the functions ξ^α according to (5.9.2). To check that $x \mapsto L_x$ is of class C^r, it is necessary and sufficient to check that the coordinates of L_x are C^r functions of those of x, or, what clearly amounts to the same, that the *functions* $L\xi^\alpha$ *are of class* C^r on U for any multi-exponent α such that $|\alpha| \leq s$.

Then, to show that Lf is of class C^{t-s} if f is of class C^t, it suffices to apply (5.5.5) to f, as this obviously holds for every function of class at least C^s on U. This gives

$$Lf(x) = L_x(f) = \sum_{|\alpha| \leq s} D_\xi^\alpha f(x) \cdot L_x[(\xi - \xi(x))^\alpha]/\alpha!$$

$$= \sum_{|\alpha|, |\gamma|, |\delta| \leq s} c_{\alpha\gamma\delta} D_\xi^\alpha f(x) \cdot \xi^\gamma(x) \cdot L_x(\xi^\delta). \tag{5.9.5}$$

This can also be written

$$Lf = \sum_{|\alpha|,|\gamma|,|\delta| \leq s} c_{\alpha\gamma\delta} \xi^\gamma L(\xi^\delta) D^\alpha_\xi(f), \tag{5.9.6}$$

where $D^\alpha_\xi(f)$ stands for $D^\alpha_\xi f$, and $L(\xi^\delta)$ for $L\xi^\delta$ to avoid confusion, so that the general term in the above sum is the product function of the constant $c_{\alpha\gamma\delta}$ and of the functions $x \mapsto \xi^\gamma(x)$, $x \mapsto D^\alpha_\xi f(x)$ and $x \mapsto L_x(\xi^\delta)$. The fact that Lf is of class C^{t-s} when f is of class C^t is now clear. At the same time, *in the chart (U, ξ) of X, any differential operator admits an expression of the form*

$$Lf(x) = \sum_{|\alpha| \leq s} c_\alpha(x) D^\alpha_\xi f(x) \tag{5.9.7}$$

with functions $c_\alpha(x)$ of class C^r on U, which of course depend on the choice of the chart. As it is always possible to find a function f whose derivatives $D^\alpha_\xi f$ of order $\leq s$ have given values at x, the punctual distributions $f \mapsto D^\alpha_\xi f(x)$, for fixed x and variable α, are linearly independent, and so expression (5.9.7) for L is unique.

Conversely, fix a map $f \mapsto Lf$ which, to every function f defined and of class C^r on an open subset U of X associates a function of the same kind, and assume that there is an integer s such that, in every chart, L admits an expression of the form (5.9.7). The linear form $f \mapsto L_x(f) = Lf(x)$ is then a distribution of order $\leq s$ at x, for all $x \in X$, so that, by (5.9.2), the operator $f \mapsto Lf$ can indeed be obtained from a section of $P^{(s)}(X)$. It is of class C^r since, if this is the case of Lf for every C^r function f, then the coefficients $c_\alpha(x)$ appearing in (5.9.7) are themselves necessarily of class C^r, which the reader will verify as an exercise.

A differential operator of order 0 is of type

$$Lf(x) = c(x)f(x) \tag{5.9.8}$$

with a C^r function c which determines L; besides, c is just the function obtained by applying L to the function equal to 1 everywhere.

In a chart (U, ξ), a differential operator of order ≤ 1 is of the form:

$$Lf(x) = c_o(x)f(x) + \sum c_i(x) D^i_\xi f(x); \tag{5.9.9}$$

the most important case corresponds to $c_o(x) = 0$ for all x (the question can be reduced to it by subtracting a 0th order operator from L). Then L_x is a tangent vector to X at every point, and L may be regarded as a *section of the manifold $T(X)$ of tangent vectors* to X, i.e. a C^r map $L : X \to T(X)$ such that $\pi \circ L = \mathrm{id}$, where $\pi : T(X) \to X$ is the obvious projection. Such an operator is also called a *vector field* on X. It can be shown that vector fields are the differential operators characterized by

$$L(fg) = L(f)g + f \cdot L(g) \tag{5.9.10}$$

for all C^r functions f and g—this follows from the analogous characterization for tangent vectors.[6]

Exercise. If L is of order $\leq s$, then, for any function g, the map $f \mapsto L(fg) - L(f)g$ is a differential operator of order $\leq s - 1$. Converse?

If L and M are pth and qth order differential operators, then the map $f \mapsto LMf = L(M(f))$ is obviously also a differential operator of order $\leq p + q$; this can be readily seen from expressions (5.9.7) for L and M in a local chart. Generally, $LM \neq ML$, but *if L and M are of order p and q, then $LM - ML$ is of order at most $p + q - 1$*, as can be once again seen using a chart and (5.9.7). In particular, *if L and M are vector fields, so is the operator*

$$[L, M] = LM - ML. \tag{5.9.11}$$

If equalities

$$L = \sum a_i D_\xi^i, \quad M = \sum b_i D_\xi^i \tag{5.9.12}$$

hold in some chart (U, ξ), where a_i, b_i are functions, then clearly,

$$[L, M] = \sum c_i D_\xi^i \quad \text{with} \quad c_i = \sum_j (a_j D_\xi^j (b_i) - b_j D_\xi^j (a_i)). \tag{5.9.13}$$

5.10 Invariant Differential Operators on a Lie Group

The simplest differential operators on the space $\mathbb{R}^n$ are those that can be written using polynomials with *constant* coefficients with respect to the usual operators $D_i = \partial/\partial x_i$, where the x_i are the coordinate functions relative to the canonical basis. In the last twenty years, dramatic progress has been made in the study of these operators, directly connected to the Fourier transform and to distribution theory

---

[6]It is a matter of showing that a distribution μ at a point a of a manifold X is a tangent vector if and only if

$$\mu(fg) = \mu(f)g(a) + f(a)\mu(g)$$

for all $f, g \in \mathscr{F}_a$. The condition is clearly necessary due to (5.2.8). Conversely, if the previous condition holds, then first of all $\mu(1) = 0$—set $f = g = 1$ in the formula—, then $\mu(f) = 0$ if $f \in \mathscr{F}_a$ is a product, or more generally a finite sum of products of C^r functions vanishing at a, and it amounts to showing that this is also the case (for $r = \infty$ or $r = \omega$) if f admits at least a zero of order 2 at a. The analytic case follows readily from the local power series expansion, and the infinitely differentiable case from the formula

$$f(x) = f(0) + \sum x_i \int_0^1 D^i f(tx) \cdot dt,$$

satisfied by all functions of class at least C^1 in the neighbourhood of 0 in $\mathbb{R}^n$. These arguments easily extend to C^r manifolds with finite $r \geq 2$.

(Ehrenpreis, Hörmander, Malgrange etc.), and we are reaching a position where these results can be extended to "invariant operators" on certain classes of Lie groups or homogeneous spaces (Helgason, for the symmetric space case).

The invariance property which characterizes operators with constant coefficients on $\mathbb{R}^n$ can be easily formulated using translations $f_s : x \mapsto f(x + s)$ of any function f defined on $\mathbb{R}^n$ or more generally on any open subset U of $\mathbb{R}^n$. Indeed, a trivial calculation then shows that differential operators with constant coefficients are then characterized by

$$L(f_s) = (Lf)_s \qquad\qquad (5.10.1)$$

for all $s \in \mathbb{R}^n$ and C^∞ functions f on an open subset U of $\mathbb{R}^n$.

More generally, this prompts us to say that a differential operator L on a C^r manifold X ($r = \infty$ or $r = \omega$) acted on by a group G (if G is an "abstract" group, the maps $x \mapsto sx$ are naturally assumed to agree with the manifold structure of X) is G-*invariant* if it satisfies (5.10.1), where f_s now denotes the function $x \mapsto f(sx)$, defined on the open set $s^{-1}(U)$ if f is defined on the open subset U of X.

Exercise. Take $X = \mathbb{R}^n$ and for G the group of displacements $x \mapsto Ax + b$, where $A \in O(n)$. Invariant differential operators have constant coefficients and must more-over be invariant under the group of rotations. Show that these are polynomials with constant coefficients in the Laplacian Δ.

Exercise. There are no non-trivial differential operators on the line $\mathbb{R}$ invariant under the group of similitudes $x \mapsto ax + b$.

Exercise. On the upper half-plane $P : \mathrm{Im}\,(z) > 0$, the differential operator $y^2 \Delta$, where Δ is the usual Laplacian, is invariant under the group $SL_2(\mathbb{R})$. (Note that the transformation group of P defined by $SL_2(\mathbb{R})$ is generated by the horizontal translations $z \mapsto z + a$, with real a, the homotheties $z \mapsto tz$ ($t > 0$), and by $z \mapsto -1/z$). Are there first-order invariant operators?

In the context of the theory of Lie groups, the simplest case is when X is a Lie group and G acts on itself, either by right translations $g \mapsto gs$, or left translations $g \mapsto sg$; this gives *right (resp. left) invariant differential operators* on G. We next show that they can be readily determined from the *distributions at e* on G.

First consider a *left* invariant differential operator R on G. To each point x of G, there corresponds a distribution at x, namely the linear form $\phi \mapsto R\phi(x)$ on the C^∞ functions in the neighbourhood of x. Let $\mu : \phi \mapsto R\phi(e)$ be the distribution corresponding to R at e. To be consistent with the general notation f_s introduced above, let $\phi_s(x) = \phi(sx)$; then $\phi_s(e) = \phi(s)$ and so

$$R\phi(g) = (R\phi)_g(e) = R(\phi_g)(e) \quad \text{since} \quad (R\phi)_g = R(\phi_g)$$
$$= \mu(\phi_g) = \int \phi_g(x)d\mu(x) = \int \phi(gx)d\mu(x). \qquad (5.10.2)$$

The formula obtained shows that R *is completely determined by* μ. Besides, if a distribution μ is arbitrarily chosen at e on G, for every C^∞ function ϕ on an open subset U of G, $R\phi$ can be *defined* on U by the previous formula (g has be assumed to be in U so that the function $x \mapsto \phi(gx)$ being "integrated" with respect to μ can be defined in the neighbourhood of e), and we thereby get a left invariant differential operator on G. We prove this in two steps. First, left invariance follows from a trivial computation:

$$R(\phi_s)(g) = \int \phi_s(gx)d\mu(x) = \int \phi[s(gx)]d\mu(x)$$

$$= \int \phi[(sg)x]d\mu(x) = R\phi(sg) = (R\phi)_s(g). \tag{5.10.3}$$

We still need to show that R is a C^r differential operator if G is of class C^r. First, for each $g \in G$, the map $\phi \mapsto R\phi(g)$ is clearly a distribution at g, namely the direct image of μ under the map $x \mapsto gx$ from G to itself. It is obviously of order s for all g if μ is of order s (transforming a punctual distribution under a manifold isomorphism does not change its order), so that R indeed arises from the fibered manifold $P^{(s)}(G)$, which is consistent with the preceding section. The fact that this section is of class C^r will be established once R is shown to transform every C^r function ϕ on an open subset U of G into a function also of class C^r. But if the local coordinates $\xi_1, \ldots, \xi_n$ are chosen in the neighbourhood of the origin e in G, with $\xi_i(e) = 0$, then

$$\mu(\phi) = \sum_{|\alpha| \le s} c_\alpha D_\xi^\alpha \phi(0) \tag{5.10.4}$$

for all C^r functions in the neighbourhood of e. It then amounts to showing that the partial derivatives at e (in the chosen chart) of the function $x \mapsto \phi(gx)$ are C^r functions of g; but this is clear since the map $(g, x) \mapsto gx$ is of class C^r.

From now on, $\mathsf{U}(\mathfrak{g})$ will denote the set of distributions at e on the group G; it is an infinite-dimensional vector space containing the Lie algebra $\mathfrak{g} = T_e(G)$ of G. It can moreover be endowed with an obviously linear composition law (5.7.22) which will be shortly seen to be associative.[7] The construction of the associative algebra $\mathsf{U}(\mathfrak{g})$ will also be shortly seen to follow once $\mathfrak{g}$ is given—or at least that of an algebra canonically isomorphic to $\mathsf{U}(\mathfrak{g})$—without needing to involve a Lie group G. This explains why the notation adopted here does not indicate any dependence of $\mathsf{U}(\mathfrak{g})$ on G. Thus, (5.10.2) implies that it is possible to associate to each $\mu \in \mathsf{U}(\mathfrak{g})$ the differential operator R_μ given by

$$\boxed{R_\mu\phi(g) = \int \phi(gx)d\mu(x)} \tag{5.10.5}$$

---

[7]Using (5.4.15) to define the tensor product $\lambda \otimes \mu$ of two punctual distributions on G, by (5.7.22) $\lambda * \mu$ is clearly just the direct image of $\lambda \otimes \mu$ under the map $(x, y) \mapsto xy$ from $G \times G$ to G. Associativity (including for the convolution of distributions with "compact support") follows directly from very easy formal calculations.

for every function ϕ sufficiently differentiable on an open subset of G. The map $\mu \mapsto R_\mu$ is already known to be an isomorphism from the vector space $U(\mathfrak{g})$ onto the space of left invariant differential operators, and in fact it is even an algebra isomorphism—in other words,

$$\boxed{R_\lambda R_\mu = R_{\lambda * \mu}} \tag{5.10.6}$$

for all $\lambda, \mu \in U(\mathfrak{g})$. Indeed, both sides are clearly left invariant differential operators. Hence it suffices to show that, at the origin, $R_\lambda R_\mu$ defines the distribution $\lambda * \mu$, in other words that $\lambda * \mu(\phi) = R_\lambda R_\mu \phi(e)$. But the right-hand side is equal to $\lambda(R_\mu \phi) = \int R_\mu \phi(x) d\lambda(x) = \int d\lambda(x) \int \phi(xy) d\mu(y)$, which reduces (5.10.6) to the definition (5.7.22) of a convolution. This, by the way, implies the associativity of the convolution product in $U(\mathfrak{g})$.

Similar results obviously hold for *right* invariant differential operators; besides, as the automorphism $x \mapsto x^{-1}$ of the manifold G transforms right translations into left ones, it is not difficult to predict the results! We are now led to associate to each distribution $\mu \in U(\mathfrak{g})$ the differential operator L_μ defined by

$$\boxed{L_\mu \phi(g) = \int \phi(x^{-1}g) d\mu(x)} = \text{(definition)} \quad \mu * \phi(g). \tag{5.10.7}$$

Here too we get an isomorphism of the algebra $U(\mathfrak{g})$ onto the algebra of right invariant differential operators, with in particular the equality

$$\boxed{L_\lambda L_\mu = L_{\lambda * \mu}} \tag{5.10.8}$$

analogous to (5.10.6). Concerning the notation $\mu * \phi$ just introduced, note that

$$\phi * \mu(g) := \int \phi(gx^{-1}) d\mu(x); \tag{5.10.9}$$

and so

$$R_\mu \phi = \phi * \check{\mu}, \text{ where } \check{\mu} \text{ is the image of } \mu \text{ under } x \mapsto x^{-1}. \tag{5.10.10}$$

Exercise. $\overline{(l * \mu)} = \check{\mu} * \check{\lambda}$ for all $\lambda, \mu \in U(\mathfrak{g})$. The simplest invariant operators— apart from scalars—are those that correspond to the elements of the Lie algebra $\mathfrak{g}$. This prompts us to associate to each $X \in \mathfrak{g}$ the differential operators

$$\boxed{\begin{aligned} L_X \phi(g) &= \int \phi(x^{-1}g) dX(x) = \frac{d}{dt} \phi[\gamma(t)^{-1}g]_{t=0}, \\ R_X \phi(g) &= \int \phi(gx) dX(x) = \frac{d}{dt} \phi[g\gamma(t)]_{t=0}, \end{aligned}} \tag{10.11'} \tag{10.11''}$$

where $t \mapsto \gamma(t)$ is a tangent curve to X in G at $t = 0$. The distribution defining L_X at g is evidently the direct image of the distribution $-X$ under the map $x \mapsto xg$ and as $-X$ is a tangent vector to G at e, this means that the distribution in question is a tangent vector to G at g, namely the image of $-X$ by the tangent linear map to $x \mapsto xg$ at e. Hence, because of the way in which the action of G on tangent vectors has been defined in (6.7), L_X *is a differential operator defined by the vector field*

$$g \mapsto -Xg \quad \text{(heed the sign!)}. \tag{5.10.12}$$

Similarly R_X *is clearly a differential operator defined by the vector field*

$$g \mapsto +gX \quad \text{(heed the sign!)}. \tag{5.10.13}$$

As explained at the end of the last section, identifying "homogeneous first-order" differential operators with vector fields, i.e. to sections of the fibered manifold $T(G)$, operated on the left and on the right by G (since $T(G)$ has previously been endowed with a Lie group structure and since G is its subgroup), clearly identifies right invariant "homogeneous first-order" differential operators with vector fields $L : G \mapsto T(G)$ such that

$$L(xg) = L(x)g, \tag{5.10.14'}$$

while left invariant homogeneous first-order differential operators are identified with vector fields $R : G \mapsto T(G)$ such that

$$R(gx) = gR(x). \tag{5.10.14''}$$

Besides, each of the four categories of objects just considered is, as a vector space, canonically isomorphic to the Lie algebra $\mathfrak{g}$ of G. In particular, fixing a basis (X_i) for $\mathfrak{g}$ enables us to construct a basis for the space of left (resp. right) invariant first-order homogeneous differential operators ("derivations"): it suffices to consider the R_{X_i} (resp. L_{X_i}).

The correspondence between elements of $\mathfrak{g}$ and invariant differential operators also enables us to define the Lie algebra structure of $\mathfrak{g}$ by using these. For this it suffices to combine computation (5.7.24) of the bracket in $\mathfrak{g}$ involving convolutions with the general formulas (5.10.6) and (5.10.8). We obviously get

$$\boxed{L_{[X,Y]} = L_X L_Y - L_Y L_X, \quad R_{[X,Y]} = R_X R_Y - R_Y R_X} \tag{5.10.15}$$

for all $X, Y \in \mathfrak{g}$; the products of differential operators on the right-hand sides are to be taken in the most literal sense. The late Sophus Lie, who was not acquainted with distributions, nor with convolutions, nor even with manifolds, had to be contented with knowing that if D' and D'' are two derivations, then so is $D'D'' - D''D'$, which already in his days was obvious from (5.9.13)...

We next show how to compute invariant differential operators of arbitrary order in terms of operators associated to the elements of $\mathfrak{g}$:

Poincaré–Birkoff–Witt Theorem *Let G be a Lie group and $L_1, \ldots, L_n$ be a basis of the vector field of left invariant vectors on G. Then the ordered (resp. symmetric) monomials in $L_1, \ldots, L_n$ form a basis of the space of left invariant differential operators on G.*

By *ordered monomials*, we mean expressions $L_1^{p_1} \ldots L_n^{p_n}$, with arbitrary exponents $p_i \geq 0$, and by *symmetric monomials* expressions

$$\sum_{s \in \mathfrak{S}_p} L_{i_{s(1)}} \ldots L_{i_{s(p)}} \tag{5.10.16}$$

for all integers $p \geq 0$ and all increasing sequences $(i_1 \leq i_2 \leq \ldots \leq i_p)$ of indices with values between 1 and n.

By the way, note that because of (5.10.9), the above theorem can also be interpreted as a statement on the algebra $\mathsf{U}(\mathfrak{g})$ of distributions with support $\{e\}$, endowed with the convolution product (5.7.22): choosing a basis $X_1, \ldots, X_n$ of $\mathfrak{g}$, (for example) the symmetric monomials

$$\sum_{s \in \mathfrak{S}_p} X_{i_{s(1)}} * \ldots * X_{i_{s(p)}}, \quad (1 \leq i_1 \leq i_2 \leq \ldots \leq i_p \leq n) \tag{5.10.17}$$

in X_i form a basis of $\mathsf{U}(\mathfrak{g})$.

The PBW theorem is itself a particular case of a result for C^∞ and C^ω manifolds. *Let X be an n-dimensional manifold and $L_1, \ldots, L_n$ C^r vector fields on X, i.e. first-order homogeneous differential operators. Suppose that at each point x of X the vectors $L_i(x)$ form a basis of the tangent space $T_x(X)$, i.e. are linearly independent. Then, for every sth-order differential operator L on X, there is a uniquely defined symmetric family of functions $a_{i_1 \ldots i_q} \in C^r(X)$, with $l \leq i_1, \ldots, i_q \leq n$ and $q \leq s$:*

$$a_{i_{s(1)} \ldots i_{s(q)}} = a_{i_1 \ldots i_q} \quad \text{for every permutation } s \in \mathfrak{S}_q, \tag{5.10.18}$$

and such that

$$L = \sum_{q \leq s} \sum_{i_1, \ldots, i_q = 1}^{n} a_{i_1 \ldots i_q} L_{i_1} \ldots L_{i_q}. \tag{5.10.19}$$

Similarly there is a unique family of functions $b_{j_1 \ldots j_n} \in C^r(X)$ with $j_1 + \ldots + j_n \leq s$ such that

$$L = \sum_{j_1, \ldots, j_n} b_{j_1 \ldots j_n} L_1^{j_1} \ldots L_n^{j_n}. \tag{5.10.20}$$

The result about right invariant operators on a Lie group G will be obtained by observing that, if L and all L_i are right invariant, then the uniqueness of expansions

(5.10.19) and (5.10.20) shows that the functions $a_{i_1 \ldots i_q}$ and $b_{j_1 \ldots j_n}$ must be right invariant and hence *constant*.

For the proofs, we need only investigate what happens in a local chart of X. Indeed, if the desired expansions are shown to exist and be unique in every sufficiently small open subset of X, then uniqueness will show that the functions $a_{\ldots}$ and $b_{\ldots}$ associated to these open sets can be "glued" together and hence globally define the desired coefficients.

So we may assume that X is an open subset of $\mathbb{R}^n$. Then, in addition to the vector fields L_i there are the usual operators

$$D_i = \partial / \partial x_i \tag{5.10.21}$$

on X, and there are C^r coefficients b' on X such that

$$L = \sum_{\substack{j_1, \ldots, j_n \\ j_1 + \ldots j_n \leq s}} b'_{j_1 \ldots j_n} D_1^{j_1} \ldots D_n^{j_n}. \tag{5.10.22}$$

At every point of X, the $L_i(x)$ form a basis of $T_x(X)$. So

$$D_k(x) = \sum_h u_{kh}(x) L_h(x) \tag{5.10.23}$$

with C^r functions u_{kh} on X (write the L_i in terms of the D_j and compute the inverse of the coefficient matrix). Now, given two vector fields L' and L'' and two functions u' and u'', for every function f,

$$u' L' u'' L'' f = u' u'' L' L'' f + u L'(u'') L'' f, \tag{5.10.24}$$

where uL denotes the differential operator $f \mapsto u.L(f)$. Hence we get a more general equality of the form

$$D_{k_1} \ldots D_{k_q} = \sum_{h_1, \ldots, h_q} u_{k_1 h_1} \ldots u_{k_q h_q} L_{h_1} \ldots L_{h_q} + M \tag{5.10.25}$$

with order $(M) < q$.

If the left-hand side is symmetrized by summing over all expressions derivable from it by index permutation (expressions that are anyhow all equal since, possibly unlike the L_j, the D_i mutually commute), we obviously get, up to a differential operator of order $<q$, a polynomial (5.10.19) with symmetric coefficients in L_j. Applying this result to each term of total degree s in expression (5.10.22) for L, we conclude that, up to a differential operator of order $<s$, L indeed admits a symmetric expression; induction on the order s of L then implies the existence of an expansion (5.10.19). Since moreover, as has already been seen,

$$\text{order}(MN - NM) \leq \text{order}(M) + \text{order}(N) - 1 \qquad (5.10.26)$$

for all differential operators M and N, it is possible to permute the terms in the expression $L_{i_1} \ldots L_{i_q}$ where arbitrary indices appear, in such a way as to obtain an ordered monomial; the error made is an operator of order $< q$. An induction argument on the order of L starting from (5.10.19) then shows the existence of an expansion (5.10.20).

The uniqueness of the expansion remains to be shown in both cases. By induction, it again suffices to study the terms of maximal order s. The desired result would be obvious if $L_i = D_i$ held for all i. Besides, for the function

$$f(x) = \exp\left(\sum t_i x_i\right) \qquad (5.10.27)$$

where the t_i are parameters, $D_i f = t_i f$ for all i, and so

$$L = \sum c_{i_1} \ldots c_{i_g} D_{i_1} \ldots D_{i_q} \Rightarrow Lf = P_L(t)f$$
$$\text{where } P_L(t) = \sum c_{i_1} \ldots c_{i_q} t_{i_1} \ldots t_{i_q}. \qquad (5.10.28)$$

However, the polynomial function $t \mapsto P_L(t)$ is fully determined by equality $Lf = P_L(t)f$, and since a polynomial can only be written uniquely in terms either of ordinary monomials, or symmetric monomials, the uniqueness of expansions (5.10.19) and (5.10.20) of L is clear in this case.

In the general case, instead of (5.10.23), set $L_i = \sum v_{ij} D_j$ and observe that

$$L_{i_1} \ldots L_{i_q} = \sum v_{i_1 j_1} \ldots v_{i_q j_q} D_{j_1} \ldots D_{j_q} \qquad (5.10.29)$$
$$\text{plus an operator of order } < q.$$

Substituting in (5.10.19), we get

$$L = \sum a_{i_1 \ldots i_s} v_{i_1 j_1} \ldots v_{i_s j_s} D_{j_1} \ldots D_{j_s} \qquad (5.10.30)$$
$$\text{plus an operator or order } < s,$$

and as the coefficients

$$a'_{j_1 \ldots j_s} = \sum_{i_1, \ldots, i_s} v_{i_1 j_1} \ldots v_{i_s j_s} a_{i_1 \ldots i_s}, \qquad (5.10.31)$$

like the coefficients a, are symmetric on the indices, the uniqueness of the symmetric expression of L in terms of the D_i shows that the same coefficients (5.10.31) occur in every symmetric expression of L in terms of the L_i. But the matrices (u_{ij}) and (v_{ij}) being mutual inverses, (5.10.31) is clearly equivalent to

$$a_{i_1...i_s} = \sum_{j_1,...,j_s} u_{i_1 j_1} \ldots u_{i_s j_s} a'_{j_1...j_s}, \tag{5.10.32}$$

and the uniqueness of the coefficients $a_{i_1...i_s}$ follows; the uniqueness of the others can be deduced by induction on s.

The uniqueness of expansion (5.10.20) is obtained likewise or, if preferred, from the uniqueness of expansion (5.10.19) as follows. First observe that

$$\begin{aligned} L_1^{j_1} \ldots L_n^{j_n} &= L_1 \ldots L_1 \ldots L_n \ldots L_n \\ &= L_{i_1} \ldots L_{i_N} \quad (N = j_1 + \ldots + j_n) \\ &= \frac{1}{N!} \sum_{\sigma \in \mathfrak{S}_N} L_{i_{\sigma(1)}} \ldots L_{i_{\sigma(n)}} + \ldots, \end{aligned} \tag{5.10.33}$$

the suspension points representing an operator of order $< N$. If this manipulation is performed on the terms of maximal degree s in (5.10.20), we find, up to an operator of order $< s$, an expression for L of type (5.10.19), whose terms all have total degree s, with

$$a\underbrace{1 \ldots 1}_{j_1}\underbrace{n \ldots n}_{j_n} = \frac{1}{s!} b_{j_1...j_n} \quad (j_1 \ldots + j_n = s). \tag{5.10.34}$$

This implies the uniqueness of the terms of maximal degree in (5.10.20), and the rest follows by induction on s. The results claimed are now fully proved.

In the case of left invariant operators on a Lie group G, the symmetric expression (5.10.19) can be given an intrinsic form, independent of the choice of basis for $\mathfrak{g}$, by introducing the dual vector space $\mathfrak{g}^*$ of $\mathfrak{g}$ and the algebra $S(\mathfrak{g})$ of polynomial functions on $\mathfrak{g}^*$. Indeed, the choice of a basis (X_i) of $\mathfrak{g}$ gives a corresponding basis of $\mathfrak{g}^*$, and the coordinates of an element $\omega \in \mathfrak{g}^*$ with respect to it are the numbers $x_i = \omega(X_i)$. As a polynomial $p \in S(\mathfrak{g})$ can be uniquely written as

$$p(\omega) = \sum_{q \leq s, 1 \leq i_1 \leq ... i_q \leq n} a_{i_1...i_q} x_{i_1} \ldots x_{i_q} \tag{5.10.35}$$

with symmetric coefficients $a \ldots$, it is necessary to associate to each $p \in S(\mathfrak{g})$ the element

$$\beta(p) = \sum a_{i_1...i_q} X_{i_1} * \ldots * X_{i_q} \tag{5.10.36}$$

of $U(\mathfrak{g})$ or, if arguments in terms of differential operators are preferred, the operator

$$L_p = L_{\beta(p)} = \sum a_{i_1...i_q} L_{i_1} \ldots L_{i_q}. \tag{5.10.37}$$

The existence and uniqueness of the symmetric expression of an invariant operator in terms of L_i show that the map $p \mapsto L_p$ (resp. $p \mapsto \beta(p)$) is an isomorphism from the vector space (and not from the algebra—there are unfortunately non-commutative

Lie groups) $S(\mathfrak{g})$ onto the vector space of invariant operators (resp. onto $U(\mathfrak{g})$). The advantage of this map is that it does not depend on the choice of the basis (X_i) of $\mathfrak{g}$. Indeed, if it is replaced by another basis (Y_i), then

$$Y_i = a_i^j X_j, \quad X_i = b_i^j Y_j \tag{5.10.38}$$

and the coordinate change formulas from $x_i = \omega(X_i)$ to $y_i = \omega(Y_i)$ in $\mathfrak{g}^*$ are

$$x_i = b_i^j y_j, \quad y_i = a_i^j x_j. \tag{5.10.39}$$

The symmetric expression

$$p(\omega) = \sum b_{i_1\ldots i_q} y_{i_1} \cdots y_{i_q} \tag{5.10.40}$$

for $p \in S(\mathfrak{g})$ in the new coordinate system is obtained by substituting (5.10.39) in (5.10.35). This evidently gives[8]

$$b_{i_1\ldots i_q} = \sum a_{j_1\ldots j_q} b_{j_1}^{i_1} \cdots b_{j_q}^{i_q}, \tag{5.10.41}$$

since the right-hand side is effectively a symmetric function of the indices $i_1, \ldots, i_q$. It then remains to check that (5.10.41) implies

$$\sum a_{j_1\ldots j_q} X_{j_1} * \ldots * X_{j_q} = \sum b_{i_1\ldots i_q} Y_{i_1} * \ldots * Y_{i_q}; \tag{5.10.42}$$

this obviously follows from (5.10.38).

As pointed out above, the isomorphism $\beta : S(\mathfrak{g}) \to U(\mathfrak{g})$ is certainly not compatible with the multiplicative structures of $S(\mathfrak{g})$ and $U(\mathfrak{g})$, except possibly when G is commutative, in which case β is effectively an algebra isomorphism, which the reader will check as an exercise by noting that in this case, β transforms *ordered* monomials in $x_i = \omega(X_i)$ into *ordered* monomials in X_i. If the *degree* (or *order*) of an element of $U(\mathfrak{g})$ is defined in the obvious manner by considering its symmetric

---

[8] Equalities such as (5.10.41) or (5.10.35) would have certainly resulted in protests from the founders of tensor calculus given that the "covariant" indices are not distinguished from the "contravariant" indices, which traditional conventions put in an upper and lower position, respectively. If (X_i) is a basis of $\mathfrak{g}$, in theory, the upper indices should denote the coordinates of an element $X \in \mathfrak{g}$, and the lower indices—as has been done by miracle—those of an element $\omega \in \mathfrak{g}^*$, and the coefficients of a polynomial function on $\mathfrak{g}$ (resp. $\mathfrak{g}^*$) must then be endowed with lower (resp. upper) indices in such a way that *each summation index appears* once *in a lower position* and once *in an upper position* in the formulas. These conventions make it in principle possible to guess which formulas have a geometric meaning, i.e. are independent from the coordinate system (which naturally explains their success in Physics and differential Geometry ever since Einstein's general Relativity), and this is what justifies them. In the case at hand, observing these conventions would immediately show without any further ado that (i) there is a "natural" correspondence between elements of $U(\mathfrak{g})$ and elements of $S(\mathfrak{g})$, i.e. polynomials on the dual of $\mathfrak{g}$, (ii) there is no "natural" correspondence between elements of $U(\mathfrak{g})$ and polynomial functions on $\mathfrak{g}$, i.e. between $U(\mathfrak{g})$ and $S(\mathfrak{g}^*)$.

(or ordered) expression with respect to an arbitrary basis of $\mathfrak{g}$, then on the one hand β transforms each polynomial p into a distribution $\beta(p) \in U(\mathfrak{g})$ of the same degree as p, and on the other

$$d^\circ[\beta(pq) - \beta(p)\beta(q)] \leq d^\circ(p) + d^\circ(q) - 1 \qquad (5.10.43)$$

for all $p, q \in S(\mathfrak{g})$, as implied by (5.10.26).

To conclude, let us state that it is possible to define the associative algebra $U(\mathfrak{g})$ purely algebraically from an arbitrary Lie algebra $\mathfrak{g}$ (*universal enveloping algebra of* $\mathfrak{g}$), i.e. without using the theory of Lie groups. We will return to this point later.

5.11 Some Concrete Examples

It is difficult to understand the considerations of the previous section without computing invariant operators in particular cases.

Let us first consider the group $G = GL_n(\mathbb{R})$, and compute the operator L_X associated to a matrix $X \in \mathfrak{g} = M_n(\mathbb{R})$. Formula (10.11') needs to be applied with $\gamma(t)$ chosen to be as simple as possible, namely by setting

$$\gamma(t) = 1 + tX, \qquad (5.11.1)$$

the right-hand side being effectively invertible for sufficiently small $|t|$. Hence,

$$L_X\phi(g) = -\int \phi(xg)dX(x) = -\frac{d}{dt}\phi[(1 + tX)g]_{t=0}$$
$$= -\frac{d}{dt}\phi(g + tXg)_{t=0}. \qquad (5.11.2)$$

Denoting by (e_i) the canonical basis of $\mathbb{R}^n$ and setting

$$g(e_i) = g_i^j e_j, \quad X(e_i) = \xi_i^j e_j \qquad (5.11.3)$$

in order to highlight the coefficients of the matrices g and X, we get

$$Xg(e_i) = g_i^k \xi_k^j e_j \qquad (5.11.4)$$

and so the matrix $g + tXg$ is given by

$$(g + tXg)_i^j = g_i^j + t\xi_k^j g_i^k. \qquad (5.11.5)$$

The simplest computation rules on "functions of functions" then show that

$$L_X\phi(g) = -\xi_k^j g_i^k D_j^i \phi(g) \qquad (5.11.6)$$

where we have introduced the (non-invariant) operators

$$D^i_j = \frac{\partial}{\partial g^j_i}. \tag{5.11.7}$$

In conclusion, *the operators*

$$L^j_i = \sum_\alpha g^\alpha_i \partial/\partial g^\alpha_i = g^\alpha_i D^j_\alpha \tag{5.11.8}$$

form a basis of the space of right invariant vector fields on $GL_n(\mathbb{R})$.

Let us next investigate right invariant arbitrary operators. They are linear combinations of monomials in L^j_i. So in degree 2, we already find operators

$$
\begin{aligned}
L^j_i L^h_k &= g^\alpha_i D^j_\alpha g^\beta_k D^h_\beta \\
&= g^\alpha_i D^j_\alpha(g^\beta_k) \cdot D^h_\beta + g^\alpha_i g^\beta_k D^j_\alpha D^k_\beta \\
&= g^\alpha_i \delta^\beta_\alpha \delta^j_k D^h_\beta + g^\alpha_i g^\beta_k D^j_\alpha D^h_\beta \\
&= \delta^j_k g^\beta_i D^h_\beta + g^\alpha_i g^\beta_k D^j_\alpha D^h_\beta,
\end{aligned}
$$

and so

$$L^j_i L^h_k = \delta^j_k L^h_i + g^\alpha_i g^\beta_k D^j_\alpha D^h_\beta \tag{5.11.9}$$

where δ^j_i denotes the Kronecker delta:

$$\delta^j_i = \begin{cases} 1 & \text{if } i = j, \\ 0 & \text{if } i \neq j. \end{cases} \tag{5.11.10}$$

In the result obtained, the first term is in $\mathfrak{g}$, from which it follows that the operators $g^\alpha_i g^\beta_k D^j_\alpha D^h_\beta$ are left invariant and that every left invariant operator of order ≤ 2 is a linear combination of the unit operator, of the $g_i D^i$ and of the $g^\alpha_i g^\beta_k D^j_\alpha D^h_\beta$.

Let us now consider the monomials of degree 3 in the operators (5.11.8). We need to compute the expressions

$$
\begin{aligned}
L^j_i L^h_k L^q_p &= L^j_i(\delta^h_p L^q_k + g^\beta_k g^\gamma_p D^h_\beta D^q_\gamma) \\
&= \delta^h_p L^j_i L^q_k + g^\alpha_i D^j_\alpha(g^\beta_k g^\gamma_p)D^h_\beta D^q_\gamma + g^\alpha_i g^\beta_k g^\gamma_p D^j_\alpha D^h_\beta D^q_\gamma \\
&= \delta^h_p(\delta^j_k L^q_i + g^\alpha_i g^\beta_k D^j_\alpha D^q_\beta) + g^\alpha_i \delta^j_k \delta^\beta_\alpha g^\gamma_p D^h_\beta D^q_\gamma \\
&\quad + g^\alpha_i g^\beta_k \delta^j_p \delta^\gamma_\alpha D^h_\beta D^q_\gamma + g^\alpha_i g^\beta_k g^\gamma_p D^j_\alpha D^h_\beta D^q_\gamma.
\end{aligned}
$$

We find

$$L_i^j L_k^h L_p^q = \delta_k^j \delta_p^h L_i^q + \delta_p^h g_i^\alpha g_k^\beta D_\alpha^j D_\beta^q + \delta_k^j g_i^\alpha g_p^\gamma D_\beta^h D_\alpha^q$$
$$+ \delta_p^j g_i^\alpha g_k^\gamma D_\alpha^h D_\gamma^q + g_i^\alpha g_k^\beta g_p^\gamma D_\alpha^j D_\beta^h D_\gamma^q. \tag{5.11.11}$$

This implies that every right invariant operator of order ≤ 3 is the linear combination of operators

$$1, \quad g_i^\alpha D_\alpha^j, \quad g_i^\alpha g_k^\beta D_\alpha^j D_\beta^h, \quad g_i^\alpha g_k^\beta g_p^\gamma D_\alpha^j D_\beta^h D_\gamma^q. \tag{5.11.12}$$

The reader will easily be able to pursue these computations (or try to do without...) and show that the operators of type

$$g_{i_1}^{\alpha_1} g_{i_2}^{\alpha_2} \cdots g_{i_r}^{\alpha_r} D_{\alpha_1}^{j_1} D_{\alpha_2}^{j_2} \cdots D_{\alpha_r}^{j_r}, \tag{5.11.13}$$

where summation is over $\alpha_1, \ldots, \alpha_r$ and where r takes all possible integral values and the indices $i_1, \ldots, j_r$ every value between 1 and n, form a *basis* of the space of right invariant operators.

Another example is provided by the Heisenberg group of Chap. 3 consisting of real matrices of the form

$$g = \begin{pmatrix} 1 & x & z \\ 0 & 1 & y \\ 0 & 0 & 1 \end{pmatrix} \tag{5.11.14}$$

and whose Lie algebra, identified with its image in $M_3(\mathbb{R})$, admits as basis the three matrices

$$P = \begin{pmatrix} 0 & 1 & 0 \\ 0 & 0 & 0 \\ 0 & 0 & 0 \end{pmatrix}, \quad Q = \begin{pmatrix} 0 & 0 & 0 \\ 0 & 0 & 1 \\ 0 & 0 & 0 \end{pmatrix}, \quad R = \begin{pmatrix} 0 & 0 & 1 \\ 0 & 0 & 0 \\ 0 & 0 & 0 \end{pmatrix}. \tag{5.11.15}$$

To compute the operators L_P, L_Q and L_R, we apply $(10.11')$ and identify the functions on the group with functions in three variables x, y and z. For example,

$$L_P \phi(x, y, z) = -\frac{d}{dt} \phi \left[\begin{pmatrix} 1 & t & 0 \\ 0 & 1 & 0 \\ 0 & 0 & 1 \end{pmatrix} \begin{pmatrix} 1 & x & z \\ 0 & 1 & y \\ 0 & 0 & 1 \end{pmatrix} \right]_{t=0}$$

$$= -\frac{d}{dt} \phi \begin{pmatrix} 1 & x+t & z+ty \\ 0 & 1 & y \\ 0 & 0 & 1 \end{pmatrix}_{t=0} \tag{5.11.16}$$

from which

$$L_P = -\frac{\partial}{\partial x} - y\frac{\partial}{\partial z} \tag{5.11.17}$$

immediately follows; similar computations would show that

$$L_Q = -\frac{\partial}{\partial y}, \quad L_R = -\frac{\partial}{\partial z}. \tag{5.11.18}$$

To recover the operators x and d/dx of Quantum Mechanics, first note that given a Lie group G, a subgroup H of G and a homomorphism π from H to $\mathbb{C}^*$, the set of functions ϕ defined on G and satisfying

$$\phi(gh) = \pi(h)^{-1}\phi(g) \tag{5.11.19}$$

is invariant under *left* convolution operators L_μ for every punctual distribution μ on G—this trivial computation rests uniquely on the associativity of the multiplication in G. In what precedes, ϕ may even be allowed to take values in a finite-dimensional vector space F and π to be a linear representation of H in F.

When G is the Heisenberg group, H can be taken to be the commutative and normal subgroup consisting of matrices with $x = 0$, and π defined by

$$\pi\begin{pmatrix} 1 & 0 & z \\ 0 & 1 & y \\ 0 & 0 & 1 \end{pmatrix} = e^{by+cz}, \tag{5.11.20}$$

where b and c are complex constants. Then, (5.11.19) states that

$$\phi(x, y+\eta, z+\eta x + \xi) = \phi(x, y, z)e^{-b\xi - c}, \tag{5.11.21}$$

which immediately implies that

$$\phi(x, y, z) = f(x)e^{-by - c(z-xy)} \tag{5.11.22}$$

with an "arbitrary" function f, i.e. C^∞ on $\mathbb{R}$ if we wish ϕ to be C^∞ on G. The application of the differential operators (5.11.17) and (5.11.18) to the functions (5.11.22) must result in functions of the same kind, and hence be rendered by operators on the corresponding functions $f(x)$, and this is indeed what occurs. For example L_P takes us from the function (5.11.22) to the function

$$- e^{-by - c(z.xy)}[f'(x) + cyf(x) - cyf(x)] = -f'(x)e^{-by - c(z-xy)}, \tag{5.11.23}$$

whereas L_Q and L_R multiply the function (5.11.22) by $b - cx$ and c respectively. The effect on functions $f(x)$ is therefore given by the three operators

$$p = -\frac{d}{dx}, \quad q = -cx + b, \quad r = c \tag{5.11.24}$$

and it is easy and reassuring to check that they satisfy the same commutator rules $[p, r] = [q, r] = 0$, $[p, q] = r$ as the basis matrices P, Q and R of the Lie algebra of the group considered. We would recover $pq - qp = 1$ by taking, for example, $b = 0, c = 1$.

Exercise. Let U be the associative algebra over $\mathbb{C}$ which can be derived from the algebra $U(\mathfrak{g})$ of the Heisenberg group by extending the scalar field to $\mathbb{C}$. Hence the set of monomials $P^i Q^j R^k$ is a basis of U, with commutator formulas $[P, R] = [Q, R] = 0$ and $[P, Q] = R$. The purpose here is to determine all the maximal two-sided ideals of U.

(1) Show that, in U,

$$[P, Q^i] = i Q^{i-1} R, \quad [Q, P^i] = -i P^{i-1} R. \tag{5.11.25}$$

(2) Write each $u \in U$ as $u = \sum u_{ij} P^i Q^j$ where the u_{ij} are polynomials in R with complex coefficients. Using (5.11.25), show that the centre Z of U consists of polynomials in R.

(3) Let I be a maximal two-sided ideal of U and z an element of Z not in I; show that if $zu \in I$, then $u \in I$; deduce that $I_o = I \cap Z$ is a prime ideal of Z, namely the ideal of Z generated by $R - c$ for some well-determined $c \in \mathbb{C}$.

(4) Suppose $c \neq 0$, i.e. $R \notin I$; using (5.11.25), show that $\sum u_{ij} P^i Q^j \in I$ if and only if $u_{ij} \in I_o$ for all i, j and that I is the two-sided ideal of U generated by I_o, i.e. by $R - c$.

(5) Show that the map $I \mapsto I \cap Z$ from the set of maximal two-sided ideals of U not containing R to the set of maximal ideals of Z not containing R is bijective.

(6) Next suppose that $c = 0$, i.e. $R \in I$; show that U/I is a commutative field isomorphic to the quotient of the polynomial ring $\mathbb{C}[X, Y]$ by a maximal ideal, and that as a result $U/I = \mathbb{C}$ (it is also possible to directly check that every $x \in U/I$ is algebraic over $\mathbb{C}$, because otherwise the elements $(x - \lambda)^{-1}$ of U/I, for $\lambda \in \mathbb{C}$, would be linearly independent over $\mathbb{C}$, contradicting the fact that the dimension of U/I over $\mathbb{C}$ is at most countable).

(7) Let a and b be two complex numbers; show that there is a unique homomorphism χ from the algebra U to $\mathbb{C}$ satisfying $\chi(P) = a$, $\chi(Q) = b$, $\chi(R) = 0$; we thereby get all such homomorphisms. Show that the map which associates to each homomorphism $\chi : U \to \mathbb{C}$ the two-sided ideal $I = \mathrm{Ker}(\chi)$ is a bijection onto the set of maximal two-sided ideals of U containing R. [For further information on these questions, see J. Dixmier, *Représentations irréductibles des algébres de Lie nilpotente*, Anais da Acad. Brasil. de Ciências, *35* (1963), pp. 491–519).]

Let us now move on to the group $G = SL_2(\mathbb{R})$ to compute the effect of its right invariant operators on functions ϕ such that $\phi(gu) = \phi(g)$, where u varies in the subgroup U of matrices $\begin{pmatrix} 1 & * \\ 0 & 1 \end{pmatrix}$, i.e. the stabilizer in G of the first basis vector of $\mathbb{R}^2$; these functions can obviously be identified with the functions $\phi \begin{pmatrix} x \\ y \end{pmatrix}$ on the *pointed plane* $\mathbb{R}^2 - \{0\}$. The following matrices form a basis of the Lie algebra $\mathfrak{g} \subset M_2(\mathbb{R})$ of G:

$$X = \begin{pmatrix} 0 & 1 \\ 0 & 0 \end{pmatrix}, \quad H = \begin{pmatrix} 1 & 0 \\ 0 & -1 \end{pmatrix}, \quad Y = \begin{pmatrix} 0 & 0 \\ 1 & 0 \end{pmatrix} \tag{5.11.26}$$

with

$$[H, X] = 2X, \quad [H, Y] = -2Y, \quad [X, Y] = H. \tag{5.11.27}$$

To compute $X * \phi$, apply (10.11'), which now implies

$$X * \phi(g) = \frac{d}{dt} \phi[\exp(-tX)g]_{t=0} = \frac{d}{dt} \phi\left[\begin{pmatrix} 1 & -t \\ 0 & 1 \end{pmatrix} \begin{pmatrix} x \\ y \end{pmatrix} \right]_{t=0}$$

$$= \frac{d}{dt} \phi \begin{pmatrix} x - ty \\ y \end{pmatrix}_{t=0}. \tag{5.11.28}$$

Hence, it is immediate that the functions considered on the pointed plane satisfy

$$L_X = -y \frac{\partial}{\partial x}. \tag{5.11.29}$$

Similarly,

$$L_Y = -x \frac{\partial}{\partial y}. \tag{5.11.30}$$

Finally,

$$H * \phi(g) = \frac{d}{dt} \phi[\exp(-tH)g]_{t=0} = \phi\left[\begin{pmatrix} e^{-t} & 0 \\ 0 & e^t \end{pmatrix} g \right]_{t=0} \tag{5.11.31}$$

and as

$$\begin{pmatrix} e^{-t} & 0 \\ 0 & e^t \end{pmatrix} \begin{pmatrix} x \\ y \end{pmatrix} = \begin{pmatrix} xe^{-t} \\ ye^t \end{pmatrix}, \tag{5.11.32}$$

it is clear that

$$L_H = -x \frac{\partial}{\partial x} + y \frac{\partial}{\partial y}. \tag{5.11.33}$$

We strongly urge the reader to check that these differential operators on $\mathbb{R}^2$ do satisfy the same commutator formulas (5.11.27) as X, Y and H, for this type of verification can bring to light possible calculation mistakes.

Note that, for all n, L_X, L_Y and L_H preserve the vector space P_n formed by the *homogeneous polynomials of degree n* over $\mathbb{R}^2$. It admits as basis the $n + 1$ functions

$$e_p \begin{pmatrix} x \\ y \end{pmatrix} = (-1)^p \binom{n}{p} x^p y^{n-p}, \quad 0 \le p \le n, \tag{5.11.34}$$

and it can be readily checked that

$$L_H e_p = (n - 2p)e_p, \quad L_X e_p = (n - p + 1)e_{p-1}, \quad L_Y e_p = (p + 1)e_{p+1}.$$
$$(5.11.35)$$

Confining ourselves to this space P_n gives us a linear representation of $\mathfrak{g}$ in P_n. It, by the way, arises from a linear representation of the group G itself, namely that of Chap. 3 which, to every $g \in G$, associates the operator

$$\pi(g) : \{x \mapsto \phi(x)\} \mapsto \{x \mapsto \phi(g^{-1}x)\}. \tag{5.11.36}$$

Returning to the notation $\phi \begin{pmatrix} x \\ y \end{pmatrix}$,

$$\pi \begin{pmatrix} a & b \\ c & d \end{pmatrix}^{-1} \phi \begin{pmatrix} x \\ y \end{pmatrix} = \phi \left[\begin{pmatrix} a & b \\ c & d \end{pmatrix} \begin{pmatrix} x \\ y \end{pmatrix} \right] = \phi \begin{pmatrix} ax + by \\ cx + dy \end{pmatrix}. \tag{5.11.37}$$

This is evidently an operation which indeed transforms a homogeneous polynomial of degree n in $\begin{pmatrix} x \\ y \end{pmatrix}$ into another one. To find the corresponding representation π' of the Lie algebra $\mathfrak{g}$ of G, first choose an arbitrary $U \in \mathfrak{g}$, then a curve $\gamma(t)$ in G tangent to U at $t = 0$, for example (sic) the curve $t \mapsto \exp(tU)$, then replace $\begin{pmatrix} a & b \\ c & d \end{pmatrix}^{-1}$ in (5.11.37) by this curve and finally differentiate the function obtained at $t = 0$; it obviously leads to the operators computed above for $U = X, Y$ or H. It is possible to show that the linear representations of G (or of $\mathfrak{g}$) in the spaces P_n thus obtained are irreducible and that there are no others (in finite dimensions). See MA XII, Sect. 28.

5.12 Differential Operators with Complex Coefficients. The Case of $SL_2(\mathbb{R})$

Until now we have merely defined and used differential operators whose expression (5.9.7) in a chart only involves real-valued coefficients $c_\alpha(x)$. But obviously one might as well—and in practice it is essential to do so—define differential operators with *complex coefficients* (possibly even with coefficients in an arbitrary finite-dimensional real vector space F). Such an operator on a C^r manifold X, with $r = \infty$ or $r = \omega$, transforms every function $f \in C^r(U)$ into a complex (or F-) valued C^r function Lf defined on U, for any open subset U of X. Finally, L is given in every local chart of X by a formula of type (5.9.7) where the C^r coefficients c_α are now complex or F-valued.

By the way, note that if the manifold X and the operator L are of class C^r and if L is of (finite) order s in the obvious sense, that Lf can in fact be defined for every function f defined and at least of class C^s on an open subset of X. In particular, every operator with analytic coefficients on an analytic manifold can be applied to

C^∞ functions. It is of course also possible to define operators of class C^p ($p \leq r$) on any C^r manifold, but it is not very useful to insist on these trivial versions of the definition, which will sometimes be used without further ado.

The reader should also note that if differential operators with coefficients in $\mathbb{C}$ or F can be defined, then it should be similarly possible to define distributions with values in $\mathbb{C}$ or F. At a point a of a manifold X, hence where the pseudo-algebra $\mathscr{F}_a$ of functions defined and of class C^r in the neighbourhood of a is at our disposal, a distribution at a with values in F will be a linear map μ from $\mathscr{F}_a$ to F which, moreover, vanishes at functions having a zero of sufficiently high order at a. This has already been stated in Sect. 5.1 in the case $F = \mathbb{R}$. Choosing a basis (a_i) of F, $\mu(f) = \sum \mu_i(f)a_i$, where the μ_i are now real distributions at a. In particular, if $F = \mathbb{C}$, then $\mathrm{Re}(\mu)$ and $\mathrm{Im}(\mu)$ can be defined in an obvious manner.

It is equally clear that complex distributions can be applied to complex functions by requiring that, for complex f, the number $\mu(f)$ should be a $\mathbb{C}$-linear function of f; this amounts to setting $\mu(g + ih) = \mu(g) + i \cdot \mu(h)$ for real g and h.

If G is a Lie group, then not only is it possible to define the Lie algebra $\mathfrak{g}$ of "real" tangent vectors to G at e, but also the set $\mathfrak{g}_{\mathbb{C}}$ of complex tangent vectors to G at e. Such a vector is written uniquely as $X + iY$ with $X, Y \in \mathfrak{g}$, so that $\mathfrak{g}_{\mathbb{C}}$ is just the complex vector space arising from $\mathfrak{g}$ by extending the scalars to $\mathbb{C}$. As the convolution $\lambda * \mu$ of two complex distributions on G are, thanks to (5.7.22), defined as that of two real distributions, we get a complex Lie algebra structure on $\mathfrak{g}_{\mathbb{C}}$ by setting $[X, Y] = X * Y - Y * X$ as in the case of $\mathfrak{g}$. The map $(X, Y) \mapsto [X, Y]$, which is $\mathbb{C}$-bilinear can then also be derived from the analogous map relative to $\mathfrak{g}$ by extending the scalars. For example, if $G = SL_2(\mathbb{R})$, in which case $\mathfrak{g}$ is canonically identified with the Lie algebra of 2×2 *real* matrices having zero trace, $\mathfrak{g}_{\mathbb{C}}$ becomes clearly identified with the Lie algebra of 2×2 *complex* matrices having zero trace (i.e., surprisingly, with the real Lie algebra of the group $SL_2(\mathbb{C})$, except for the fact that the latter is obtained by considering the Lie algebra $\mathfrak{g}_{\mathbb{C}}$ in question as a 6-dimensional Lie algebra over $\mathbb{R}$, and not as a complex Lie algebra...).

Considering the set of complex distributions on G at e, endowed with the obvious algebraic structures, we get the complex *infinitesimal algebra* $\mathsf{U}_{\mathbb{C}}(\mathfrak{g})$ or $\mathsf{U}(\mathfrak{g}_{\mathbb{C}})$ of the group G; here too, it arises from $\mathsf{U}(\mathfrak{g}) = \mathsf{U}_{\mathbb{R}}(\mathfrak{g})$ by extending the scalars, and the Poincaré–Birkhoff–Witt Theorem continues to hold in this context provided a complex basis of $\mathfrak{g}_{\mathbb{C}}$ is considered. For example, for $SL_2(\mathbb{R})$, the monomials $Y^q H^r X^p$ form a basis of $\mathsf{U}_{\mathbb{C}}(\mathfrak{g})$ over $\mathbb{C}$ given that X, Y and H form a real basis of $\mathfrak{g}$ and hence a complex basis of $\mathfrak{g}_{\mathbb{C}}$.

To highlight the usefulness of these trivial considerations, let us consider the group $G = SL_2(\mathbb{R})$, the compact subgroup $K = SO_2(\mathbb{R})$ of G consisting of matrices of type $\begin{pmatrix} a & b \\ -b & a \end{pmatrix}$ with $a^2 + b^2 = 1$, whence $a = \cos\theta$ and $b = \sin\theta$, and for some fixed integer r, let F_r denote the space of C^∞ functions on G satisfying

$$\phi(kg) = \chi_r(k)\phi(g) \quad \text{where} \quad \chi_r\begin{pmatrix} \cos\theta & -\sin\theta \\ \sin\theta & \cos\theta \end{pmatrix} = e^{ri\theta}. \tag{5.12.1}$$

Our purpose is to evaluate the effect of differential operators on these functions, and notably of *right* invariant vector fields on G. The difficulty stems from the fact that if μ is a distribution at the origin, then there is no reason why the function

$$L_\mu \phi(g) = \mu * \phi(g) = \int \phi(x^{-1}g)d\mu(x) \tag{5.12.2}$$

should still satisfy (5.12.1). Indeed, setting $\mu * \phi = \psi$,

$$\psi(kg) = \int \phi(x^{-1}kg)d\mu(x) = \chi_r(k) \int \phi(k^{-1}x^{-1}kg)d\mu(x)$$

$$= \chi_r(k) \int \phi[(k^{-1}xk)^{-1}g]d\mu(x) \tag{5.12.3}$$

$$= \chi_r(k) \int \phi(y^{-1}g)d\mu_k(y) = \chi_r(k)\mu_k * \phi(g)$$

where μ_k is the direct image of μ under the automorphism $x \mapsto k^{-1}xk$ of G—see (5.3.10). The problem here is that although left translations commute with right translations (associativity!), left translations do not mutually commute.

Before going any further, as an aside note that if μ is a distribution at the origin on G, the image of μ under the automorphism $x \mapsto gxg^{-1}$ of G is again a distribution of the same type; for $\mu \in \mathfrak{g}$, by (5.6.14) and (5.6.15), it is just the distribution $\mathrm{ad}\,(g)\mu$. Despite confusions, albeit inoffensive, that might follow, it is convenient to define

$$\mathrm{ad}\,(g)\mu = \text{direct image of } \mu \text{ under } x \mapsto gxg^{-1} \tag{5.12.4}$$

for any distribution $\mu \in \mathsf{U}_\mathbb{C}(\mathfrak{g})$ and $g \in G$. Although $\mathsf{U}_\mathbb{C}(\mathfrak{g})$ is infinite-dimensional, for every integer s, the distributions of order $\leq s$ form a finite-dimensional subspace of $\mathsf{U}_\mathbb{C}(\mathfrak{g})$, and it is obviously invariant under the operators $\mathrm{ad}\,(g)$ which therefore define, for all s, a linear representation of G in this subspace. These considerations apply to all Lie groups G, not just $SL_2(\mathbb{R})$.

Exercise. Let G be a Lie group, U the algebra $\mathsf{U}_\mathbb{C}(\mathfrak{g})$ and, for every integer s, U_s the subspace of distributions of order $\leq s$, so that the U_s form an increasing sequence of finite-dimensional subspaces, having U as their union. For any $g \in G$, let $\mathrm{ad}_s(g)$ denote the restriction of $\mathrm{ad}\,(g)$ to U_s. This gives a linear representation of class C^∞ of G in U_s. Differentiation at the origin associates to it a linear representation $X \mapsto \mathrm{ad}_s'(X)$ from $\mathfrak{g}$ to the vector space U_s. The aim here is to compute it. In the following, the elements of U will be denoted by letters such as L, M, U, V, etc. and multiplication in U by $(L, M) \mapsto LM$. (i) Let $t \mapsto \gamma(t)$ be a curve in G tangent to the vector $X \in \mathfrak{g}$ at $t = 0$; hence

$$\mathrm{ad}_s'(X)L = \frac{d}{dt}\{\mathrm{ad}\,(\gamma(t))L\}_{t=0} \tag{5.12.5}$$

for all $L \in \mathsf{U}_s$; deduce that $\mathrm{ad}\,'_s(X)$ is the restriction to U_s of the operator $\mathrm{ad}\,'(X)$ defined on the whole of U by the right-hand side of (5.12.5). (ii) Taking into account the fact that, for every $g \in G$, the operator $\mathrm{ad}\,(g)$ is an automorphism of the algebra U, show that $\mathrm{ad}\,'(X)$ is a *derivation* of U. (iii) Show that

$$\mathrm{ad}\,'(X)L = XL - LX \tag{5.12.6}$$

(show that $L \mapsto XL - LX$ is the only derivation of U which reduces to $\mathrm{ad}\,(X)$ on $\mathfrak{g}$).

Returning to $SL_2(\mathbb{R})$ and to (5.12.3), where we now have

$$\mu_k = \mathrm{ad}\,(k)^{-1}\mu, \tag{5.12.7}$$

if the distribution μ itself satisfies an equality such as

$$\mathrm{ad}\,(k)\mu = \chi_m(k)\mu \tag{5.12.8}$$

for some integer m, then (5.12.3) clearly becomes

$$\psi(kg) = \chi_r(k)\chi_m(k)^{-1}\psi(g) = \chi_{r-m}(k)\psi(g) \tag{5.12.9}$$

and so (5.12.28) shows that

$$\phi \in F_r \Rightarrow \mu * \phi \in F_{r-m}. \tag{5.12.10}$$

Hence the distributions μ for which $L\mu$ preserve F_r are those for which $m = 0$ ("K-invariant" distributions).

Not every distribution $\mu \in \mathsf{U}_\mathbb{C}(\mathfrak{g}) = \mathsf{U}$ satisfies an equality (5.12.8), but it is nevertheless possible to find a basis of the vector space U consisting of such distributions. Let us first consider the Lie algebra $\mathfrak{g}_\mathbb{C}$ of 2×2 complex matrices having trace zero. The elements of K are the matrices

$$k = k(\theta) = \begin{pmatrix} \cos\theta & -\sin\theta \\ \sin\theta & \cos\theta \end{pmatrix} = \exp(\theta U) \quad \text{where} \quad U = \begin{pmatrix} 0 & -1 \\ 1 & 0 \end{pmatrix}, \tag{5.12.11}$$

so that

$$\mathrm{ad}\,(k)X = \exp(\theta \cdot \mathrm{ad}\,(U))X \tag{5.12.12}$$

for all matrices $X \in \mathfrak{g}$, and hence also for all $X \in \mathfrak{g}_\mathbb{C}$. The equality

$$\exp(\theta \cdot \mathrm{ad}\,(U))X = e^{mi\theta}X \tag{5.12.13}$$

is then evidently equivalent to

$$\mathrm{ad}\,(U)X = miX, \tag{5.12.14}$$

and it is then a matter of diagonalizing ad (U) in $\mathfrak{g}_\mathbb{C}$. It is then immediate that the matrices

$$W = \begin{pmatrix} 0 & -i \\ i & 0 \end{pmatrix} = iU, \quad Z = \frac{1}{2}\begin{pmatrix} 1 & i \\ i & -1 \end{pmatrix}, \quad \bar{Z} = \frac{1}{2}\begin{pmatrix} 1 & -i \\ -i & 1 \end{pmatrix} \qquad (5.12.15)$$

form a basis of $\mathfrak{g}_\mathbb{C}$ and satisfy

$$\mathrm{ad}\,(U)W = 0, \quad \mathrm{ad}\,(U)Z = -2iZ, \quad \mathrm{ad}\,(U)\bar{Z} = 2i\bar{Z} \qquad (5.12.16)$$

on the one hand, and the commutator formulas

$$[W, Z] = 2Z, \quad [W, \bar{Z}] = -2\bar{Z}, \quad [Z, \bar{Z}] = W \qquad (5.12.17)$$

on the other. As will be remembered, these formulas are also satisfied by the matrices X, Y and H introduced in the previous sections (which explains why U was replaced by $W = iU$). But then the monomials $W^r Z^p \bar{Z}^q$ form a basis of U over $\mathbb{C}$, and as equalities (5.12.16) also state that

$$\mathrm{ad}\,(k)W = W = \chi_0(k)W, \quad \mathrm{ad}\,(k)Z = \chi_{-2}(k)Z, \quad \mathrm{ad}\,(k)\bar{Z} = \chi_2(k)\bar{Z}, \quad (5.12.18)$$

it follows that

$$\mathrm{ad}\,(k)W^r Z^p \bar{Z}^q = \chi_{2q-2p}(k)W^r Z^p \bar{Z}^q; \qquad (5.12.19)$$

whence the desired basis of U. Its construction would not have been possible in $\mathsf{U}_\mathbb{R}(\mathfrak{g})$.

We next compute the effect of the operators L_W, L_Z and $L_{\bar{Z}}$ on the solution of (5.12.1). Equalities (5.12.16) show that this will give maps of the following type:

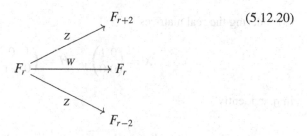

$$(5.12.20)$$

and to compute them we replace the functions $\phi(g)$ of F_r by those corresponding to them in the *upper half-plane* P as in Chap. 1, Sect. 1.4, except for some trivial differences. Indeed, (5.12.1) merely says that for $g = \begin{pmatrix} a & b \\ c & d \end{pmatrix}$,

$$(ci + d)^r \phi(g^{-1}) = f(z) \qquad (5.12.21)$$

depends solely on $z = g(i) = \frac{ai+b}{ci+d}$, as shown by an easy computation, and for all $\gamma \in G$, the operation (which preserves F_r) which takes us from the function $\phi(g)$ to the function $\phi(g\gamma^{-1})$ then becomes the operation that takes us from a function $f(z)$ in P to a function

$$(uz + v)^{-r} f(\gamma(z)) \quad \text{if} \quad \gamma = \begin{pmatrix} * & * \\ u & v \end{pmatrix}. \tag{5.12.22}$$

This the reader can check as a computation exercise. Now the problem consists in expressing the effect of the operators L_W, L_Z and $L_{\bar{Z}}$ on the function $f(z)$ corresponding to a given function $\phi(g)$ in F_r. In other words, the functions of z corresponding to the transforms of ϕ under these three operators need to be computed.

The case of L_W is particularly simple. $L_W = iL_U$ and

$$L_U \phi(g) = \frac{d}{d\theta} \phi[\exp(-\theta U)g]_{\theta=0} = \frac{d}{d\theta} \{e^{-ri\theta} \phi(g)\}_{\theta=0} = -ir\phi(g), \tag{5.12.23}$$

and so

$$L_W = r \text{ on } F_r. \tag{5.12.24}$$

To find L_Z, set $\phi' = L_Z\phi \in F_{r+2}$ and let f and f' be the functions corresponding to ϕ and ϕ' on the upper half-plane. Applying (5.12.21) to a triangular matrix readily shows that

$$y^{r/2} f(x + iy) = \phi \begin{pmatrix} y^{-1/2} & -y^{-1/2}x \\ 0 & y^{1/2} \end{pmatrix} \tag{5.12.25}$$

for real x and positive y, and as $\phi' \in F_{r+2}$, similarly

$$y^{1+r/2} f'(x + iy) = \phi' \begin{pmatrix} y^{-1/2} & -y^{-1/2}x \\ 0 & y^{1/2} \end{pmatrix}. \tag{5.12.26}$$

Considering the real matrices

$$X = \begin{pmatrix} 0 & 1 \\ 0 & 0 \end{pmatrix}, \quad H = \begin{pmatrix} 1 & 0 \\ 0 & -1 \end{pmatrix} \tag{5.12.27}$$

in $\mathfrak{g}$, evidently

$$Z = \frac{1}{2}W + iX + \frac{1}{2}H, \quad \bar{Z} = -\frac{1}{2}W - iX + \frac{1}{2}H \tag{5.12.28}$$

and it is therefore a matter of computing $X * \phi$ and $H * \phi$ for the triangular matrix appearing in (5.12.25). First,

$$X * \phi \begin{pmatrix} y^{-1/2} & -y^{-1/2}x \\ 0 & y^{1/2} \end{pmatrix}$$

$$= \frac{d}{dt}\phi\left[\begin{pmatrix} 1 & -t \\ 0 & 1 \end{pmatrix}\begin{pmatrix} y^{-1/2} & -y^{-1/2}x \\ 0 & y^{1/2} \end{pmatrix}\right]_{t=0}$$

$$= \frac{d}{dt}\phi\begin{pmatrix} y^{-1/2} & -y^{-1/2}(x+ty) \\ 0 & y^{1/2} \end{pmatrix}_{t=0} \tag{5.12.29}$$

$$= \frac{d}{dt}\{y^{r/2}f(x+ty+iy)\}_{t=0} = y^{1+r/2}D_xf(x+iy);$$

similarly

$$H * \phi \begin{pmatrix} y^{-1/2} & -y^{-1/2}x \\ 0 & y^{1/2} \end{pmatrix}$$

$$= \frac{d}{dt}\phi\left[\begin{pmatrix} e^{-t} & 0 \\ 0 & e^t \end{pmatrix}\begin{pmatrix} y^{-1/2} & -y^{-1/2}x \\ 0 & y^{1/2} \end{pmatrix}\right]_{t=0} \tag{5.12.30}$$

$$= \frac{d}{dt}\{y^{r/2}e^{rt}f(x+ie^{2t}y)\}_{t=0}$$

$$= ry^{r/2}f(x+iy) + 2y^{1+r/2}D_yf((x+iy),$$

where D_x and D_y are the derivatives with respect to x and y. As $W\phi = r\phi$, using (5.12.28), it follows that

$$Z * \phi \begin{pmatrix} y^{-1/2} & -y^{-1/2}x \\ 0 & y^{1/2} \end{pmatrix}$$

$$= y^{1+r/2}\left[\frac{r}{y}f(x+iy) + i(D_x - iD_y)f(x+iy)\right], \tag{5.12.31}$$

$$\bar{Z} * \phi \begin{pmatrix} y^{-1/2} & -y^{-1/2}x \\ 0 & y^{1/2} \end{pmatrix} = -iy^{1+r/2}(D_x + iD_y)f(x+iy). \tag{5.12.32}$$

Using (5.12.26) and the well-known operators

$$\frac{\partial}{\partial z} = \frac{1}{2}\left(\frac{\partial}{\partial x} - i\frac{\partial}{\partial y}\right), \quad \frac{\partial}{\partial \bar{z}} = \frac{1}{2}\left(\frac{\partial}{\partial x} + i\frac{\partial}{\partial y}\right) \tag{5.12.33}$$

shows that the function f' corresponding to $\phi' = Z * \phi$ is given by

$$\boxed{f'(z) = \frac{r}{y}f(z) + 2i \cdot \partial f/\partial z.} \tag{5.12.34}$$

To find f'', which corresponds to $\phi'' = \bar{Z} * \phi$, apply (5.12.26) with f' and ϕ' replaced by f'' and ϕ'' and the exponent $1 + r/2$ by $-1 + r/2$ since now $\phi'' \in F_{r-2}$; hence the right-hand side of (5.12.32) has to be divided by $y^{-1+r/2}$, and so

$$\boxed{f''(z) = -2iy^2 \partial f/\partial \bar{z}.}$$
(5.12.35)

Exercise. Let Γ be a discrete subgroup of G, and define $F_r(\Gamma)$ to be the set of C^∞ functions f on the upper half-plane satisfying

$$f(\gamma(z)) = (cz + d)^r f(z) \quad \text{for all} \quad \gamma = \begin{pmatrix} a & b \\ c & d \end{pmatrix} \in \Gamma.$$
(5.12.36)

Show that operators (5.12.34) and (5.12.35) take $F_r(\Gamma)$ to $F_{r+2}(\Gamma)$ and to $F_{r-2}(\Gamma)$ respectively. (The desired result should be obvious without any calculation, and besides, easy to check by explicit computation.)

Exercise. Holomorphic solutions of (5.12.36) correspond to functions ϕ of class C^∞ on G satisfying

$$\phi(kg\gamma) = \chi_r(k)\phi(g), \quad \bar{Z} * \phi = 0.$$
(5.12.37)

Exercise. Consider a C^∞ function ϕ on G satisfying

$$\phi(kg) = \chi_r(k)\phi(g), \quad \bar{Z} * \phi = 0,$$
(5.12.38)

and denote by $V(\phi)$ the vector space generated over $\mathbb{C}$ by the functions

$$\phi_{r+2n} = L_Z^n \phi = Z * \ldots * Z * \phi, \quad (n \text{ factors})$$
(5.12.39)

for $n = 0, 1, 2, \ldots$. Suppose that $r \geq 1$. (i) Show that

$$W * \phi_{r+2n} = (r + 2n)\phi_{r+2n}, \quad Z * \phi_{r+2n} = \phi_{r+2n+2},$$
$$\bar{Z} * \phi_{r+2n} = -n(r + n - 1)\phi_{r+2n-2} \quad (= 0 \text{ if } n = 0).$$
(5.12.40)

(ii) Show that the ϕ_{r+2n}, $(n \in \mathbb{N})$ form a basis of $V(\phi)$ and that every L_W-invariant subspace of $V(\phi)$ is generated by the vectors ϕ_{r+2n} contained in it. (iii) Using the convolution $\mu * \phi$ of a punctual distribution and a function, consider $V(\phi)$ as a left module over the ring $U_\mathbb{C}(\mathfrak{g})$. Show that it is a *simple* (or *irreducible*) module, i.e. that $V(\phi)$ does not contain any non-trivial submodules (or what comes to the same, any non-trivial L_X-invariant subspaces, for $X \in \mathfrak{g}$). (iv) What happens when $r \leq 0$? (See MA XII, Sect. 28.)

Exercise. Consider once again the subgroup U of the preceding section consisting of matrices $\begin{pmatrix} 1 & * \\ 0 & 1 \end{pmatrix}$ and denote by H the subgroup of matrices $\begin{pmatrix} t & 0 \\ 0 & 1/t \end{pmatrix}$ with $t \neq 0$. (i) Show that $G = KHU$ and that $K \cap HU$ consists of the matrices 1 and -1. (ii) For all $s \in \mathbb{C}$, denote by V_s^+ the space of functions ϕ on G satisfying the following two conditions:

$$\phi(ghu) = \phi(g)|t|^{-2s} \quad \text{for} \quad u \in U \text{ and } h = \begin{pmatrix} t & 0 \\ 0 & 1/t \end{pmatrix} \in H. \tag{5.12.41}$$

On the other hand, the restriction of ϕ to K is a trigonometric polynomial (a finite linear combination of functions $\chi_r(k)$). Show that V_s^+ is a left $U_{\mathbb{C}}(\mathfrak{g})$-module, i.e. invariant under operators L_W, L_Z and $L_{\overline{Z}}$. (iii) Show that the functions

$$\phi_{r,s} \begin{pmatrix} a & b \\ c & d \end{pmatrix} = (a + ic)^r |a + ic|^{-r-2s}, \tag{5.12.42}$$

where r is an even rational integer, form a basis for V_s^+ and that

$$\begin{cases} W * \phi_{r,s} = r\phi_{r,s} \\ Z * \phi_{r,s} = (s + \frac{r}{2})\phi_{r+2,s} \\ \overline{Z} * \phi_{r,s} = (s - \frac{r}{2})\phi_{r-2,s} \end{cases} \tag{5.12.43}$$

(iv) Show that V_s^+ is a simple module if and only if $s \notin \mathbb{Z}$. (v) Consider the same questions with (5.12.41) replaced by

$$\phi(ghu) = \phi(g)|t|^{-2s}\text{sign}(t). \tag{5.12.44}$$

(Formulas such as (5.12.43) obtainable from (5.12.44) are typical examples of infinite-dimensional irreducible representations of the Lie algebra $\mathfrak{g}$.)

5.13 Representations of G and Representations of $U(\mathfrak{g})$

We have already shown that (see (5.8.8)), if π is a C^∞ homomorphism from a Lie group G to a Lie group H, then the map $\mu \mapsto \pi_*(\mu)$ transforming each distribution at a point $a \in G$ into a distribution at $b = \pi(a) \in H$ is compatible with the convolution. Since it is linear, it is therefore a homomorphism from the associative algebra of distributions with punctual support on G to the analogous algebra relative to H. Now,

$$\pi_*(\varepsilon_g) = \varepsilon_{\pi(g)} \quad \text{for all} \quad g \in G, \tag{5.13.1}$$

where

$$\varepsilon_g : \phi \mapsto \phi(g) \tag{5.13.2}$$

is the "Dirac measure" at g. So the homomorphism π of the *group* G can conversely be extracted from the homomorphism π_* of the *algebra* of distributions with punctual support on G. On the other hand, every distribution μ on G with support $\{a\}$ is obtained by applying the translation $x \mapsto ax$ on an element of $U(\mathfrak{g})$, namely on a distribution with support $\{e\}$, i.e. on the convolution

$$\varepsilon_a * \lambda(\phi) = \int \int \phi(xy) d\varepsilon_a(x) d\lambda(y) = \int \phi(ay) d\lambda(y) \tag{5.13.3}$$

with measure ε_a. Thus to find the effect of π_* on punctual distributions (apart from the homomorphism π from G to H), it suffices to know the effect of π_* on the infinitesimal algebra $\mathsf{U}(\mathfrak{g})$, and the latter is obviously a homomorphism from $\mathsf{U}(\mathfrak{g})$ to $\mathsf{U}(\mathfrak{h})$.

As an aside, note that if it is merely a matter of defining the homomorphism

$$\pi_* : \mathsf{U}(\mathfrak{g}) \to \mathsf{U}(\mathfrak{h}), \tag{5.13.4}$$

then the "global" homomorphism π can be replaced by a local one, i.e. by an infinitely differentiable map $\pi : V \to H$, where V is an open neighbourhood of e in G and where

$$\pi(xy) = \pi(x)\pi(y) \quad \text{if} \quad x, y, xy \in V \tag{5.13.5}$$

is assumed to hold—this will prove essential in the study of the locally linear groups of Chap. 3. Indeed, if g is a C^∞ function defined in the neighbourhood of e in H, the composite function $g \circ \pi$ is again C^∞ and defined in the neighbourhood of e in G. This enables us to define $\nu(g) = \mu(g \circ \pi)$ for any distribution μ with support $\{e\}$ on G. Setting $\nu = \pi_*(\mu)$ defines the map (5.13.4), and calculation (5.8.10) used to show that π_* is compatible with the convolution continues to hold here without any changes. It goes without saying that if π is a *local isomorphism* (i.e. if π induces an isomorphism from a neighbourhood of e in G to a neighbourhood of e in H), then π_* is an algebra isomorphism. Finally,

$$\pi_*(X) = \pi'(X) \quad \text{for all } X \in \mathfrak{g} \tag{5.13.6}$$

since, for a manifold representation or more generally morphism, the tangent map $\pi'(e)$ is just the restriction to particular distributions, namely tangent vectors to G at e, of the operation consisting in taking the direct image of a distribution under π.

The above considerations apply notably to the case where H is the group $GL(E)$ of automorphisms of a finite-dimensional vector space E over $\mathbb{R}$ or $\mathbb{C}$; π is then the *local linear representation* of G in E. Since $\mathfrak{h} = \mathscr{L}(E)$ now, as already mentioned several times, it first follows that the tangent linear map $\pi' = \pi'(e)$ can be identified with a homomorphism from the Lie algebra $\mathfrak{g}$ of G to the Lie algebra $\mathscr{L}(E)$, i.e. with a linear representation of the Lie algebra of G in E (if E is complex, this representation can be extended to the complexification $\mathfrak{g}_\mathbb{C}$ of $\mathfrak{g}$). It will be seen that π' can in turn be extended to a homomorphism from the infinitesimal algebra $\mathsf{U}(\mathfrak{g})$ to the algebra $\mathscr{L}(E)$, which should *not be* confused with the homomorphism π_* from $\mathsf{U}(\mathfrak{g})$ to $\mathsf{U}(GL(E))$ which follows directly from the general theory:

Theorem 1 *Let π be a local linear representation of a Lie group G in a finite-dimensional real (resp. complex) vector space E. Then there exists a unique homomorphism from the real (resp. complex) associative algebra $\mathsf{U}(\mathfrak{g})$ (resp. $\mathsf{U}_\mathbb{C}(\mathfrak{g})$) to the algebra $\mathscr{L}(E)$ which, for all $X \in \mathfrak{g}$, reduces to $\pi'(X)$.*

We first give an elementary proof of this result. Let E^* denote the dual of E, and $\langle a, b \rangle$ be the canonical bilinear form on $E \times E^*$. If $t \mapsto \lambda(t)$ is a curve drawn in $GL(E)$ and such that $\lambda(0) = e$, then, as has been shown far back, the element u of $\mathscr{L}(E)$ with which the tangent vector to λ at $t = 0$ can be identified is just the operator

$$u = \frac{d}{dt}\lambda(t) \qquad\qquad (5.13.7)$$

where a vector-valued function—to be more precise with values in $\mathscr{L}(E)$—is differentiated at $t = 0$. As a result, for $a \in E$ and $b \in E^*$,

$$\langle u(a), b \rangle = \frac{d}{dt}\langle \lambda(t)a, b \rangle_{t=0} \qquad\qquad (5.13.8)$$

(differentiating a matrix-valued function consists in differentiating its entries). This formula enables us to compute the operator $\pi'(X)$ corresponding to an element $X \in \mathfrak{g}$. For this choose a curve $t \mapsto \gamma(t)$ in G tangent to X at $t = 0$, take its composition with π, which gives the curve $t \mapsto \pi(\gamma(t))$ in $GL(E)$, and once this is done, $\pi'(X)$ is its tangent vector at $t = 0$. Hence, identifying it with the element of $\mathscr{L}(E)$ corresponding to it gives

$$\langle \pi'(X)a, b \rangle = \frac{d}{dt}\langle \pi(\gamma(t))a, b \rangle_{t=0}. \qquad\qquad (5.13.9)$$

Since in general,

$$\int f(x)dX(x) = \frac{d}{dt}f(\gamma(t))_{t=0} \qquad\qquad (5.13.10)$$

for every C^∞ function f in the neighbourhood of e on G, we finally get

$$\langle \pi'(X)a, b \rangle = \int \langle \pi(x)a, b \rangle dX(x) \qquad\qquad (5.13.11)$$

for all $a \in E$ and $b \in E^*$. This formula determines $\pi'(X)$ as it enables us to compute its coefficients with respect to a basis of E.

Then, replacing X by an arbitrary punctual distribution μ, it is natural to associate an operator $\pi(\mu) \in (E)$ to μ given by[9]

---

[9]The notation $\pi(\mu)$ used here contradicts the notation $\pi'(\mu)$ used so far for $\mu \in \mathfrak{g}$. Hence one more "unimportant contradiction" in the notation used in this theory. Besides, knowing how to define directly the effect of a distribution on a function with values in a finite-dimensional space enables us to restate definition (5.13.8) as

$$\pi(\mu) = \int \pi(x)d\mu(x).$$

This would naturally result in the notation $\mu(\pi)$ instead of the operator $\pi(\mu)$ obtained (value of the distribution μ on the function $\pi : x \mapsto \pi(x)$, with values in $\mathscr{L}(E)$). When G is the additive group $\mathbb{R}$, π may be chosen to be a one-dimensional representation, i.e. such as $\pi(x) = \exp(2\pi i u x)$ with a parameter $u \in \mathbb{C}$ characterizing π

$$\langle \pi(\mu)a, b \rangle = \int \langle \pi(x)a, b \rangle d\mu(x) \qquad (5.13.12)$$

for all $a \in E$ and $b \in E^*$. The existence of $\pi(\mu)$ follows by observing that the right-hand side is a bilinear form on $E \times E^*$, and then that, as suggested by (5.13.12), *every* bilinear form on $E \times E^*$ corresponds to an operator on E. Besides, the coefficients of $\pi(\mu)$ with respect to a fixed basis of E are clearly obtained by merely applying the distribution μ to the coefficients of the function $x \mapsto \pi(x)$, which are C^r functions on G.

That $\mu \mapsto \pi(\mu)$ is indeed an algebra homomorphism from $\mathsf{U}(\mathfrak{g})$ to $\mathscr{L}(E)$ has yet to be shown. Linearity being obvious, it suffices to prove that

$$\pi(\lambda * \mu) = \pi(\lambda)\pi(\mu) \qquad (5.13.13)$$

when λ and μ are punctual distributions. But if $\nu = \lambda * \mu$, then

$$
\begin{aligned}
\langle \pi(\nu)a, b \rangle &= \int \langle \pi(x)a, b \rangle d\nu(x) = \int\int \langle \pi(xy)a, b \rangle d\lambda(x)d\mu(y) \\
&= \int\int \langle \pi(x)\pi(y)a, b \rangle d\lambda(x)d\mu(y) \\
&= \int \langle \pi(\lambda)\pi(y)a, b \rangle d\mu(y) = \int \langle \pi(y)a, {}^t\pi(\lambda)b \rangle d\mu(y) \\
&= \langle \pi(\mu)a, {}^t\pi(\lambda)b \rangle = \langle \pi(\lambda)\pi(\mu)a, b \rangle
\end{aligned}
\qquad (5.13.14)
$$

as claimed.

(Footnote 9 continued)
(on the right-hand side, $\pi = 3.14159\ldots$). Then $\pi(\mu)$ or $\mu(\pi)$ is the scalar operator

$$\pi(\mu) = \int \exp(2\pi iux)d\mu(x),$$

which most authors write as $\hat{\mu}(u)$, namely the value at u of the complex Fourier transform of the distribution μ. For real u, i.e. when the operators $\pi(x)$ are unitary, the same value at u is found for the Fourier transform of μ. Hence in the general case, writing $\hat{\mu}(\pi)$ for what one hesitates to write as $\mu(\pi)$ instead of $\pi(\mu)$ is somewhat justified. In reality, the usual convention for defining the Fourier transform is known to consist in setting

$$\hat{\mu}(u) = \int \exp(-2\pi iux)d\mu(x),$$

so that in the general case, it would be more natural to agree to

$$\hat{\mu}(\pi) = \int \pi(x)^{-1}d\mu(x).$$

However, this then gives

$$\mu * \nu(\pi) = \hat{\nu}(\pi)\hat{\mu}(\pi).$$

This is not important when the group is commutative but in the general case involves hazards that are better avoided. Etc...

Exercise. Let G be a locally linear group and π a local embedding of G into a group $GL_n(\mathbb{R})$, in the sense of Chap. 3, Sect. 3.5. Explain how the homomorphism π' from the Lie algebra of G to $M_n(\mathbb{R})$ which follows from the previous construction can be identified with the homomorphism described in Sects. 3.5 and 3.6 of Chap. 3 (the reader should first show how $T_e(G)$ can be identified with the Lie algebra of G in the sense of Chap. 3, Sect. 3.6).

Exercise. Let E be a finite-dimensional real vector space; denote by $U(GL(E))$ the algebra of distributions on $GL(E)$ at e. To each $\mu \in U(GL(E))$ associate the operator $j(\mu) \in \mathscr{L}(E)$ given by

$$\langle j(\mu)a, b \rangle = \int \langle g(a), b \rangle d\mu(g) \tag{5.13.15}$$

for $a \in E$ and $b \in E^*$. Let π be a local homomorphism from a Lie group G to $GL(E)$. Show that $\pi : U(\mathfrak{g}) \to \mathscr{L}(E)$ is obtained by taking the composition of $\pi_* : U(\mathfrak{g}) \to U(GL(E))$ with j.

5.14 Invariant Subspaces. Schur's Lemma and Burnside's Theorem

Let G be a Lie group, $\mathfrak{g}$ its Lie algebra, E a finite-dimensional real vector space and $\pi : G \to GL(E)$ a linear representation of G in E. As seen in the previous section, formula (5.13.12) enables us to associate a homomorphism to π, also written π, from the associative algebra $U(\mathfrak{g})$ to the algebra $\mathscr{L}(E)$ of endomorphisms of E. As $U(\mathfrak{g})$ is generated by the elements $X \in \mathfrak{g}$, the image of $U(\mathfrak{g})$ under π is an associative subalgebra of $\mathscr{L}(E)$ generated by the operators $\pi(X)$, $X \in \mathfrak{g}$. In particular, the vector subspaces of E (resp. the endomorphisms of E) invariant under (resp. commuting with) the operators $\pi(X)$ are also invariant under (resp. commuting with) the operators $\pi(\mu)$, and conversely. These subspaces (resp. endomorphisms) obviously include those that are invariant under (resp. commuting with) the operators $\pi(g)$, $g \in G$. In the latter case, this can for example be seen by differentiating the equality $\pi(g) \circ u = u \circ \pi(g)$ at the origin. The converse holds provided G is connected:

Theorem 2 *Let $\pi : G \to GL(E)$ be a linear representation from a connected Lie group G in a finite-dimensional vector space E. A vector $a \in E$ is $\pi(g)$-invariant for all $g \in G$ if and only if it is a zero of the maps $\pi'(X)$, for all $X \in \mathfrak{g}$. A vector subspace E' of E is invariant under all $\pi(g)$ if and only if it is invariant under all $\pi'(X)$. An endomorphism u of E commutes with all $\pi(g)$ if and only if it commutes with all $\pi'(X)$.*

Recall first that when a Lie group G acts on a manifold M, the map $g \mapsto g(m)$ is of *constant rank* for all $m \in M$ as seen in Chap. 4, Sect. 4.8. If its tangent linear map *at the origin* is the null map, so will it be everywhere else, and the standard form for the subimmersions (Chap. 4, Theorem 1) then shows that the map considered is locally constant. If G is connected it is then constant, and $g(m) = m$ for all $g \in G$, proving the first statement of the theorem.

To show the third one, take M to be the vector space $\mathscr{L}(E)$ and make G act on it by setting

$$g(u) = \pi(g) \circ u \circ \pi(g)^{-1} \tag{5.14.1}$$

for $g \in G$ and $u \in \mathscr{L}(E)$. The differential at the origin is obviously the map

$$X \mapsto \pi'(X) \circ u - u \circ \pi'(X) \tag{5.14.2}$$

and saying that it is the null map means that u commutes with all $\pi'(X)$; hence the theorem in the case of endomorphisms commuting with all $\pi'(X)$.

The case of a vector subspace E' of E could be tackled in the same way by considering the manifold M of subspaces of dimension $r = \dim(E')$ in E, defined at the end of Chap. 4, Sect. 4.9, but the following arguments also hold. Let P be the subgroup of $u \in GL(E)$ such that $u(E') = E'$. It is a Lie subgroup of $GL(E)$: choosing a basis of E whose first r vectors generate E' shows that P is clearly the intersection of the open submanifold $GL(E)$ of $\mathscr{L}(E)$ and the vector subspace of matrices in which the last $n - r$ entries of each of its first r columns are zero. The elements $g \in G$ such that $\pi(g)E' = E'$ form the inverse image subgroup of P under π, and from one of the exercises of Sect. 8, we know that the subalgebra of $\mathfrak{g}$ corresponding to this subgroup is the inverse image under $\pi'(e) = \pi'$ of that of P, the latter obviously consisting of the elements $X \in \mathscr{L}(E)$ such that $X(E') \subset E'$. This can notably be seen by arguing in terms of matrices. Hence if $\pi'(X)E' \subset E'$ for *all* $X \in \mathfrak{g}$, it follows that the Lie subgroup of G defined as the inverse image of P under π has the same Lie algebra as G, hence of the same dimension as G, and so is open in G, and if G is connected, it is necessarily the whole of G, qed. (In Chap. 6, the exponential map will provide a much more direct proof of these results.)

Corollary *Let π be a linear representation of a connected Lie group G. For all $g \in G$, there exists a $\mu \in \mathsf{U}(\mathfrak{g})$ such that $\pi(g) = \pi(\mu)$. In other words, if (X_i) is a basis of the Lie algebra of G, then every operator $\pi(g)$ is a polynomial in $\pi(X_i)$.*

Let E be the representation space; set $F = \mathscr{L}(E)$ and consider the representation ρ of G in F given by $\rho(g)u = \pi(g) \circ u$ for all $u \in F$. Its extension to $\mathsf{U}(\mathfrak{g})$ is obviously (exercise!) given by $\rho(\mu)u = \pi(\mu) \circ u$. Let F' be the subspace of elements $\pi(\mu)$, $\mu \in \mathsf{U}(\mathfrak{g})$; it is $\rho(\mu)$-invariant for all $\mu \in \mathsf{U}(\mathfrak{g})$, and so (Theorem 1) under all $\rho(g)$. As $1 \in F'$, $\rho(g)1 = \pi(g) \in F'$ for all g, qed.

Note that the corollary sheds somewhat of a harsh light on the statement of Theorem 2.

Exercise. Let π be a representation of a connected Lie group G in a finite-dimensional vector space E and E' a subspace of E. Let $a \in E'$. Show that $\pi(g)a \in E'$ for all g if and only if $\pi(\mu)a \in E'$ for all distributions $\mu \in \mathsf{U}(\mathfrak{g})$.

Exercise. Find the gap in the following proof. The purpose is to show that $\pi(g)a \in E'$ for all $a \in E'$. By duality, it suffices to show that if there exists a $b \in E^*$ orthogonal to E', then it is orthogonal to $\pi(g)a$. For every distribution $\mu \in U(\mathfrak{g})$, $\int \langle \pi(g)a, b \rangle d\mu(g) = \langle \pi(\mu)a, b \rangle = 0$ since by assumption E' contains the vector $\pi(\mu)a$ for all μ. As a result, the function $g \mapsto \langle \pi(g)a, b \rangle$ vanishes at $g = e$ and so do its successive derivatives. As G is connected, it follows that it is null everywhere, qed.

Exercise. We keep the same assumptions. A bilinear form B on $E \times E$ is said to be π-invariant if $B(\pi(g)x, \pi(g)y) = B(x, y)$ for all $x, y \in E$. Show that this condition is equivalent to

$$B(\pi'(X)a, b) + B(a, \pi'(X)b) = 0 \quad \text{for all } X \in \mathfrak{g}, \, a, b \in E. \tag{5.14.3}$$

As a side note, we next show some well-known properties of linear representations. Theorem 2 suggests that the group G should be replaced by the ring $A = U(\mathfrak{g})$ or $U_{\mathbb{C}}(\mathfrak{g})$, depending on the case. A linear representation π of G in a finite-dimensional vector space E then becomes a *left A-module* (see, for example, the author's *Cours d'Algèbre*, §10) provided the product of a "scalar" $\lambda \in A$ and a vector $x \in E$ is defined by

$$\lambda x = \pi(\lambda)x \tag{5.14.4}$$

where the extension to $A = U(\mathfrak{g})$ or $U_{\mathbb{C}}(\mathfrak{g})$ of the given representation π of the Lie group G appears on the right-hand side. Naturally, there exist much more complicated A-modules than those thus obtained from representations of G (for example the ring A is an obvious left A-module, and being infinite-dimensional as a vector space over $\mathbb{R}$ or $\mathbb{C}$, cannot be derived from a representation of G by the indicated method; similarly for the modules V_s^+ constructed in the last exercise of Sect. 12). Nonetheless, even in this case, some of the general properties of modules over a ring give rise to interesting results about Lie group representations.

Let E be a left module over the ring A. E is said to be *simple* or *irreducible* if it only contains 0 and E itself as submodules. If A is a field, this means that E is a 1-dimensional vector space over A (even if A is not commutative...); in the general case, this means that, if a vector $x \neq 0$ is chosen in E, then *every* element of E is of the form λx for some (not in general unique) $\lambda \in A$.

An A-module is *semi-simple* or *completely reducible* if every submodule E' of E admits a complement E'' which is also a submodule:

$$E = E' + E'', \quad E' \cap E'' = 0. \tag{5.14.5}$$

This is always the case if A is a field, but not so in the general case, including that of rings of type $A = U(\mathfrak{g})$. The adjoint representation of the group $ax + b$ is an immediate counterexample. Indeed, this group G can be identified with the set of matrices of the form

$$g = \begin{pmatrix} a & b \\ 0 & 1 \end{pmatrix} \tag{5.14.6}$$

with $a \in \mathbb{R}^*$, $b \in \mathbb{R}$. Its Lie algebra admits as a basis the two matrices $H = \begin{pmatrix} 1 & 0 \\ 0 & 0 \end{pmatrix}$
and $X = \begin{pmatrix} 0 & 1 \\ 0 & 0 \end{pmatrix}$, with $[H, X] = X$. The adjoint representation of G in its Lie algebra
$\mathfrak{g}$ enables us to consider $\mathfrak{g}$ as a left $\mathsf{U}(\mathfrak{g})$-module, the generators H and X of $\mathsf{U}(\mathfrak{g})$
acting on $\mathfrak{g}$ via the operators ad (H) and ad (X). Here the submodules are the vector
subspaces (the scalar operators are in $\mathsf{U}(\mathfrak{g})$ since $\mathsf{U}(\mathfrak{g})$ is an algebra with unit element
over $\mathbb{R}$) of $\mathfrak{g}$ invariant under ad (H) and ad (X). But trivial calculations show that the
only subspace of $\mathfrak{g}$, other than $\{0\}$ and $\mathfrak{g}$, answering this question is the line through
X. Semisimplicity therefore fails to hold.[10]

Theorem 3 *A left module E over a ring A is semisimple if and only if it is a direct
sum of simple submodules.*

By a simple submodule of E we mean a submodule $E' \neq 0$ and containing no
other submodules of E except $\{0\}$ and E' itself. In the preceding statement, the
"direct sum" may well be infinite: E is said to be the direct sum of a family $(E_i)_{i \in I}$
of submodules if, for every $x \in E$, there is a unique family of vectors $x_i, i \in I$ having
the following properties: $x_i \in E_i$ for all i, $x_i \neq 0$ only for finitely many indices $i \in I$,
and $x = \sum x_i$.

In order not to bore the reader with a completely pointless use of Zorn's Lemma in
the applications we have in mind, we will only prove the theorem for an algebra with
unity A over a commutative field k, so that the A-modules will also be k-vector spaces,
and show that all *finite-dimensional* semisimple A-modules E over k are direct sums
of submodules (hence of vector subspaces). For this, consider a submodule $E' \neq E$
of maximal dimension (over k). Let E'' be a complement submodule of E'. Then
E'' is simple because otherwise the dimension of E' could be increased by adding
a non-trivial submodule of E'' to E'. But arguing by induction on the dimension of

---

[10]This argument suggests the definition of a particular class of Lie groups. If the Lie algebra $\mathfrak{g}$ of a
Lie group G is considered a module over $\mathsf{U}(\mathfrak{g})$ by extending the adjoint representation to $\mathsf{U}(\mathfrak{g})$:

$$\text{ad}\,(X_1 \ldots X_p)Y = \text{ad}\,(X_1) \ldots \text{ad}\,(X_p)Y,$$

the submodules of $\mathfrak{g}$ are just the *ideals* of the Lie algebra $\mathfrak{g}$ in the usual sense, i.e. the vector subspaces
$\mathfrak{a}$ such that $[X, Y] \in \mathfrak{a}$ for all $X \in \mathfrak{g}$ and $Y \in \mathfrak{a}$. Semisimplicity of the $\mathsf{U}(\mathfrak{g})$-module $\mathfrak{g}$ then means
that *every ideal of $\mathfrak{g}$ has a complement which is also an ideal of $\mathfrak{g}$*. The Lie algebras having this
property are called *reductive*, and so are the corresponding Lie groups (it can be shown that this is
the case for all classical groups over $\mathbb{R}$ or $\mathbb{C}$). More particularly, a Lie algebra $\mathfrak{g}$ is called *semisimple*
if it is reductive and if its centre, the set of $Z \in \mathfrak{g}$ such that ad $(Z) = 0$, is trivial. For example, the
Lie algebra of $SL(n, \mathbb{R})$ is semisimple.

Every linear representation of a semisimple Lie algebra in a *finite-dimensional* vector space can
be shown to be completely reducible (H. Weyl, 1927). Proofs of this result are widely found; the
most recent reference is J. Dixmier, *Enveloping Algebras* (North-Holland, 1977), pp. 4–25, which
addresses this question from scratch and devotes 300 pages to algebras of type $\mathsf{U}(\mathfrak{g})$. See also
J.E. Humphreys, *Introduction to Lie Algebras and Representation Theory* (Springer, 1972) or V.S.
Varadarajan, *Lie Groups, Lie Algebras and their Representations*, (Prentice-Hall, 1974), chap. 3,
or, for a thorough treatment, chapters on Lie algebras in N. Bourbaki.

E (the initial case, where $\dim(E) = 0$, being trivial), E' may be assumed to be the direct sum of simple submodules. The result is then immediate for E.

To prove the converse, it obviously suffices to show that *if E' and E'' are two semisimple A-modules, so is $E' \times E''$*. In this case the proof is elementary even if we do not assume that E' and E'' are finite-dimensional. Indeed, consider a submodule F of $E' \times E''$. If it contains E', then $F = E' \oplus F''$ where $F'' = F \cap E''$. As E'' is semisimple, there is a submodule F' of E'' such that $E'' = F'' \oplus F'$; then $E = F \oplus F'$ in this case. If, however, F does not contain E', then E' is the direct sum of $F \cap E'$ and a submodule M intersecting F trivially. As $E' \subset F \oplus M$, we are back in the previous case, and there is a submodule F' such that $E' \times E'' = (F \oplus M) \oplus F' = F \oplus (M + F')$, qed.

Exercise. Let G be an "abstract" group and K a commutative ring. The set of "linear combinations of elements of G with coefficients in K" is called the *group algebra of G over K*. In other words, the desired algebra $A = K[G]$ is a free K-module admitting a basis $(\varepsilon_g)_{g \in G}$ indexed by G, multiplication in A being defined by the distributivity condition with respect to K and by the "multiplication table"

$$\varepsilon_x \varepsilon_y = \varepsilon_{xy}; \qquad (5.14.7)$$

the resemblance with (5.7.23) is no coincidence. Let π be a linear representation of G in a K-module E, i.e. a homomorphism from the group G to the group of automorphisms of the K-module E. E can be turned into a left A-module by setting

$$\lambda x = \sum_{g \in G} \lambda(g) \pi(g) x \quad \text{if} \quad \lambda = \sum_{g \in G} \lambda(g) \varepsilon_g \in A, \quad \lambda(g) \in K, \qquad (5.14.8)$$

for all $x \in E$. The A-submodules of E are then just the K-submodules invariant under the representation of G. The exercise consists in restating Theorem 3 in representation theory language for the group G.

Let us return to a left module E over an arbitrary ring A. The *commutant* of the A-module E is the set of endomorphisms u of the *additive group* E commuting with the operations of A:

$$u(x + y) = u(x) + u(y), \quad u(\lambda x) = \lambda u(x); \qquad (5.14.9)$$

namely the endomorphisms of the A-module E. Hence their set should be written $\mathscr{L}_A(E)$.

Theorem 4 (Schur's Lemma) *Let E be a simple A-module over a ring A. Then the ring $\mathscr{L}_A(E)$ of endomorphisms of E (commutant of the A-module E) is a field. If A is an algebra over an algebraically closed commutative field k and if E is a finite-dimensional vector space over k, then $\mathscr{L}_A(E) = k$.*

Let u be an endomorphism of E. Equalities (5.14.9) show that $\mathrm{Ker}(u)$ and $\mathrm{Im}(u)$ are submodules of E. If E is simple and $u \neq 0$, i.e. if $\mathrm{Ker}(u) \neq E$, then $\mathrm{Ker}(u) = \{0\}$, and $\mathrm{Im}(u) = E$ since $\mathrm{Im}(u) \neq 0$. Hence u is bijective and its inverse u^{-1} is obviously also in $\mathscr{L}_A(E)$, and the first assertion follows.

To prove the second one, observe that $\mathscr{L}_A(E)$ contains k [note that, A being an algebra with unity,[11] each A-module is inevitably a vector space over k, and as the definition of an algebra implies that for $\lambda \in k$, the elements $\lambda 1$ of A are in the *centre* of A, the homotheties $x \mapsto \lambda x$, $\lambda \in k$, of E are in $\mathscr{L}_A(E)$]. On the other hand, $\mathscr{L}_A(E)$ is obviously contained in $\mathscr{L}_k(E)$, which is finite-dimensional over k if this is the case for E. The assumptions in the statement therefore imply that $\mathscr{L}_A(E)$ is a (not necessarily commutative) field and a finite extension of k. But if k is algebraically closed the only possibility is $\mathscr{L}_A(E) = k$, qed.

Exercise. Rewrite Theorem 4 in terms of group representations.

Exercise. Let π be a representation of a ("abstract") group G in a finite-dimensional *complex* vector space E. Assume π is irreducible. Let u be an endomorphism of E commuting with all $\pi(g)$. Show directly that u is a scalar using the existence of at least one eigenvalue of u and considering the corresponding eigenspace. (This is the traditional version of "Schur's Lemma".)

Exercise. Find counterexamples to the second part of the statement for $k = \mathbb{R}$. (Possible commutants are then $\mathbb{R}$, $\mathbb{C}$ and the quaternion field because these are the only finite degree extensions of $\mathbb{R}$, regardless of commutativity. The reader is advised to look for an $\mathbb{R}$-algebra A and a finite-dimensional simple A-module E over $\mathbb{R}$ admitting the quaternion field as commutant.)

The following result aims to show that when an algebra A over an algebraically closed commutative field k acts irreducibly on a finite-dimensional vector space E over k, then *every* endomorphism of the vector space E is of the form $x \mapsto \lambda x$ for some properly chosen $\lambda \in A$. This follows from a slightly more general statement.

Before stating it, one remark: given a simple A-module E (for an arbitrary ring A), E can be considered as a vector space over the field $K = \mathscr{L}_A(E)$ and more precisely a left vector space over K, by agreeing to set

$$ux = u(x) \text{ for } u \in K = \mathscr{L}_A(E) \text{ and } x \in E. \qquad (5.14.10)$$

For all $\lambda \in A$, the map

$$\rho(\lambda) : x \mapsto \lambda x \qquad (5.14.11)$$

from E to E is then compatible with the vector space structure over K for obvious reasons. Denoting by $\mathscr{L}_K(E)$ the ring (which *is not* an algebra over K if K is not commutative) of endomorphisms of E regarded as a vector space over K, it is possible to consider the map ρ defined in (5.14.10) as a homomorphism from the ring A to the ring $\mathscr{L}_K(E)$. Thus:

---

[11] Some authors do not baulk at the thought of elaborating a theory of rings without unit element, as though it were not always possible to add a unit to rings lacking one.

Theorem 5 (Burnside) *Let E be a simple module over a ring A. Suppose that E is finite-dimensional as a vector space over its commutant K. Then the map ρ from A to $\mathscr{L}_K(E)$ given by (5.14.9) is surjective.*

To prove this, we start by fixing a basis $(e_i)_{1 \le i \le r}$ of E over K. It is a matter of showing that for all $a_1, \ldots, a_r \in E$, there exists a $\lambda \in A$ such that $\rho(\lambda)$ maps each e_i onto a_i, in other words satisfies

$$\lambda e_i = a_i \quad \text{for} \quad 1 \le i \le r. \tag{5.14.12}$$

For this, consider the product module E^r; in E^r (5.14.12) becomes

$$\lambda(e_i, \ldots, e_r) = (a_1, \ldots, a_r) \tag{5.14.13}$$

and so it amounts to showing that the submodule F of E^r generated by the vector $(e_1, \ldots, e_r)$ is the whole of E^r.

As E is simple, E^r is semisimple (Theorem 3) and hence F admits a complement A-module F' in E^r. Let $P : E^r \to F'$ be the projection operator corresponding to the decomposition $E^r = F \oplus F'$. As F and F' are submodules, P clearly belongs to the commutant of the A-module E^r or, to put it another way, is an endomorphism of this A-module. But an endomorphism of a Cartesian product is also represented by a "matrix". More precisely, given an endomorphism u of the A-module E^r, there are endomorphisms u_i^j of the A-module E such that $u(x_1, \ldots, x_r) = (y_1, \ldots, y_r)$ with

$$y_i = \sum_j u_i^j(x_j) \quad \text{for all } j. \tag{5.14.14}$$

As $P(e_1, \ldots, e_r) = 0$ since P is the projection onto the complement F' of the submodule F generated by $(e_1, \ldots, e_r)$, here $\sum P_i^j(e_j) = 0$ for all i. But let us now regard E as a vector space over $K = \mathscr{L}_A(E)$; the P_i^j become scalars, and we get linear relations between the basis vectors e_j. Hence $P_i^j = 0$ for all i and j, and so $P = 0$, thus $F' = 0$, and in turn $F = E^r$, proving the theorem.

Corollary *Let π be an irreducible representation of a group G in a finite-dimensional complex vector space E. Then every endomorphism of E is a linear combination of operators π(g). Or equivalently, if n = dim(E), then n^2 linearly independent operators can be extracted from the family of π(g), g ∈ G.*

Apply Theorem 5 to the algebra A consisting of the linear combinations of all $\pi(g)$. Clearly, E is a simple A-module and its commutant reduces to $\mathbb{C}$ by Schur's Lemma.

A similar result obviously holds for *connected* Lie groups: if π is a (C^∞ or analytic) representation of G in a finite-dimensional complex vector space E and if π is irreducible, then the image of the infinitesimal algebra $U_{\mathbb{C}}(\mathfrak{g})$ under π is the whole of $\mathscr{L}(E)$.

5.15 Invariant Central Distributions of the Co-adjoint Representation

Let G be a *connected* Lie group and π a representation of G in a finite-dimensional real or complex vector space E. In this section, our intention is to find out if the infinitesimal algebra $U(\mathfrak{g})$ contains distributions μ such that the operator $\pi(\mu)$ commutes with all $\pi(g)$ or, what comes to the same since G is connected, with all $\pi(X)$, $X \in \mathfrak{g}$. Due to (5.13.13) and (5.12.4), the former condition can also be formulated as

$$\pi(\mu) = \pi(g)\pi(\mu)\pi(g^{-1}) = \pi(\varepsilon_g)\pi(\mu)\pi(\varepsilon_{g^{-1}})$$
$$= \pi(\varepsilon_g * \mu * \varepsilon_{g^{-1}}) = \pi(\mathrm{ad}\,(g)\mu), \tag{5.15.1}$$

while the latter can be written as $\pi(X * \mu) = \pi(\mu * X)$. Hence μ will be the answer to the question if it satisfies any one of the following relations

$$\mathrm{ad}\,(g)\mu = \mu \quad \text{for all } g \in G, \quad X * \mu = \mu * X \quad \text{for all } X \in \mathfrak{g}. \tag{5.15.2}$$

Since $\mathfrak{g}$ generates the algebra $U(\mathfrak{g})$, the latter equality says that μ *belongs to the centre* $Z(\mathfrak{g})$ *of the infinitesimal algebra* $U(\mathfrak{g})$, and the former one, as we are about to see, is equivalent to it.

Indeed, for every natural integer s, let $U_s(\mathfrak{g})$ be the finite-dimensional subspace of $U(\mathfrak{g})$ generated by the monomials of degree at most s in the elements of $\mathfrak{g}$ (left invariant differential operators on G of order at most s). As already remarked, the restrictions of the ad (g) to $U_s(\mathfrak{g})$ define a linear representation (of class C^∞ or C^ω depending on the case) of G in $U_s(\mathfrak{g})$. The coefficients of this representation with respect to a basis of $U_s(\mathfrak{g})$ are obviously polynomials of degree $\le s$ in the coefficients of the adjoint representation in $\mathfrak{g}$. Hence Theorem 2 of Sect. 14 can be applied. This enables us to conclude that

$$\mathrm{ad}\,(g)\mu = \mu \quad \text{for all } g \in G \tag{5.15.3}$$

is equivalent to the relation stemming from it by differentiating at the origin. It then remains to observe that choosing a tangent curve $t \mapsto \lambda(t)$ to a fixed element X of $\mathfrak{g}$ at $t = 0$ in G, we get

$$\frac{d}{dt}\mathrm{ad}\,(\gamma(t))\mu_{t=0} = X * \mu - \mu * X \tag{5.15.4}$$

for all $X \in \mathfrak{g}$ [since (5.15.3) is then conveyed by a null right-hand side of (5.15.4) for all X, i.e. by the second condition (5.15.2)]. But this follows from the fact that the map

$$\mathrm{ad}\,'(X) : \mu \mapsto \frac{d}{dt}\mathrm{ad}\,(\lambda(t))\mu_{t=0} \tag{5.15.5}$$

is a derivation of the algebra $U(\mathfrak{g})$. As (5.15.4) holds for all $\mu \in \mathfrak{g}$ generating $U(\mathfrak{g})$, it extends to all distributions μ in $U(\mathfrak{g})$, and the equivalence of the two conditions (5.15.2) finally follows.

We next present a method which, theoretically, enables us to determine all the distributions in $\mu \in Z(\mathfrak{g})$ (we will call them *central distributions* in $U(\mathfrak{g})$) from the adjoint representation of G in $\mathfrak{g}$. Indeed, it was shown in Sect. 10 that if $S(\mathfrak{g})$ denotes the algebra of polynomial functions on the dual $\mathfrak{g}^*$ of the vector space $\mathfrak{g}$, then there is a "natural" isomorphism of vector spaces

$$\beta : S(\mathfrak{g}) \to U(\mathfrak{g}). \tag{5.15.6}$$

Choosing a basis (X_i) of $\mathfrak{g}$, every polynomial p on $\mathfrak{g}^*$ can be uniquely written as

$$p(\omega) = \sum_{r \leq s} a^{i_1 \dots i_r} \omega(X_{i_1}) \dots \omega(X_{i_r}) \tag{5.15.7}$$

with *symmetric* coefficients. Omitting the useless signs $*$, this gives

$$\beta(\rho) = \sum_r a^{i_1 \dots i_r} X_{i_1} \dots X_{i_r}, \tag{5.15.8}$$

the result being independent of the choice of basis. We now compute the effect of the operator ad (g) on $u = \beta(p)$. As ad (g) is an automorphism of the algebra $U(\mathfrak{g})$,

$$\mathrm{ad}\,(g)u = \sum_r a^{i_1 \dots i_r} \mathrm{ad}\,(g)X_{i_1} \dots \mathrm{ad}\,(g)X_{i_r}. \tag{5.15.9}$$

In order to highlight the matrix of the operator ad (g) in the Lie algebra $\mathfrak{g}$ with respect to the basis considered, let us set

$$\mathrm{ad}\,(g)X_i = c_i^j(g)X_j. \tag{5.15.10}$$

Then,

$$\mathrm{ad}\,(g)u = u^{i_1 \dots i_r} c_{i_1}^{j_1}(g) \dots c_{i_r}^{j_r}(g)X_{j_1} \dots X_{j_r} \tag{5.15.11}$$

with the usual summation conventions. However, the expressions

$$a^{j_1 \dots j_r}(g) = c_{i_1}^{j_1}(g) \dots c_{i_r}^{j_r}(g)a^{i_1 \dots i_r} \tag{5.15.12}$$

appearing in these formulas are, like the initial coefficients $a \dots$, symmetric functions of their indices $j_1, \dots, j_r$. Hence

$$\mathrm{ad}\,(g)u = \mathrm{ad}\,(g)\beta(p) = \beta(p_g) \tag{5.15.13}$$

where
$$p_g(\omega) = a^{j_1 \dots j_r}(g)\omega(X_{j_1})\dots\omega(X_{j_r})$$
$$= a^{i_1 \dots i_r} c_{i_1}^{j_1}(g)\dots c_{i_r}^{j_r}(g)\omega(X_{j_1})\dots\omega(X_{j_r}), \tag{5.15.14}$$

and as
$$c_i^j(g)\omega(X_j) = \omega(c_i^j(g)X_j) = \omega(\mathrm{ad}\,(g)X_i), \tag{5.15.15}$$

the remainder is

$$p_g(\omega) = a^{i_1 \dots i_r}\omega(\mathrm{ad}\,(g)X_{i_1})\dots\omega(\mathrm{ad}\,(g)X_{i_r}) \tag{5.15.16}$$

with the same coefficients $a \dots$ as p, but with respect to the basis arising from (X_i) by applying ad (g).

However, the operator ad $(g) \in GL(\mathfrak{g})$ admits a transpose acting on $\mathfrak{g}^*$; by definition, it transforms a linear form ω on $\mathfrak{g}$ into the linear form $X \mapsto \omega(\mathrm{ad}\,(g)X)$. Comparing (5.15.16) and (5.15.7), (5.15.16) can be seen to follow from (5.15.7) by replacing ω by ${}^t\mathrm{ad}\,(g)\omega$ in it. In other words, we finally get

$$p_g(\omega) = p({}^t\mathrm{ad}\,(g)\omega) \tag{5.15.17}$$

and therefore, on $\mathfrak{g}^*$, the desired $\mu = \beta(p)$ correspond to polynomials p satisfying

$$p({}^t\mathrm{ad}\,(g)\omega) = p(\omega) \tag{5.15.18}$$

for all $\omega \in \mathfrak{g}^*$; they are called *the invariants of the co-adjoint representation* of G (the latter being, by definition, the representation $g \mapsto {}^t\mathrm{ad}\,(g)^{-1}$ of G in $\mathfrak{g}^*$).

Exercise. Let E be a finite-dimensional real vector space, $S = S(E^*)$ the algebra of polynomial functions on E and π a representation of a connected Lie group G in E. (i) Let $t \mapsto \gamma(t)$ be a tangent curve to some fixed $X \in \mathfrak{g}$ at $t = 0$ in G. Associate to each $p \in S$ the polynomial $D_X p \in S$ given by

$$D_X p(u) = \frac{d}{dt} p(\mu(\gamma(t)^{-1}u)_{t=0};$$

show that D_X only depends on X and that it is the only *derivation* of the algebra S transforming every linear form $f \in E^*$ on E into the linear form $-f \circ \pi'(X)$. (ii) Let I be the set of *invariants of* π, i.e. of $p \in S$ such that $p(\pi(g)u) = p(u)$ for all u and g. Show that the elements of I are characterized by the condition $D_X p = 0$ for all $X \in \mathfrak{g}$. (iii) Let E be the dual vector space $\mathfrak{g}^*$ of $\mathfrak{g}$ and π the co-adjoint representation so that $E^* = \mathfrak{g}$ and $\mathfrak{g}$ embeds into S and generates it. Show that for all $X \in \mathfrak{g}$, there is a unique derivation of S which on $\mathfrak{g}$ reduces to the operator ad (X), and that the invariants of the co-adjoint representation are the zeros of the derivations belonging to S. (iv) Into what does the map $\beta : S(\mathfrak{g}) \to U(\mathfrak{g})$ transform the derivation of $S(\mathfrak{g})$

extending ad (X)? Deduce from this a new proof of the fact that β transforms the algebra of invariants of the co-adjoint representation into the centre of $U(\mathfrak{g})$.

Let us, for example, take the *Heisenberg group*. Then the three matrices P, Q, R form a basis of $\mathfrak{g}$ and G is generated by the three one-parameter subgroups $\exp(tP)$, $\exp(tQ)$ and $\exp(tR)$. The equality ad $(\exp(X)) = \exp(\mathrm{ad}\,(X))$ of Chap. 3 shows that the group ad (G) of automorphisms of $\mathfrak{g}$ is generated by the operators

$$\exp(t \cdot \mathrm{ad}\,(P)) = 1 + t \cdot \mathrm{ad}\,(P) \quad \text{and} \quad \exp(t \cdot \mathrm{ad}\,(Q)) = 1 + t \cdot \mathrm{ad}\,(Q). \tag{5.15.19}$$

The subsequent terms of the exponential series vanish since ad $(P)^2 = \mathrm{ad}\,(Q)^2 = 0$. On the other hand, the one-parameter subgroup $\exp(tR)$, being in the centre of G, is the kernel of the adjoint representation. Hence it is a matter of finding the polynomial functions on $\mathfrak{g}^*$ invariant under the transposes of operators (5.15.19). Denoting by

$$u = \omega(P), \quad v = \omega(Q), \quad w = \omega(R) \tag{5.15.20}$$

the coordinates of an element $\omega \in \mathfrak{g}$ with respect to the basis P, Q, R of $\mathfrak{g}$, the transpose of $1 + t \cdot \mathrm{ad}\,(P)$ is readily seen to be the map $(u, v, w) \mapsto (u, v + tw, w)$ and the transpose of $1 + t.\mathrm{ad}\,(Q)$ to be $(u, v, w) \mapsto (u - tw, v, w)$. The desired polynomials must therefore satisfy

$$p(u + xw, v, w) = p(u, v + yw, w) = p(u, v, w) \tag{5.15.21}$$

for all x and y, and so obviously (exercise) $p(u, v, w) = q(w)$. Hence

$$p(\omega) = q(\omega(R)) \tag{5.15.22}$$

with a polynomial q in one indeterminate. The image of p under the Birkhoff–Witt map β remains to be found. It suffices to do so when $p(\omega) = \omega(R)^n$ for some integer n. But in this case, the *symmetric* expansion of $p(\omega)$ with respect to the coordinates $\omega(P), \omega(Q)$ and $\omega(R)$ is $p(\omega) = \omega(R) \ldots \omega(R)$, with n factors, all its other coefficients being zero, so that (5.15.8) gives $\beta(p) = R \ldots R$ with n factors. Hence, finally, we get the *polynomials in the indeterminate R* in $U(\mathfrak{g})$. The resemblance between this result and that of question (2) of the *exercise* on page 191 is no coincidence!

Next, let G be the *group* $\begin{pmatrix} a & b \\ 0 & 1 \end{pmatrix}$. The two matrices $X = \begin{pmatrix} 1 & 0 \\ 0 & 0 \end{pmatrix}$ and $Y = \begin{pmatrix} 0 & 1 \\ 0 & 0 \end{pmatrix}$ form a basis of its Lie algebra, and trivial calculations show that for $g = \begin{pmatrix} a & b \\ 0 & 1 \end{pmatrix}$,

$$\mathrm{ad}\,(g)X = X - bY, \quad \mathrm{ad}\,(g)Y = aY. \tag{5.15.23}$$

Setting $u = \omega(X)$ and $v = \omega(Y)$ for all $\omega \in \mathfrak{g}^*$, the invariants of the co-adjoint representation are seen to be polynomials $p(u, v)$ such that

$$p(u - bv, av) = p(u, v) \tag{5.15.24}$$

for all $a \neq 0$ and b. Even if this condition is imposed only for $a > 0$ (so that G may be replaced by its connected component), the only solutions found are the constant polynomials. In other words, for the group considered, the only central distribution having support $\{e\}$ is, up to a constant factor, the Dirac measure at $\{e\}$; as will be shortly seen, this means that the centre of the algebra $U(\mathfrak{g})$ reduces to scalars.

We next consider the *hyperbolic group* $G = SL_2(\mathbb{R})$, whose Lie algebra consists of trace zero matrices in $M_2(\mathbb{R})$. The adjoint representation of G in $\mathfrak{g}$ has dimension 3 and is obviously irreducible; and hence so is the co-adjoint representation. However, as seen in Chap. 3, the group $SL_2(\mathbb{R})$ admits a unique irreducible representation of dimension n for each n. This suggests the existence of an isomorphism from the vector space $\mathfrak{g}$ onto its dual $\mathfrak{g}^*$, which for all $g \in G$, transforms the operator ad (g) into the contragredient operator. But, as is well known, an isomorphism from a vector space onto its dual is obtained from a non-degenerate bilinear form. Hence we are led to search for a non-degenerate bilinear form on $\mathfrak{g}$ *invariant under the adjoint representation* (so that the isomorphism from $\mathfrak{g}$ to $\mathfrak{g}^*$ defined by it transforms ad (g) into ad (g) for all $g \in G$). But replacing X by ad $(g)X$ replaces ad (X) by ad (g)ad (X)ad $(g)^{-1}$. Setting

$$K(X, Y) = \text{Tr}(ad(X)ad(Y)) \tag{5.15.25}$$

we get a bilinear form on $\mathfrak{g}$ which, by the way, is symmetric, and

$$\begin{aligned} K(\text{ad } (g)X, \text{ad } (g)Y) &= \text{Tr}(\text{ad } (g)\text{ad } (X)\text{ad } (g)^{-1}\text{ad } (g)\text{ad } (Y)ad(g)^{-1}) \\ &= \text{Tr}(\text{ad } (g)\text{ad } (X)\text{ad } (Y)\text{ad } (g)^{-1}) \tag{5.15.26} \\ &= \text{Tr}(\text{ad } (X)\text{ad } (Y)) = K(X, Y) \end{aligned}$$

for all X and Y in $\mathfrak{g}$ and g in G; this argument obviously applies to *all* Lie groups and gives what is known as the *Killing form* of $\mathfrak{g}$, and which most authors write as $B(X, Y)$.

Denoting the usual basis of $\mathfrak{g}$ by X, Y, H, where

$$\text{ad } (H)X = 2X, \quad \text{ad } (H)Y = -2Y, \quad \text{ad } (X)Y = H, \tag{5.15.27}$$

easy calculations show that the matrix of K with respect to this basis of $\mathfrak{g}$ is equal to

$$\begin{pmatrix} 8 & 0 & 0 \\ 0 & 0 & 4 \\ 0 & 4 & 0 \end{pmatrix}, \tag{5.15.28}$$

and as a result, the Killing form is non-degenerate in this case. For example, the entry 8 means that $\text{Tr}(\text{ad } (H)^2) = 8$, which follows from the fact that ad $(H)^2$ multiplies X and Y by 4 and vanishes at H.

Returning to our problem, thanks to the use of the Killing form, the search for the invariants of the co-adjoint representation is reduced to a search for the invariants of the adjoint representation, i.e. for the polynomial functions p on $\mathfrak{g}$ such that

$$p(\mathrm{ad}\,(g)U) = p(U) \quad \text{for all } U \in \mathfrak{g} \text{ and } g \in G, \tag{5.15.29}$$

which can also be written as

$$p(gUg^{-1}) = p(U) \tag{5.15.30}$$

since g and U are 2×2 matrices. Solutions are then readily found: functions $U \mapsto \mathrm{Tr}(U^n)$ and more generally expressions of the form

$$p(U) = \sum a_n \mathrm{Tr}(U^n) \tag{5.15.31}$$

with finitely many non-trivial coefficients (note that $\mathrm{Tr}(U) = 0$ for $U \in \mathfrak{g}$, so that the term $n = 1$ disappears from the result). We next show that these are *all* the solutions of (5.15.30).

Let $\mathfrak{g}^+$ denote the set of $U \in \mathfrak{g}$ whose two eigenvalues (necessarily equal up to sign since $\mathrm{Tr}(U) = 0$) are *real and non-trivial*. If $U = \begin{pmatrix} a & b \\ c & d \end{pmatrix}$ with $a + d = 0$, a trivial calculation shows that the condition for U to be in $\mathfrak{g}^+$ is

$$ad - bc < 0. \tag{5.15.32}$$

It follows that $\mathfrak{g}^+$ is *open* in the real vector space $\mathfrak{g}$, obviously invariant under the operators $U \mapsto gUg^{-1}$. The matrices

$$tH = \begin{pmatrix} t & 0 \\ 0 & -t \end{pmatrix} \tag{5.15.33}$$

with $t \neq 0$ are in $\mathfrak{g}^+$. We show that all $U \in \mathfrak{g}^+$ can be transformed into a matrix of type (5.15.33) by a properly chosen operator ad (g). Indeed, the eigenvalues of U being real and distinct, U is diagonalizable over $\mathbb{R}$. Hence there exists a $g \in GL_2(\mathbb{R})$ such that gUg^{-1} is of type (5.15.33). If need be multiplying g on the left by the matrix $\begin{pmatrix} 1 & 0 \\ 0 & -1 \end{pmatrix}$, we may assume that $\det(g) > 0$. Then g is the product of a real scalar matrix, namely $\det(g)^{1/2}$, and an element of $SL_2(\mathbb{R})$, and as gUg^{-1} remains unchanged when g is divided by the scalar $\det(g)^{1/2}$, our claim follows.

Hence $\mathfrak{g}^+$ is the union of the images under ad (g) of the line (5.15.33) of $\mathfrak{g}$, except for the point 0 temporarily excluded from the argument. A function defined on $\mathfrak{g}^+$ and invariant under the operators ad (g) is therefore determined by its restriction to the non-trivial matrices (5.15.33). However, for the element $w = \begin{pmatrix} 0 & -1 \\ 1 & 0 \end{pmatrix}$ of G,

$$\operatorname{ad}(w)H = -H; \tag{5.15.34}$$

hence the invariant polynomial function p must satisfy

$$p(tH) = p(-tH) \tag{5.15.35}$$

where t is a non-trivial real. Hence it is an even polynomial in t, implying that

$$p(tH) = \sum c_n t^{2n}. \tag{5.15.36}$$

But,

$$\operatorname{Tr}(t^n H^n) = \begin{cases} 0 & \text{if } n \text{ is odd,} \\ 2t^n & \text{if } n \text{ is even.} \end{cases} \tag{5.15.37}$$

It follows that the desired formula (5.15.31) holds in $\mathfrak{g}^+$, hence also in the whole of $\mathfrak{g}$ because if two polynomials are equal in a non-trivial open subset of a vector space, then they are equal in the whole space (extension principle of algebraic identities). It also follows that it suffices to consider even degree terms in (5.15.31)—this could have been guessed beforehand.

The corresponding elements of $U(\mathfrak{g})$ are still to be constructed. For clarity of presentation, we first do so in a more general context. Consider a Lie algebra $\mathfrak{g}$ whose Killing form K is non-degenerate (this condition happens to provide an alternative characterization of *semisimple* Lie algebras). Choose a basis (X_i) of $\mathfrak{g}$; identifying $\mathfrak{g}^*$ with $\mathfrak{g}$ using K, the dual basis of $\mathfrak{g}^*$ becomes the basis (X^i) of $\mathfrak{g}$ defined by

$$K(X_i, X^j) = \delta_i^j. \tag{5.15.38}$$

The Birkhoff–Witt map β then transforms each polynomial function p defined on $\mathfrak{g}$ (and no longer on $\mathfrak{g}^*$) into an element $\beta(p)$ of $U(\mathfrak{g})$. To find the latter, consider a "variable" element X of $\mathfrak{g}$. Identify it with an element ω of $\mathfrak{g}^*$, namely with the linear form $Y \mapsto K(X, Y)$ on $\mathfrak{g}$. Assign it coordinates

$$\xi_i = \omega(X_i) = K(X, X_i), \tag{5.15.39}$$

so that

$$X = \xi_i X^i \tag{5.15.40}$$

with the usual Einstein convention; write p as

$$p(X) = \sum_r a^{i_1 \dots i_r} \xi_{i_1} \dots \xi_{i_r} \tag{5.15.41}$$

with *symmetric* coefficients $a \dots$. We then get

$$\beta(p) = \sum a^{i_1 \dots i_r} X_{i_1} \dots X_{i_r}. \tag{5.15.42}$$

Notational conventions in classical tensor calculus prove to be particularly useful here since the nature of the operations is automatically suggested by the double co- and contravariant indices. In particular, note that (5.15.42) is written in the basis X_i, whereas (5.15.41) gives p with respect to the coordinates of X in the "contragredient" basis X^i.

This construction can be explicitly written down notably for the polynomial $p(U) = \mathrm{Tr}(U^n)$ on the Lie algebra of $SL_2(\mathbb{R})$. Start from the basis consisting of vectors H, X and Y. The matrix (5.15.28) of the Killing form shows that the contragredient matrix consists of the elements

$$H' = H/8, \quad X' = Y/4, \quad Y' = X/4. \tag{5.15.43}$$

Modifying the notation, (5.15.40) becomes

$$U = zH' + uX' + vY'. \tag{5.15.44}$$

The polynomial $\mathrm{Tr}(U^n)$ needs to be computed using variables z, u and v; but, given the definition of the matrices H, X and Y, (5.15.44) implies

$$U = \frac{1}{8} \begin{pmatrix} z & 2v \\ 2u & -z \end{pmatrix}. \tag{5.15.45}$$

Brute force calculation of the trace of the matrix U^n holds little attraction (still...) and it is better to observe that the eigenvalues of the matrix $\begin{pmatrix} a & b \\ c & -a \end{pmatrix}$ are given by $t^2 = a^2 + bc$; hence

$$\mathrm{Tr}\left[\begin{pmatrix} a & b \\ c & -a \end{pmatrix}^{2n} \right] = 2(a^2 + bc)^n. \tag{5.15.46}$$

Invariant polynomials on $\mathfrak{g}$ are therefore all the functions of type

$$p(U) = \sum c_n (z^2 + 4uv)^n. \tag{5.15.47}$$

To find $\beta(p)$, it still remains to write $p(U)$ in a *symmetric* form in terms of monomials in u, v and z, then to replace u, v and z, by X, Y and H in the result.

For the polynomial $z^2 + 4uv = z^2 + 2uv + 2vu$, the symmetric form is obvious; hence, writing p_n for the polynomial $(z^2 + 4uv)^n$,

$$\begin{aligned} \beta(p_1) &= H^2 + 2XY + 2YX = H^2 + 2H + 4YX \\ &= H^2 - 2H + 4XY = Z. \end{aligned} \tag{5.15.48}$$

Next consider

$$p_2(U) = (z^2 + 4uv)^2 = z^4 + 8z^2uv + 16u^2v^2. \tag{5.15.49}$$

This expression has to be written symmetrically in terms of monomials in u, v and z. The result is

$$
\begin{aligned}
p_2(U) =& zzzz + \frac{8}{12}(zzuv + zzvu + zuzv + zvzu + zuvz + zvuz \\
& + uzzv + uzvz + uvzz + vzzu + vzuz + vuzz) \\
& + \frac{16}{6}(uuvv + uvuv + uvvu + vuuv + vuvu + vvuu),
\end{aligned} \tag{5.15.50}
$$

and so

$$
\begin{aligned}
\beta(p_2) =& H^4 + \frac{2}{3}(H^2XY + H^2YX + HXHY + HYHX \\
& + HXYH + HYXH + XH^2Y + XHYH + XYH^2 \\
& + YH^2X + YHXH + YXH^2) + \frac{8}{3}(X^2Y^2 + XYXY \\
& + XY^2X + YX^2Y + YXYX + Y^2X^2),
\end{aligned} \tag{5.15.51}
$$

a result that does not prompt us to push our investigations further, even though physicists have carried on such computations every day ever since the early 1930s.

In fact, in the case at hand, the elements $\beta(p)$ of $U(\mathfrak{g})$ obtained from invariant polynomials on $\mathfrak{g}$ or $\mathfrak{g}^*$ are just the *polynomials in the element* $\beta(p_1) = H^2 + 2XY + 2YX$ of $U(\mathfrak{g})$ found above. This would be obvious if we had $\beta(p_n) = \beta(p_1)^n$ but, as already remarked in Sect. 10, the map $\beta : S(\mathfrak{g}) \to U(\mathfrak{g})$ is not an algebra homomorphism, even when restricted to polynomials invariant under the co-adjoint representation.[12] Nonetheless, as p_1 is of degree 2, (5.10.43) shows that

---

[12] As already shown, the image $Z(\mathfrak{g})$ of the algebra $J(\mathfrak{g})$ of invariants of the co-adjoint representation under β is in every case the centre of $U(\mathfrak{g})$, hence a commutative subalgebra of $U(\mathfrak{g})$. So the idea that

$$\beta : J(\mathfrak{g}) \to Z(\mathfrak{g})$$

could be an algebra isomorphism and not simply a vector space isomorphism seems reasonable. Although this is false, this conjecture is not too far removed from reality. First of all, there is a large class of Lie algebras satisfying it—the "nilpotent" algebras (J. Dixmier, 1959). In the general case, β is not compatible with the multiplicative structures. Nonetheless, it is possible to find an *algebra* isomorphism from $J(\mathfrak{g})$ onto $Z(\mathfrak{g})$ using significantly more complicated constructions than those of Poincaré–Birkhoff–Witt. In Dixmier's book on enveloping algebras mentioned earlier, this result (M. Duflo, 1971) of a seemingly inoffensive nature requires the application of every known result in this field and to be exact is the last proof given by Dixmier (Theorem 10.4.5, p. 345). It is obviously "scandalous" that the proof of such a simple statement should require so much effort and technique when the proof is easy in the nilpotent case and not so difficult in the semisimple case, which was expounded by Harish-Chandra around 1950 using the theory of finite-dimensional representations.

$$\deg[\beta(p_1^n) - \beta(-p_1)^n] \leq 2n - 1. \qquad (5.15.52)$$

This enables us to prove the claimed result by arguing by induction on n. It is a matter of showing that $\beta(p_1^n) = \beta(p_n)$ is a polynomial of degree $\leq n$ in $\beta(p_1) = Z$. It suffices to show it for $\beta(p_1^n) - \beta(p_1)^n = \beta(p_n) - Z^n$. Now, this element of $U(\mathfrak{g})$ is obviously invariant under the operators ad (g) since so are $\beta(p_n)$ and Z. It is therefore of type $\beta(q)$ for some polynomial q over $\mathfrak{g}$ invariant under the adjoint representation. As β preserves the degree, the polynomial q is of degree at most $2n - 1$ by (5.15.52); hence it is a linear combination of $p_1, \ldots, p_{n-1}$; in other words, $\beta(p_n)$ is a linear combination of $\beta(p_1), \ldots, \beta(p_{n-1})$ and of Z^n, qed.

Exercise. Check that (5.15.51) is a polynomial of the second degree in Z.

Exercise. Let $\mathfrak{g}$ be a Lie algebra. Suppose that, as an algebra, $J(\mathfrak{g})$ is generated by the homogeneous polynomials $p_1, \ldots, p_n$; show that the $\beta(p_i)$ then generate $Z(\mathfrak{g})$. If the p_i are algebraically independent (i.e. if $J(\mathfrak{g})$ is an algebra of polynomials in n indeterminates), then so are the $\beta(p_i)$. (Use the same arguments as in the case of the hyperbolic group by systematically applying (5.10.43).) By the way, there exist (solvable) Lie algebras for which $Z(\mathfrak{g})$ is not an algebra of finite type over $\mathbb{R}$ (for an example of dimension 45, see J. Dixmier's book, *Enveloping Algebras*, 4.9.20, p. 165). On the contrary, (Harish-Chandra) if $\mathfrak{g}$ is reductive, then $Z(\mathfrak{g})$ can be shown to be generated by finitely many algebraically independent generators. The case of the group GL_n is the subject of the next exercise.

Exercise. Let $\mathfrak{g} = M_n(\mathbb{R})$ be the Lie algebra of the group $G = GL_n(\mathbb{R})$. (i) Show that the symmetric bilinear form

$$H(X, Y) = \text{Tr}(XY) \qquad (5.15.53)$$

is invariant under the adjoint representation and non-degenerate. Let (X_i^j) be the obvious basis of $\mathfrak{g}$; find the "contragredient" basis with respect to the form (5.15.53). (ii) Let $\mathfrak{h}$ be the set of diagonal $X \in \mathfrak{g}$. Show that if the eigenvalues of some $X \in \mathfrak{g}$ are real and distinct, then there exists a $g \in G$ such that ad $(g)X \in \mathfrak{h}$. Show that the set of these matrices is an open subset $\mathfrak{g}^+$ of $\mathfrak{g}$. (iii) Let p be an invariant of the adjoint representation; show that p is fully determined by its restriction to $\mathfrak{h}$, and that it is a *symmetric* polynomial whose indeterminates are the eigenvalues of the variable $H \in \mathfrak{h}$ (note that $GL_n(\mathbb{R})$ contains elements, the permutation matrices, inducing every possible permutation of the eigenvalues in $\mathfrak{h}$). (iv) Using the fact (Newton's formulas) that every symmetric polynomial in the indeterminates $x_1, \ldots, x_n$ can be polynomially written in terms of $x_1^k + \ldots + x_n^k$, $(1 \leq k \leq n)$, show that the algebra of invariants of the adjoint representation is generated by n algebraically independent elements, namely the functions $X \to \text{Tr}(X^k)$ for $1 \leq k \leq n$. (v) Let (X_i^j) be the obvious basis of $\mathfrak{g}$. Show that in $U(\mathfrak{g})$, the n elements

$$X_i^i, X_j^j X_j^i, \ldots, X_{i_1}^{i_2} X_{i_2}^{i_3} \ldots X_{i_n}^{i_1} \qquad (5.15.54)$$

generate $Z(\mathfrak{g})$ and are algebraically independent. (vi) How does replacing GL_n by SL_n change these results?

Exercise. Let $\mathfrak{g}$ be a Lie algebra whose Killing form $K(X, Y) = \mathrm{Tr}(ad(X)ad(Y))$ is non-degenerate (i.e. a semisimple Lie algebra!). Choose a basis (X_i) of $\mathfrak{g}$ and consider the "contragredient" basis (X^i) given by $K(X_i, X^j) = \delta_i^j$. Show that the element $X_i X^i$ of $U(\mathfrak{g})$ is independent of the choice of basis and that it belongs to $Z(\mathfrak{g})$. It is called the *Casimir operator* of $\mathfrak{g}$ after H.B.G. Casimir, who discovered it for the Lorentz group in the early thirties and who then, in collaboration with B.L. van der Waerden, used it to give the first algebraic proof of the complete reducibility of finite-dimensional representations of semisimple Lie algebras, after which Casimir disappeared from the sight of mathematicians.[13]

Exercise. Take $G = SL(2, \mathbb{C})$, so that $\mathfrak{g}$ is the Lie algebra of 2×2 complex matrices with trace zero regarded as an algebra of dimension 6 over $\mathbb{R}$. Consider the subalgebras of dimension 3 of $\mathfrak{g}$ corresponding to the subgroups $SL_2(\mathbb{R})$ and $SU(2)$ of G and the Casimir operators of these subalgebras. (Note that thanks to the Poincaré–Birkhoff–Witt theorem, the algebra $U(\mathfrak{h})$ of a subalgebra $\mathfrak{h}$ of a Lie algebra $\mathfrak{g}$ can always be identified with the subalgebra of $U(\mathfrak{g})$ generated by $\mathfrak{h}$.) Show that they are algebraically independent and generate the algebra $Z(\mathfrak{g})$. [The reader may observe that, as a Lie algebra over $\mathbb{C}$, $\mathfrak{g}$ can be canonically identified with the complexification of each of the two subalgebras considered.]

The usefulness of central distributions on a Lie group G becomes clear once it is remembered that they were defined so that, if $\mu \in Z(\mathfrak{g})$, then the corresponding operator $\pi(\mu)$ commutes with all $\pi(g)$ for all finite-dimensional representations π of G (at least in the most important case when G is connected). Hence if π is an *irreducible* representation in a finite-dimensional *complex* vector space V, Schur's lemma (Sect. 14, Theorem 4) shows that $\pi(\mu)$ is a scalar for all $\mu \in Z(\mathfrak{g})$. So considering an arbitrary coefficient

$$f(x) = \langle \pi(x)a, b \rangle \quad (a \in V, b \in V^*) \tag{5.15.55}$$

of the representation π, in notation (5.10.5), we get

$$R_\mu f(x) = \int f(xy)d\mu(y) = \int \langle \pi(x)\pi(y)a, b \rangle d\mu(y) \tag{5.15.56}$$
$$= \langle \pi(x)\pi(\mu)a, b \rangle = \chi_\pi(\mu)f(x)$$

where $\chi_\pi(\mu)$ denotes the scalar that the operator $\pi(\mu)$ reduces to. Note that $\mu \mapsto \chi_\pi(\mu)$ is necessarily an associative algebra homomorphism from $Z(\mathfrak{g})$ to $\mathbb{C}$. In other words, *the coefficients of the irreducible representations of G are eigenfunctions of the invariant operators R_μ associated to $\mu \in Z(\mathfrak{g})$.* This sometimes enables us to compute them by solving differential equations, at least for groups of small dimensions. Besides, this method extends to infinite-dimensional representations, where it

---

[13] He recently retired from Philips (Netherlands), where he was head of their research laboratories.

plays a far more important role than for finite-dimensional representations. Consequently, the coefficients of these representations can be expressed using practically all known "special functions" (see Y. Vilenkin, *Special Functions and the Theory of Group Representations*, Moscow, 1965, eng. trans. AMS) and an infinite number of others that had never been encountered until then.

Exercise. Set $G = SL_2(\mathbb{R})$ and $K = SO(2)$, so that $G/K = P$ is the upper half-plane. (i) Identify the functions $\phi(g)$ such that $\phi(gk) = \phi(g)$ with the functions defined on P (i.e. of interest here is the function space F_o of Sect. 12). Show that the effect of the Casimir operator of G on these functions is, up to a constant factor, given in P by the operator

$$\Omega = y^2 \left(\frac{\partial^2}{\partial x^2} + \frac{\partial^2}{\partial y^2} \right) \tag{5.15.57}$$

(use (5.12.20), (5.12.30), (5.12.31) and (5.12.32) of Sect. 12). (ii) Assume furthermore that $\phi(kg) = \phi(g)$, i.e. that ϕ is both left and right invariant under the subgroup K or, what comes to the same, that the corresponding function $f(z)$ is invariant under the "rotations"

$$z \mapsto \frac{z \cdot \cos\theta - \sin\theta}{z \cdot \sin\theta + \cos\theta} \tag{5.15.58}$$

around $z = i$. Show that $\phi(g)$ depends only on $\xi = \frac{1}{2}\mathrm{Tr}(g'g) = \frac{1+z\bar{z}}{2y}$ if $z = g(i)$. (iii) Assume that ϕ is an eigenfunction of the Casimir operator of G. Setting $\phi(g) = f(z) = P(\xi)$ and taking (5.15.57) into account, show that P satisfies the Legendre differential equation

$$\frac{d}{d\xi}(\xi^2 - 1)\frac{dP}{d\xi} = \lambda P. \tag{5.15.59}$$

[The classical proof is to show that (5.15.59) possesses, up to a constant factor, a unique solution which remains regular when $\xi > 1$ tends to 1. This is certainly the case if ϕ is of class C^∞ on G since $\xi = 1$ corresponds to $g = e$—namely

$$P(\xi) = \frac{1}{2\pi} \int_0^{2\pi} (\xi + \cos(\theta)\sqrt{\xi^2 - 1})^s d\theta \tag{5.15.60}$$

where the exponent s to be chosen depends on the eigenvalue λ.] (iv) Let V_n be the vector space of homogeneous polynomial functions with complex coefficients of degree n in $\mathbb{R}^2$, and let π_n be the obvious (irreducible) representation of G in V_n, (see (5.11.37)). Set $z = x + iy$, so that the polynomials $(x + iy)^p(x - iy)^q = z^p\bar{z}^q$, with $p + q = n$, form a basis of V_n. By computing the effect of the operators $\pi_n(k)$, $k \in K$, on these polynomials, show that V_n contains a non-trivial $\pi_n(K)$-invariant vector if and only if n is even; this vector, namely $(x^2 + y^2)^{n/2}$, is then unique up to a constant factor. (v) Let

$$u_p(x, y) = (x + iy)^p(x - iy)^{n-p}, \quad (0 \le p \le n) \tag{5.15.61}$$

be the basis of V_n in question; set

$$\pi_n(g)u_p = \sum \pi_{n;pq}(g)u_q \tag{5.15.62}$$

by highlighting the matrix of $\pi_n(g)$ with respect to basis (5.15.61). Suppose that n is even. Show that, for $p = q = n/2$, the function $\pi_{n;pq}$ is right and left K-invariant and is the eigenvalue of the Casimir operator. (vi) Write this function using Legendre polynomials

$$P_k(\xi) = \frac{(-1)^k}{2^k k!} \frac{dk}{d\xi^k} (1 - \xi^2)^k. \tag{5.15.63}$$

Exercise. Let μ be a distribution at the origin on a Lie group G, so that the differential operator R_μ commutes with left translations. Show that R_μ also commutes with right translations if and only if $\mu \in Z(\mathfrak{g})$ when G is connected. How should the statement be modified when the group is not connected?

5.16 The Universal Property of $U(\mathfrak{g})$

Before ending this chapter, let us return to the Poincaré–Birkhoff–Witt theorem to show that if G is a Lie group, with Lie algebra $\mathfrak{g}$, and π is a finite-dimensional representation of G, and hence π' a representation of $\mathfrak{g}$, then it is possible to give a purely algebraic proof of Theorem 1 of Sect. 13 in a much more general setup. Indeed, the following result holds:

Theorem 6 *Let G be a Lie group, $\mathfrak{g}$ its Lie algebra, A an associative algebra over $\mathbb{R}$ and π a linear map from $\mathfrak{g}$ to A such that*

$$\pi([X, Y]) = \pi(X)\pi(Y) - \pi(Y)\pi(X) \tag{5.16.1}$$

for all $X, Y \in \mathfrak{g}$. Then π can be uniquely extended to a homomorphism from the algebra $U(\mathfrak{g})$ to the algebra A.

Theorem 1 can be recovered by taking A to be the algebra of endomorphisms of a finite-dimensional real vector space. But note that Theorem 6 makes no assumptions on A, which may well be infinite-dimensional over $\mathbb{R}$.

The uniqueness of the extension of π to $U(\mathfrak{g})$ is obvious since $\mathfrak{g}$ generates $U(\mathfrak{g})$, so that only existence requires proving. To do so, we start from the *tensor algebra* $T(\mathfrak{g})$. It is constructed as follows. First set

$$T_0(E) = k, \quad T_1(E) = E, \quad T_2(E) = E \otimes E, \quad T_3(E) = E \otimes E \otimes E, \ldots \tag{5.16.2}$$

for every vector space E over a commutative field k. The tensor product $T_n(E)$ has the following properties: (i) there is a multilinear map

$$(x_1, \ldots, x_n) \mapsto x_1 \otimes \ldots \otimes x_n \qquad (5.16.3)$$

from $E \times \ldots \times E$ to $T_n(E)$; (ii) if $(e_i)_{i \in I}$ is a basis of E, then, as the x_p vary in the given basis, the products (5.16.3) form a basis of $T_n(E)$; (iii) for every n-linear map f from E^n to a vector space F over k, there is a unique linear map $\overline{f} : T_n(E) \to F$ such that

$$\overline{f}(x_1 \otimes \ldots \otimes x_n) = f(x_1, \ldots, x_n). \qquad (5.16.4)$$

$T_n(E)$ is (more than) fully determined by these properties; its dimension is r^n if E is of finite dimension r, and the tensor algebra $T(E)$ is the direct sum

$$T(E) = \bigoplus T_n(E) = \coprod T_n(E) \qquad (5.16.5)$$

of the $T_n(E)$ thus defined (we leave the choice of notation to the reader). Multiplication is defined in $T(E)$ in the obvious way, i.e. by requiring it to be distributive with respect to addition and to satisfy

$$(x_1 \otimes \ldots \otimes x_p).(y_1 \otimes \ldots \otimes y_q) = x_1 \otimes \ldots \otimes x_p \otimes y_1 \otimes \ldots \otimes y_q \qquad (5.16.6)$$

for all elements x_i and y_j of E. This gives $T(E)$ a (non-commutative) associative algebra structure, and the elements of E obviously generate the algebra $T(E)$. Then, the following "universal" property holds: if A is an associative algebra over k, then every linear map f from E to A can be uniquely extended to a *homomorphism* from the algebra $T(E)$ to A. Indeed, for all n, the map

$$(x_1, \ldots, x_n) \mapsto f(x_1) \ldots f(x_n) \qquad (5.16.7)$$

from E^n to A is multilinear, and so, by property (iii) of $T_n(E)$, arises from a linear map $\overline{f}$ from $T_n(E)$ to A. As this construction holds for all n, we thereby get a linear map $\overline{f}$ from $T(E)$ to A which obviously satisfies the required condition. Its uniqueness is obvious.

Let us now return to Theorem 6 and consider the tensor algebra $T(g)$, constructed without any reference to the Lie algebra structure of g. The given map π from g to A can be extended to a homomorphism $\overline{\pi}$ from $T(g)$ to A given by

$$\overline{\pi}(X_1 \otimes \ldots \otimes X_n) = \pi(X_1) \ldots \pi(X_n) \qquad (5.16.8)$$

for all $X_i \in g$. The same holds for the canonical map j from g to the algebra $U(g)$; it can be extended to a homomorphism

$$\overline{j} : T(g) \to U(g). \qquad (5.16.9)$$

Note that $\bar{j}$ is surjective since $\mathfrak{g}$ generates $U(\mathfrak{g})$. As a result, $\bar{j}$ gives an isomorphism

$$U(\mathfrak{g}) = T(\mathfrak{g})/J, \text{ where } J = \text{Ker}(\bar{j}). \tag{5.16.10}$$

If $\bar{\pi}(J) = 0$ is shown to hold, then this will imply that the homomorphism $\bar{\pi}$ is the composite of $\bar{j}$ and a homomorphism from $U(\mathfrak{g})$ to A, proving the theorem. Now, condition (5.16.1) can also be stated as

$$\bar{\pi}(X \otimes Y - Y \otimes X - [X, Y]) = 0 \quad \text{for all } X, Y \in \mathfrak{g}. \tag{5.16.11}$$

As $\bar{\pi}$ is an algebra homomorphism, its kernel (as well as J for the same reason) is a double-sided ideal of $T(\mathfrak{g})$. Since it contains the elements of type

$$X \otimes Y - Y \otimes X - [X, Y] \in T_1(\mathfrak{g}) \oplus T_2(\mathfrak{g}), \tag{5.16.12}$$

it contains the double-sided ideal they generate. Hence it suffices to show that *J is the double-sided ideal I of $T(\mathfrak{g})$ generated by the elements of type* (5.16.12).

To show that

$$I = \text{Ker}(T(\mathfrak{g}) \to U(\mathfrak{g})), \tag{5.16.13}$$

let us choose a basis (X_i) of $\mathfrak{g}$, and consider the ordered monomials

$$\underbrace{X_1 \otimes \ldots \otimes X_1}_{\alpha_1} \otimes \underbrace{X_2 \otimes \ldots \otimes X_2}_{\alpha_2} \otimes \ldots \otimes \underbrace{X_n \otimes \ldots \otimes X_n}_{\alpha_n} \tag{5.16.14}$$

in $T(\mathfrak{g})$. The image of (5.16.14) in $U(\mathfrak{g})$ is obviously the ordered monomials

$$X_1^{\alpha_1} X_2^{\alpha_2} \ldots X_n^{\alpha_n} \tag{5.16.15}$$

and as seen above they form a *basis* of $U(\mathfrak{g})$. It follows first that the expressions (5.16.14) are *linearly independent* mod I, since if a *linear combination* of the products (5.16.14) not all of whose coefficients are zero were in $I \subset J$, there would be a non-trivial relation between the expressions (5.16.15). The map from $T(\mathfrak{g})$ to $U(\mathfrak{g})$, restricted to the vector subspace V of $T(\mathfrak{g})$ generated by the expressions (5.16.14), is moreover bijective—it maps basis (5.16.14) of V onto a *basis* of $U(\mathfrak{g})$.[14] To prove (5.16.13), it therefore suffices to show that $T(g) = I + V$, and for this to show that, mod $I + V$, the unordered monomials

$$X_{i_1} \otimes \ldots \otimes X_{i_q} \tag{5.16.16}$$

can be written in terms of the ordered monomials (5.16.14). But this follows readily.

Indeed, if two consecutive factors in (5.16.16) are permuted, say X_k and X_h, then the difference between the old and the new expression is

---

[14]Due to the Poincaré–Birkhoff–Witt theorem for $U(\mathfrak{g})$.

$$u \otimes (X_k \otimes X_h - X_h \otimes X_k) \otimes v = u \otimes (X_k \otimes X_h - X_h \otimes X_k - [X_k, X_h]) \otimes v$$
$$+ u \otimes [X_k, X_h] \otimes v$$
$$\equiv u \otimes [X_k, X_h] \otimes v \quad \mod I, \qquad (5.16.17)$$

where u and v are monomials of type (5.16.16) such that the sum of their degrees equals $q - 2$. As $X_k, X_h \in \mathfrak{g}$, $u \otimes [X_k, X_h] \otimes v$ can be written linearly in terms of the monomials (5.16.16) of degree $\leq q - 1$. By induction on q, it may therefore be assumed that $u \otimes [X_k, X_h] \otimes v$ is in $I + V$. This implies that if two consecutive factors in (5.16.16) are exchanged, it remains in the same class mod $I + V$. Since every permutation is obtained by a sequence of operations of this type, the unordered monomial (5.16.16) is equal to an ordered monomial mod $I + V$, i.e. to an element of V, and so is in $I + V$. This shows that $T(\mathfrak{g}) = I + V$ as claimed, completing the proof of the theorem.

A *Lie algebra* over a field k of characteristic 0 (there is no satisfactory theory in characteristic p) is defined to be a vector space $\mathfrak{g}$ over k endowed with a bilinear composition law $(X, Y) \mapsto [X, Y]$ satisfying equalities

$$[X, Y] + [Y, X] = 0$$
$$[X, [Y, Z]] + [Y, [Z, X]] + [Z, [X, Y]] = 0 \quad \text{(Jacobi identity)}, \qquad (5.16.18)$$

which obviously hold in any associative algebra endowed with the product $XY - YX$. For such an algebra, the *universal enveloping algebra* of $\mathfrak{g}$ is by definition the associative algebra

$$U(\mathfrak{g}) = T(\mathfrak{g})/I, \qquad (5.16.19)$$

where $T(\mathfrak{g})$ is the tensor algebra of $\mathfrak{g}$, defined by (5.16.5) for every vector space, and where I is the double-sided ideal of $T(\mathfrak{g})$ generated by the elements (5.16.12). Without any further assumption on $\mathfrak{g}$, it is possible to show that, if (X_i) is a (finite or infinite) basis of $\mathfrak{g}$, then the ordered monomials in X_i form a basis of $U(\mathfrak{g})$—Poincaré–Birkhoff–Witt theorem—and that any linear map π from $\mathfrak{g}$ to an associative algebra A such that $\pi([X, Y]) = \pi(X)\pi(Y) - \pi(Y)\pi(X)$ can be extended to a homomorphism from $U(\mathfrak{g})$ to A. The arguments used in this section and in Sect. 10 have enabled us to prove these facts in a roundabout way when $\mathfrak{g}$ is the Lie algebra of a Lie group. But there are direct purely algebraic proofs of these statements. These can be found in most presentations on Lie groups (and in all those that go beyond the elementary part of the theory, the associative algebra $U(\mathfrak{g})$ being absolutely fundamental in the study of linear representations of Lie groups and algebras) and so perforce in J. Dixmier's book, *Enveloping Algebras* (North-Holland 1977) since it is fully dedicated to the subject. See also N. Bourbaki's book on Lie algebras.

Chapter 6
The Exponential Map for Lie Groups

6.1 One-Parameter Subgroups

Let G be a C^r ($r = \infty$ or ω) Lie group; a *one-parameter subgroup* of G is by definition a homomorphism $\gamma : \mathbb{R} \to G$ from the additive group $\mathbb{R}$ to G, i.e. a C^r map (it will soon become clear that assuming only continuity would be sufficient) from $\mathbb{R}$ to G such that $\gamma(s + t) = \gamma(s)\gamma(t)$ for all s, t. In Chap. 3, it was shown that if $G = GL_n(\mathbb{R})$, then these one-parameter subgroups correspond bijectively to the matrices $X \in M_n(\mathbb{R})$, i.e. to the elements of the Lie algebra of G, the subgroup γ_X corresponding to a matrix X being given by

$$\gamma_X(t) = e^{tX} = \sum t^n X^n / n! \tag{6.1.1}$$

Note that we then have

$$X = \gamma'_X(0), \tag{6.1.2}$$

so that conversely, the element of the Lie algebra which, in this case, corresponds to a one-parameter subgroup of G is its tangent vector at the origin. We show that this holds for all Lie groups G.

Theorem 1 *Let G be a Lie group. Then the map $\gamma \mapsto \gamma'(0)$ from the set of one-parameter subgroups of G to the Lie algebra of G is bijective.*

The proof of this theorem is very simple if one assumes the elementary theory of differential equations. Indeed, let us begin with a one-parameter subgroup γ of G. Set $X = \gamma'(0)$, so that

$$X(f) = df(e; X) = \frac{d}{dt} f(\gamma(t))_{t=0} \tag{6.1.3}$$

© Springer International Publishing AG 2017
R. Godement, *Introduction to the Theory of Lie Groups*,
Universitext, DOI 10.1007/978-3-319-54375-8_6

for any C^r function f in the neighbourhood of e, and differentiate

$$\gamma(s + t) = \gamma(s)\gamma(t) \tag{6.1.4}$$

with respect to t; computations from Chap. 5, Sect. 5.6, immediately give

$$\gamma'(s + t) = \gamma(s)\gamma'(t). \tag{6.1.5}$$

Taking $t = 0$,

$$\gamma'(t) = \gamma(t)X \quad \text{for all } t. \tag{6.1.6}$$

Hence, considering the vector field $R = R_X$ on G given by

$$R(g) = gX \tag{6.1.7}$$

shows that the curve γ satisfies the differential equation

$$\gamma'(t) = R(\gamma(t)) \tag{6.1.8}$$

with initial condition $\gamma(0) = e$. The uniqueness theorem from the theory of differential equations[1] immediately shows that γ is fully determined by X.

Conversely, starting from X and considering the vector field (6.1.7), the theory of differential equations enables us to construct, possibly on the whole of $\mathbb{R}$, but at least on an open interval $I \subset \mathbb{R}$ containing 0, a C^r curve γ satisfying

$$\gamma'(t) = R(\gamma(t)), \quad \gamma'(0) = e. \tag{6.1.9}$$

We now show that taking the maximal solution of (6.1.9) for γ—so that it cannot be extended beyond I—then $I = \mathbb{R}$ and γ is a one-parameter subgroup of G.

---

[1] Let R be a vector field on a manifold G; assume that both R and G are of class C^r with $r = \infty$ or ω. The *integral* of (6.1.8) is then defined to be any pair (I, γ) where I is an open interval in $\mathbb{R}$ and $\gamma : I \to G$ is of class C^r and satisfies (6.1.8) for all $t \in I$. Replacing G and R by the domain of a local chart and the restriction of R to it, locally, the search for integrals of (6.1.8) amounts to integrating a system of ordinary differential equations

$$dx_i/dt = R_i(x_1, \dots, x_n)$$

in an open subset of a Cartesian space; it follows immediately that (a) every C^1 solution of (6.1.8) is in fact C^r, (b) if two integrals (I, γ) and (J, δ) of (6.1.8) are equal at some point of $I \cap J$, they are equal in the neighbourhood of this point and so in all of $I \cap J$ since $I \cap J$ is connected, (c) every integral can be extended to a *maximal* integral, i.e. which cannot be extended to a strictly larger interval, (d) for all $a \in G$ and $s \in \mathbb{R}$, there is a unique maximal integral of (6.1.8) satisfying $\gamma(s) = a$. See M. Berger and B. Gostiaux, *Differential Geometry*, Chap. I, or H. Cartan, *Calcul Différentiel*, Chap. II (where, for the needs of our presentation, the Banach spaces can be assumed to be finite-dimensional), or J. Dieudonné, *Eléments d'Analyse*, X.4, X.5, X.7 and XVIII.1 and XVIII.2, or MA IX, Sect. 15, etc.

To show that $I = \mathbb{R}$, it suffices to show that, otherwise, γ would be extendable beyond I. For this, choose some $s \in I$ and consider the map $\gamma_s : s + I \to G$ given by $\gamma_{s(t)} = \gamma_{(s)}\gamma_{(t-s)}$. Obviously

$$\gamma'_s(t) = \gamma(s)\gamma'(t-s) = \gamma(s)\gamma(t-s)X = \gamma_s(t)X \qquad (6.1.10)$$

for all $t \in s + I$. As a result γ_s is a solution of (6.1.8) defined in $s + I$. Moreover, $\gamma_s(s) = \gamma(s)\gamma(0) = \gamma(s)$, so that the integrals γ and γ_s of (6.1.8), defined in I and $s + I$ coincide at $s \in I \cap (s + I)$, hence in the whole of the interval $I \cap (s + I)$ by the uniqueness of solutions of a differential equation. Hence, we get a solution of (6.1.9) defined in $I \cup (s + I)$ by requiring it to be equal to γ in I and to γ_s in $s + I$. As γ is maximal, it first follows that $I \cup (s + I) = I$ for all $s \in I$, and so obviously $I = \mathbb{R}$, and moreover, $\gamma_s = \gamma$, which means that $\gamma(s)\gamma(t-s) = \gamma(t)$ for all $s, t \in \mathbb{R}$; as a result, γ is indeed a one-parameter subgroup.

From now on, γ_X will denote the tangent one-parameter subgroup of G to the vector $X \in \mathfrak{g}$ at e. Since, in the case of $GL_n(\mathbb{R})$,

$$\exp(X) = \gamma_X(1), \qquad (6.1.11)$$

in the general case, it is natural to *define* a map

$$\exp : \mathfrak{g} \to G \qquad (6.1.12)$$

using (6.1.11); it is the *exponential map* from $\mathfrak{g}$ to G which will sometimes be written $\exp_G$ to avoid confusion with similar maps for other groups. Observe that, if, as above, γ_X denotes the tangent one-parameter subgroup to the vector X at $t = 0$, then, for all s, $t \mapsto \gamma_X(st)$ is *a* tangent one-parameter subgroup to the vector sX at $t = 0$; hence it is *the* one-parameter subgroup γ_{sX}. Therefore, setting $t = 1$ in equality $\gamma_X(st) = \gamma_{sX}(t)$ which follows from this, for all $X \in \mathfrak{g}$,

$$\gamma_X(s) = \exp(sX) \quad \text{for all} \quad s \in \mathbb{R}. \qquad (6.1.13)$$

Exercise. Find all the exponential maps with respect to the groups $\mathbb{R}$ and $\mathbb{R}^*$.

6.2 Elementary Properties of the Exponential Map

The most "obvious" property is the following:

Theorem 2 *Let G be a Lie group and $\mathfrak{g}$ its Lie algebra. For $X, Y \in \mathfrak{g}$,*

$$\boxed{\exp(X)\exp(Y) = \exp(X + Y) \quad \text{if} \quad [X, Y] = 0.} \qquad (6.2.1)$$

To prove this let us consider the one-parameter subgroups $\gamma_X(t) = \exp(tX)$ and $\gamma_Y(t)$ defined by X and Y. Once $\gamma_X(t)\gamma_Y(t) = \gamma_Y(t)\gamma_X(t)$ will have been shown to hold, an obvious computation will show that

$$t \mapsto \gamma(t) = \gamma_X(t)\gamma_Y(t) \tag{6.2.2}$$

is also a one-parameter subgroup. However,

$$\gamma'(0) = \gamma_X'(0)\gamma_Y(0) + \gamma_X(0)\gamma_Y'(0) = X + Y, \tag{6.2.3}$$

and so $\gamma(t) = \exp(t(X + Y))$; setting $t = 1$ in the result then gives (6.2.1).

Hence, since tX and tY commute like X and Y, it is a matter of proving that

$$\exp(X)\exp(Y) = \exp(Y)\exp(X) \quad \text{if} \quad [X, Y] = 0. \tag{6.2.4}$$

Setting $g = \exp(Y)$, this becomes $g \cdot \exp(X) \cdot g^{-1} = \exp(X)$. To establish this equality, we show that, on the one hand,

$$\boxed{g \cdot \exp(X) \cdot g^{-1} = \exp(\mathrm{ad}\,(g)X)} \quad \text{for all } g \in G \text{ and } X \in \mathfrak{g}, \tag{6.2.5}$$

where $\mathrm{ad}\,(g)X$ is defined in Chap. 5, Sect. 5.6 (it is the image of the distribution X under the automorphism $x \mapsto gxg^{-1}$ of G), and on the other, that

$$\boxed{g = \exp(Y) \Rightarrow \mathrm{ad}\,(g) = e^{\mathrm{ad}\,(Y)} = \sum \mathrm{ad}\,(Y)^n/n!} \tag{6.2.6}$$

for all $Y \in \mathfrak{g}$. As $[X, Y] = 0$ is equivalent to $\mathrm{ad}\,(Y)X = 0$ and obviously implies $\mathrm{ad}\,(Y)^n X = 0$ for all $n > 0$, so that

$$\boxed{e^{\mathrm{ad}\,(Y)}X = X \quad \text{if } [X, Y] = 0,} \tag{6.2.7}$$

the desired equality (6.2.4) readily follows by substituting (6.2.7) on the right-hand side of (6.2.5).

The proof of (6.2.5) is very easy. The curve $\gamma(t) = g \cdot \exp(tX) \cdot g^{-1}$ is obviously a one-parameter subgroup of G, obtained by taking the composition of the homomorphism $\gamma_X : \mathbb{R} \to G$ and the automorphism ϕ of G given by $\phi(x) = gxg^{-1}$. The chain rule shows that

$$\gamma'(0) = \phi'(\gamma_X(0))\gamma_X'(0) = \phi'(e)X = \mathrm{ad}\,(g)X \tag{6.2.8}$$

by definition of $\mathrm{ad}\,(g)$ (see Chap. 5, (6.6.14)). Since $\gamma_Y(t) = \exp(tY)$ for all t and Y, $g \cdot \exp(tX) \cdot g^{-1} = \exp(t \cdot \mathrm{ad}\,(g)X))$, and hence (6.2.5) follows by setting $t = 1$.

The proof of (6.2.6) is similar. As $g \mapsto \mathrm{ad}\,(g)$ is a homomorphism from the Lie group G to the Lie group $GL(\mathfrak{g})$, the map

$$t \mapsto \mathrm{ad}\,(\exp(tX)) = \gamma(t) \tag{6.2.9}$$

is a one-parameter subgroup of $GL(\mathfrak{g})$, and so $\mathrm{ad}\,(\exp(tX)) = e^{t\gamma'(0)}$ by Chap. 3. As $\gamma = \mathrm{ad}\,\circ\,\gamma_X$, the chain rule shows that

$$\gamma'(0) = \mathrm{ad}\,'(\gamma_X(0))\gamma_X'(0) = \mathrm{ad}\,'(e)X = \mathrm{ad}\,(X), \tag{6.2.10}$$

as shown in Chap. 5, Sect. 5.7; (6.2.6) is then immediate.

The arguments used to prove (6.2.5) and (6.2.6) are in fact particular cases of the following result: *if $\phi : G \to H$ is a Lie group homomorphism, then the diagram*

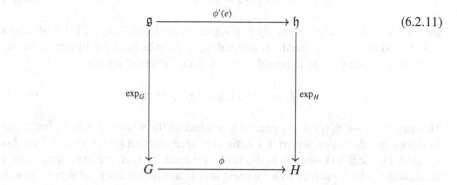

$$(6.2.11)$$

is commutative. To see this, start with a one-parameter subgroup $\gamma_X(t) = \exp(tX)$ of G and consider its image $\gamma(t) = \phi(\gamma_X(t))$ under ϕ; it is again a one-parameter subgroup of H, and as $\gamma'(0) = \phi'(e)\gamma_X'(0) = \phi'(e)X = Y$, we get $\phi(\exp(tX)) = \exp(tY)$; the desired result follows for $t = 1$.

Exercise. Recover (6.2.5) and (6.2.6) from the general result.

Exercise. Let G be a Lie group and π a linear representation of G in a finite-dimensional real vector space E, i.e. a homomorphism from G to the Lie group $GL(E)$; let $\pi' : \mathfrak{g} \to \mathscr{L}(E)$ be the tangent linear map to π at the origin. Show that

$$\pi(\exp_G(X)) = e^{\pi'(X)} = \sum \pi'(X)^n/n! \quad \text{for all } X \in \mathfrak{g}. \tag{6.2.12}$$

6.3 Regularity of the Exponential Map

We now establish the following result:

Theorem 3 *Let G be a C^r Lie group and $\mathfrak{g}$ its Lie algebra. The map $\exp : \mathfrak{g} \to G$ is then of class C^r and its tangent map at the origin:*

$$\exp'(0) : T_0(\mathfrak{g}) = \mathfrak{g} \to T_e(G) = \mathfrak{g} \tag{6.3.1}$$

is the identity map.

The proof that exp is of class C^r rests on well-known results about differential equations involving a single parameter, and which can be formulated as follows. Take two C^r manifolds U and V and a morphism

$$L : U \times V \to T(U) \tag{6.3.2}$$

from the product manifold $U \times V$ to the manifold $T(U)$ of tangent vectors to U (Chap. 5, Sect. 5.5) such that, for all $x \in U$ and $y \in V$, x is the initial point of the vector $L(x, y)$. Then, for each $y \in Y$, there clearly exists a vector field on U whose vectors L_y are given by

$$L_y(x) = L(x, y), \tag{6.3.3}$$

and so, for each y, there is a family of integral curves from L_y to U. Fix $t_o \in \mathbb{R}$ and $x_o \in U$ and consider the maximal integral γ_y of L_y which, at the instant t_o passes through x_o; hence it is the maximal solution of the differential equation

$$x'(t) = L(x(t), y), \quad x(t_o) = x_o. \tag{6.3.4}$$

The map $(t, y) \mapsto \gamma_y(t)$ is not generally defined on the whole of $\mathbb{R} \times Y$; but it can be shown that the set on which it is defined is open and that it is of class C^r on this set (MA IX, Sect. 15)—and it is the result we need. In fact, the statement is of a local nature, which enables us to assume that U and V are open subsets of spaces $\mathbb{R}^p$ and $\mathbb{R}^q$. The function L can then be identified with a C^r map from $U \times V$ to $\mathbb{R}^p$, and (6.3.4) becomes an ordinary differential equation whose right-hand side is a C^r function of x and of a parameter $y \in V$, thereby reducing the question to a well-known situation.

To apply this result to the map exp : $\mathfrak{g} \to G$, take $U = G$, $V = \mathfrak{g}$, and let L be the map from $G \times \mathfrak{g}$ to $T(\mathfrak{g})$ given by

$$L(g, X) = gX. \tag{6.3.5}$$

Choosing $t_o = 0$ and $x_o = e$, for each $Y \in V = \mathfrak{g}$, we need to integrate the equation

$$\gamma'(t) = \gamma(t)Y, \quad \gamma(0) = e, \tag{6.3.6}$$

so that the maximal integral is just the one-parameter subgroup $\gamma_Y(t) = \exp(tY)$. The general result then shows that the map $(t, Y) \mapsto \exp(tY)$ from $\mathbb{R} \times \mathfrak{g}$ to G is of class C^r. Thus this is also necessarily the case for the exponential map from $\mathfrak{g}$ to G, proving the first claim.

We next compute the image $Y = \exp'(0)X$ of an element $X \in \mathfrak{g}$ under the tangent linear map to exp at the origin. For this it suffices to choose a curve γ in $\mathfrak{g}$ such that $\gamma(0) = 0$ and $\gamma'(0) = X$, take its image $t \mapsto \exp(\gamma(t))$ under the exponential map, and find its tangent vector at $t = 0$: it will be the desired vector Y. The simplest curve γ is the line $\gamma(t) = tX$; then $\exp(\gamma(t))$ is just the tangent one-parameter subgroup γ_X to X at $t = 0$. Consequently, $Y = X$, proving the second claim:

$$\exp'(0) = \mathrm{id}. \qquad\qquad (6.3.7)$$

Corollary 1 *For any sufficient small open neighbourhood U of 0 in $\mathfrak{g}$, the exponential map induces a C^r isomorphism from U onto an open neighbourhood of e in G. If G is connected, then G is generated by the image of the exponential map.*

This is the theorem on maps with non-trivial Jacobian.

Exercise. Use the above corollary to recover Theorem 2 of Chap. 5, Sect. 5.14, and its corollary (the fact that in finite dimensions, the sum of a convergent series belongs to the subspace generated by its terms may be used).

Corollary 2 *For any point x of G sufficiently near e, there is a one-parameter subgroup of G passing through it.*

Indeed, there exists an X such that $x = \exp(X)$, whence $\gamma_X(1) = x$, and the one-parameter subgroup γ_X answers the question. (Note that condition $\exp(X) = x$ only determines X if, beforehand, X is required to be in a neighbourhood of 0 on which the exponential map is injective; in addition to "the" one-parameter subgroup "directly" connecting x to e, the existence of subgroups connecting e to x after crossing regions far away from G cannot obviously be excluded; see the one-parameter subgroups of $\mathbb{T}^n$).

Corollary 3 *For all n, there is an open neighbourhood U of e in G such that every $x \in U$ has a unique nth root in U.*

Take $U = \exp(S)$, where S is a *convex* open neighbourhood of 0 in $\mathfrak{g}$ where the exponential map is injective. To solve $x = y^n$ in U, choose $X \in S$ such that $x = \exp(X)$—which determines X—and it is then a matter of finding $Y \in S$ such that $\exp(Y)^n = \exp(X)$, i.e. such that $\exp(nY) = \exp(X)$. As S is convex, it contains X/n and this element of S answers the question, so that $x = y^n$ has at least one solution in U. The uniqueness of the solution remains to be shown when U is sufficiently small; but if $Y, Z \in S$ satisfy $\exp(Y)^n = \exp(Z)^n$, then $\exp(nY) = \exp(nZ)$ and so $Y = Z$ if nY and nZ are already known to be in S, in other words if Y and Z are in the neighbourhood $S' = S/n$. Replacing U by $U' = \exp(S')$ gives a neighbourhood answering the question (but obviously depending on n; besides if $G = \mathbb{T}$, for sufficiently large n, every neighbourhood of e contains non-trivial nth roots of unity ...).

Corollary 4 *Let $\mathfrak{a}_1, \ldots, \mathfrak{a}_n$ be vector subspaces of $\mathfrak{g}$ such that their direct sum equals $\mathfrak{g}$. Then for all i, there exist an open neighbourhood V_i of 0 in $\mathfrak{a}_i$ such that the map*

$$(X_1, \ldots, X_n) \mapsto \exp(X_1) \cdots \exp(X_n) \qquad\qquad (6.3.8)$$

induces an isomorphism from the manifold $V_1 \times \cdots \times V_n$ onto an open neighbourhood of e in G.

Let ϕ be the map considered, obviously of class C^r; it all amounts to showing that the tangent map to ϕ at $(0, \ldots, 0)$ is bijective; it is however evidently given by

$$(X_1, \ldots, X_n) \mapsto \sum \exp'(0)X_i = \sum X_i, \tag{6.3.9}$$

and as $\mathfrak{g}$ is the direct sum of the $\mathfrak{a}_i$, (6.3.9) is clearly a bijection from $\mathfrak{a}_1 \times \cdots \times \mathfrak{a}_n = T_{(0,\ldots,0)}(\mathfrak{a}_1 \times \cdots \times \mathfrak{a}_n)$ to $\mathfrak{g}$, qed.

If, for example, G contains two Lie subgroups M and N whose Lie algebras (identified with subalgebras of $\mathfrak{g}$) satisfy $\mathfrak{g} = \mathfrak{m} \oplus \mathfrak{n}$, then the above corollary can be applied; but in this case, the exponential maps from $\mathfrak{m}$ and $\mathfrak{n}$ to M and N are already known to induce isomorphisms from sufficiently small neighbourhoods of 0 in $\mathfrak{m}$ and $\mathfrak{n}$ to neighbourhoods of e in M and N (Corollary 1 above). Hence the following result follows:

Corollary 5 *Let M and N be two Lie subgroups of G and suppose that $\mathfrak{g}$ is the direct sum of the subalgebras $\mathfrak{m}$ and $\mathfrak{n}$ corresponding to M and N. Then there exist open neighbourhoods U, V and W of e in G, M and N such that the map $(m, n) \mapsto mn$ induces an isomorphism from the manifold $V \times W$ onto the manifold U. The product MN is open, and it is closed if any one of the subgroups is either compact or normal.*

The product MN is the orbit of e when the group MN is made to act on the topological space G by $(m, n)g = mgn^{-1}$; since an orbit having an interior point is always open, MN is always an open subset of G. Thanks to Lemma 3 from Chap. 1, Sect. 1.2, it is closed if M is compact. If N is normal, MN is an open *subgroup*, hence closed, of G, proving the corollary.

When G is connected, the corollary shows that $G = MN$ if M is compact or N is normal. The decomposition $g = mn$ is not in general unique, except if $M \cap N$ is trivial. In the general case, the above corollary shows that the intersection of $M \cap N$ and a sufficiently small neighbourhood of e is trivial, in other words that $M \cap N$ is discrete.

Exercise. Apply the corollary to the following cases: (1) $G = GL_n(\mathbb{R})$, $M = O(n)$, N being the subgroup of upper triangular matrices; (2) $G = GL_n(\mathbb{R})$, M is the subgroup of lower triangular matrices all of whose eigenvalues equal 1, and N the subgroup of case (1).

Corollary 6 *Every connected commutative Lie group is the product of a group $\mathbb{R}^p$ and a group $\mathbb{T}^q$.*

Suppose that G is commutative; obviously ad $(g)X = X$ for all $g \in G$ and $X \in \mathfrak{g}$, and since, as seen in Sect. 5.7 of Chap. 5, the derivative of the representation $g \mapsto$ ad (g) is the map $X \mapsto$ ad (X), ad $(X) = 0$ for all X, i.e.

$$[X, Y] = 0 \quad \text{for all } X, Y \in \mathfrak{g}; \tag{6.3.10}$$

(the Lie algebra $\mathfrak{g}$ is then said to be *commutative* or *abelian*). But then (6.2.1) shows that the exponential map is a *homomorphism* from the additive group $\mathfrak{g}$ to G, and in fact *onto* G, since $\exp(\mathfrak{g})$ is now necessarily an open subgroup as it contains a neighbourhood of e, and so, G being connected, $\exp(\mathfrak{g}) = G$. The kernel D of the exponential map is a discrete subgroup of $\mathfrak{g}$ since it is injective in the neighbourhood of 0, and so is generated by at most $\dim(\mathfrak{g})$ linearly independent vectors. The proof will be complete once G will have been shown to be isomorphic as a Lie group to $\mathfrak{g}/D$. First, the obvious bijection from $\mathfrak{g}/D$ to G is a homeomorphism, as can be seen by adapting to the present situation Theorem 1 of Chap. 1, Sect. 1.2, on homogeneous spaces of locally compact groups countable at infinity. Secondly, the map from $\mathfrak{g}$ onto $\mathfrak{g}/D$ being a submersion, and its composition with the previous map being C^r, the map in consideration from $\mathfrak{g}/D$ onto G is itself C^r ("universal property" of submersions, Theorem 4 of Chap. 4). As it is a group homomorphism, the rank of this map is constant, and as it is a homeomorphism, the standard form of subimmersions shows that it is locally, and globally, a manifold isomorphism, qed.

Corollary 7 *Let u and v be two homomorphisms from a connected Lie group G to a Lie group H, and suppose that $u'(e) = v'(e)$. Then $u = v$.*

Let $\mathfrak{g}$ and $\mathfrak{h}$ be the Lie algebras, and for $X \in \mathfrak{g}$, set $Y = u'(e)X = v'(e)X$. Then $u(\exp(X)) = \exp(Y) = v(\exp(X))$, and so $u = v$ on the subset $\exp(\mathfrak{g})$ of G. As u and v are homomorphisms, they are equal on the subgroup generated by $\exp(\mathfrak{g})$, i.e. on G by the above Corollary 1.

6.4 The Derivative of the Exponential Map

Let G be a Lie group, $\mathfrak{g}$ its Lie algebra and X an element of $\mathfrak{g}$; set $x = \exp(X)$. This gives a linear map

$$\exp'(X) : T_X(\mathfrak{g}) = \mathfrak{g} \to T_x(G) = x\mathfrak{g}. \tag{6.4.1}$$

Composition with the left translation by x^{-1} results in a map

$$\exp(X)^{-1} \circ \exp'(X) : Y \mapsto \exp(X)^{-1} \cdot \exp'(X)Y \quad \text{from } \mathfrak{g} \text{ to } \mathfrak{g}. \tag{6.4.2}$$

Theorem 4 *Let G be a Lie group and $\mathfrak{g}$ its Lie algebra. Then,*[2]

$$\boxed{\exp(X)^{-1} \circ \exp'(X)Y = \sum_1^\infty \operatorname{ad}(-X)^{n-1} Y/n!,} \tag{6.4.3}$$

for all $X, Y \in \mathfrak{g}$.

In other words,

$$\exp(X)^{-1} \circ \exp'(X) = \phi(\operatorname{ad}(X)) \quad \text{where}$$

---

[2] We let the reader recover formula (3.9.14) of Chap. 3 from this result.

$$\phi(z) = \sum (-z)^{n-1}/n! = \frac{1 - e^{-z}}{z}. \tag{6.4.4}$$

To prove (6.4.4), start from

$$\exp(sX + tX) = \exp(sX)\exp(tX) \tag{6.4.5}$$

and differentiate it with respect to X, considering s and t as constants. Setting $\exp(sX) = a$ and $\exp(tX) = b$, by the chain rule, the tangent linear map to the left-hand side in X is clearly equal to $(s + t)\exp'(sX + tX)$, which takes $\mathfrak{g}$ to $T_{ab}(G) = ab\mathfrak{g}$. The right-hand side is the composite of the map $X \mapsto (\exp(sX), \exp(tX))$ from $\mathfrak{g}$ to $G \times G$ and the map $(x, y) \mapsto xy$ from $G \times G$ to G. The tangent to the former at X is obviously the map $(s \cdot \exp'(sX), t \cdot \exp'(tX))$ from $\mathfrak{g}$ to $T_a(G) \times T_b(G)$, and the tangent map to the latter at (a, b) is the map $(h, k) \mapsto hb + ak$ from $T_a(G) \times T_b(G)$ to $T_{ab}(G)$, as seen [by definition of the composition law in $T(G)$] in Chap. 5, Sect. 5.6, (5.6.8). Hence, for all $Y \in \mathfrak{g}$,

$$(s + t)\exp'(sX + tX)Y = hb + ak = ab(\mathrm{ad}\,(b^{-1})a^{-1}h + b^{-1}k), \tag{6.4.6}$$

where

$$h = s \cdot \exp'(sX)Y, \quad k = t \cdot \exp'(tX)Y. \tag{6.4.7}$$

Setting

$$M_X(t) = \exp(tX)^{-1} \circ t \cdot \exp'(tX), \tag{6.4.8}$$

which defines a function with values in $\mathscr{L}(\mathfrak{g})$, it follows that, in (6.4.6), $a^{-1}h = M_X(s)Y$ and $b^{-1}k = M_X(t)Y$, so that (6.4.6) can now be written as

$$M_X(s + t)Y = \mathrm{ad}\,(\exp(tX)^{-1})M_X(s)Y + M_X(t)Y. \tag{6.4.9}$$

This can be seen by multiplying on the left both sides of (6.4.6) by the inverse of ab. Hence, finally,

$$M_X(s + t) = e^{-t \cdot \mathrm{ad}\,(X)} \circ M_X(s) + M_X(t), \tag{6.4.10}$$

where the sign $\circ$ is a strong reminder that these are products of endomorphisms of the vector space $\mathfrak{g}$.

Differentiating with respect to t and then putting $t = 0$, gives

$$M'_X(s) = -\mathrm{ad}\,(X) \circ M_X(s) + M'_X(0). \tag{6.4.11}$$

This linear differential equation with constant coefficients can be readily integrated; besides $M_X(0) = 0$, and on the other hand, differentiating (6.4.8) with respect to t at $t = 0$, we get

$$M'_X(0) = (\exp(tX)^{-1} \circ \exp'(tX) + t\{\cdots\})_{t=0} = 1, \tag{6.4.12}$$

since we already know that $\exp'(0) = 1$. This prompts us to integrate

$$M'_X(s) = -\text{ad}\,(X) \circ M_X(s) + 1 \quad \text{with} \quad M_X(0) = 0, \qquad (6.4.13)$$

and the only solution to the problem is then the power series

$$M_X(s) = s - \text{ad}\,(X)s^2/2! + \text{ad}\,(X)^2 s^3/3! - \cdots \qquad (6.4.14)$$

(Regardless of whether the unknown function takes numerical or vector values, the only solutions of a linear differential equation with constant coefficients are power series, and these can be formally computed by differentiating term-by-term.) The claimed formula (6.4.3) follows by setting $s = 1$.

The set of points on which $\exp'(X)$ is bijective is immediate from (6.4.3); it is a matter of expressing the invertibility of the operator (6.4.3), i.e. that its eigenvalues are non-trivial. But as ϕ is a power series convergent everywhere, for any matrix $u \in M_n(\mathbb{C})$, the eigenvalues of the operator $\phi(u)$ are clearly the numbers $\phi(z)$, where z varies over the set of eigenvalues of u (triangularize u and compute the diagonal entries of $\phi(u)$). As the zeros of the function ϕ are the numbers $2\pi i n$, where n is a *non-trivial* integer, *the map* $\exp'(X)$ *is bijective if and only if* ad (X) *does not have any eigenvalue such as* $2\pi i n$ *with* $n \neq 0$. Compare with the arguments of Chap. 3, Sect. 3.9.

6.5   The Campbell–Hausdorff Formula (Integral Form)

In what follows, in addition to the (everywhere convergent) series $\phi(z)$ of the preceding section, we will need the series

$$\psi(z) = z \sum (-1)^m (z-1)^m/(m+1) = \frac{z \cdot \log z}{z-1} \qquad (6.5.1)$$

converging in the disk $|z - 1| < 1$, which enables us to define $\psi(u)$ for all linear operators u satisfying the condition $\|u - 1\| < 1$. A trivial calculation, left for the reader to carry out in Chap. 3, Sect. 3.10, shows that

$$\phi(u)\psi(e^U) = 1 \quad \text{if} \quad \|e^U - 1\| < 1. \qquad (6.5.2)$$

Now, let us consider a Lie group G, its Lie algebra $\mathfrak{g}$ and a connected open neighbourhood U of 0 in $\mathfrak{g}$ such that exp induces an isomorphism from U onto an open neighbourhood U of e in G; log will denote the inverse isomorphism from U onto U, so that

$$\log(\exp(X)) = X \quad \text{for } X \in \mathsf{U}, \quad \exp(\log x) = x \quad \text{for } x \in U. \qquad (6.5.3)$$

The map $(X, Y) \mapsto \exp(X)\exp(Y)$ being continuous, there is a connected open neighbourhood $\mathsf{U}' \subset \mathsf{U}$ of 0 in $\mathfrak{g}$ such that $\exp(X)\exp(Y) \in U$ for $X, Y \in \mathsf{U}'$; set

$$\boxed{H(X, Y) = \log(\exp(X)\exp(Y))} \qquad (6.5.4)$$

for $X, Y \in \mathsf{U}'$. As log is an isomorphism from U onto an open subset, namely U, of a Cartesian space, the pair $(U, \log)$ is a local chart of the manifold G at e, and the function H is just *the expression in this chart of the composition law of* G. By the way, a chart such as $(U, \log)$ is frequently called a *canonical chart* of G.

To compute $H(X, Y)$, fix X and Y and set

$$H(t) = \log(\exp(tX)\exp(tY)) \quad \text{for } 0 \le t \le 1, \text{ and so } H(0) = 0. \qquad (6.5.5)$$

This expression is well defined since U' being convex and containing X and Y, it also contains tX and tY for $0 \le t \le 1$. We calculate the derivative of the function $H(t)$—obviously differentiable on $[0, 1]$ since it is clearly the restriction to $[0, 1]$ of a C^r function defined on an open interval containing 0 and 1. For this, start from

$$\exp(H(t)) = \exp(tX)\exp(tY) \qquad (6.5.6)$$

and apply the product differentiation formula; alternatively, consider the right-hand side as the composite of the maps $t \mapsto (\exp(tX), \exp(tY))$ and $(x, y) \mapsto xy$ and take the composite of the corresponding derivative maps. We thereby get

$$\exp'(H(t))H'(t) = \exp'(tX)X \cdot \exp(tY) + \exp(tX) \cdot \exp'(tY)Y. \qquad (6.5.7)$$

To make it easier to understand this equality, it may be useful to explain its meaning: H is a map from $[0, 1]$ to $\mathfrak{g}$, so that $H'(t)$ is an element of $\mathfrak{g}$, and the left-hand side of (6.5.7) is its image under the tangent map $\exp'(H(t)) : \mathfrak{g} \to T_{\exp(H(t))}(G)$ to exp at $H(t)$; on the right-hand side, $\exp'(tX))X$ is the image of $X \in \mathfrak{g}$ under the tangent map $\exp'(tX) : \mathfrak{g} \to T_{\exp(tX)}(G)$, and $\exp'(tX)X \cdot \exp(tY)$ is obtained by multiplying the result on the right and in the group $T(G)$ by $\exp(tY)$—in other words, it is the image of the vector $\exp'(tX)X$ under the tangent map to the right translation $g \mapsto g \cdot \exp(tY)$ at $\exp(tX)$. The second term on the right-hand side can be similarly interpreted.

Multiplying both sides of (6.5.7) on the left by $\exp(H(t))^{-1} = \exp(tY)^{-1}\exp(tX)^{-1}$ and taking into account the previous section, clearly gives

$$\phi(\operatorname{ad} H(t))H'(t) = \operatorname{ad}(\exp(tY))^{-1}\phi(\operatorname{ad}(tX))X + \phi(\operatorname{ad}(tY))Y. \qquad (6.5.8)$$

But in general,

$$\phi(\text{ad }(X))Y = Y - \text{ad }(X)Y/2!$$
$$+ \text{ad }(X)^2 Y/3! - \cdots = Y \quad \text{if } [X, Y] = 0 \tag{6.5.9}$$

since excepting the first one, all the terms of the series are obviously zero. Hence, instead of (6.5.8), all that remains is

$$\phi(\text{ad } H(t))H'(t) = e^{-\text{ad }(tY)}X + Y, \text{ and so } H'(0) = X + Y. \tag{6.5.10}$$

(6.5.2) may be applied to the operator $u = \text{ad }(H(t))$ provided $e^{\|u-1\|} < 1$. However,

$$e^{\text{ad } H(t)} = \text{ad }(\exp H(t)) = \text{ad }(\exp(tX)\exp(tY))$$
$$= \text{ad }(\exp(tX))\text{ad }(\exp(tY)) = e^{\text{ad }(tX)}e^{\text{ad }(tY)}, \tag{6.5.11}$$

and the condition $\|u - 1\| < 1$ will hold for $0 \le t \le 1$ if U' is sufficiently small so that (see calculation (3.10.5) of Chap. 3)

$$\|\text{ad }(X)\| < \frac{1}{2}\log 2 \quad \text{for all} \quad X \in U'. \tag{6.5.12}$$

Assuming henceforth that this condition holds, (6.5.2) takes us from (6.5.10) to

$$H'(t) = \psi(e^{\text{ad }(tX)}e^{\text{ad }(tY)})(e^{-\text{ad }(tY)}X + Y), \tag{6.5.13}$$

and as $H(0) = \log e = 0$ we finally get the desired formula needed to compute $H(X, Y) = H(1)$, namely

$$H(X, Y) = \int_0^1 \psi(e^{\text{ad }(tX)}e^{\text{ad }(tY)})(e^{-\text{ad }(tY)}X + Y)dt. \tag{6.5.14}$$

6.6 The Campbell–Hausdorff Formula (Series Expansion of Homogeneous Polynomials)

A series expansion follows from (6.5.14). It enables us to show that, even if the group G is only C^∞, then the function $H(X, Y)$ is *analytic* in the neighbourhood of 0. It can be obtained by expanding (6.5.13) into a power series in t and integrating it term-by-term.

As $e^{\text{ad }(tX)}e^{\text{ad }(tY)}$ is a factor in the expression of $\psi(e^{\text{ad }(tX)}e^{\text{ad }(tY)})$, for reasons already met before, it is first of all necessary to observe that

$$e^{\text{ad}\,(tX)}e^{\text{ad}\,(tY)}(e^{-\text{ad}\,(tY)}X + Y) = e^{\text{ad}\,(tX)}X + e^{\text{ad}\,(tX)}e^{\text{ad}\,(tY)}Y$$
$$= X + e^{\text{ad}\,(tX)}Y \tag{6.6.1}$$
$$= X + \sum t^n \text{ad}\,(X)^n Y / n!$$

Hence,

$$H'(t) = \left(\sum \frac{(-1)^m}{m+1}(e^{\text{ad}\,(tX)}e^{\text{ad}\,(tY)} - 1)^m\right) \tag{6.6.2}$$
$$\cdot \left(X + \sum t^r \text{ad}\,(X)^r Y / r!\right).$$

For the moment keeping to a formal computation, and observing that

$$(e^U e^V - 1)^m = \sum_{p_1+q_1>0} \frac{u^{p_1} v^{q_1}}{p_1! q_1!} \cdots \sum_{p_m+q_m>0} \frac{u^{p_m} v^{q_m}}{p_m! q_m!}$$
$$= \sum_{p_i+q_i>0} \frac{u^{p_1} v^{q_1} \cdots u^{p_m} v^{q_m}}{p_1! q_1! \cdots p_m! q_m!} \tag{6.6.3}$$

even and especially when the variables u and v are non-commutative operators, this clearly leads to

$$H'(t) =$$
$$\sum_{\substack{m=0 \\ p_i+q_i>0}} \frac{(-1)^m}{m+1} t^{p_1+q_1+\cdots+p_m+q_m} \frac{\text{ad}\,(X)^{p_1}\text{ad}\,(Y)^{q_1} \cdots \text{ad}\,(X)^{p_m}\text{ad}\,(Y)^{q_m} X}{p_1! q_1! \cdots p_m! q_m!}$$
$$+ \sum_{\substack{m,n\geq0 \\ p_i+q_i>0}} \frac{(-1)^m}{m+1} t^{p_1+\cdots+q_{m+n}} \frac{\text{ad}\,(X)^{p_1} \cdots \text{ad}\,(Y)^{q_m}\text{ad}\,(X)^n Y}{p_1! q_1! \cdots p_m! q_m!}. \tag{6.6.4}$$

These formal manipulations using the associativity formula for multiple series are well defined as long as the multiple series obtained is absolutely convergent. As the associativity theorem holds for all series whose terms are positive (regardless of their convergence), to check the absolute convergence of (6.6.4), the terms may be replaced by their norms, and then regrouped ingeniously—for example those that take us from (6.6.4) to (6.6.3) then to (6.6.2)—and the positive term series obtained verified to be convergent. But this is the series which follows from (6.6.2) by removing the factors $(-1)^m$ and by replacing X, Y, $\text{ad}\,(X)$ and $\text{ad}\,(Y)$ by their norms (obviously the norm chosen on $\mathfrak{g}$ is only of secondary importance for these estimations). In other words, it is all a matter of checking that

$$\sum (e^{t\|\text{ad}\,(X)\|}e^{t\|\text{ad}\,(Y)\|} - 1)^m/(m+1) < \infty \tag{6.6.5}$$

i.e. that

$$e^{t(\|\text{ad}(X)\|+\|\text{ad}(Y)\|)} < 2, \qquad (6.6.6)$$

and as $0 \leq t \leq 1$ the calculations are well defined for $X, Y \in U'$ since

$$\|\text{ad}(X)\| + \|\text{ad}(Y)\| < \log 2. \qquad (6.6.7)$$

Then (6.6.4) clearly converges normally in $[0, 1]$ with respect to t, and so can be integrated term-by-term, and substituting p_{m+1} for n in (6.6.4) now gives

$$H(X, Y) = \sum_{r+s>0} (H'_{rs}(X, Y) + H''_{rs}(X, Y)), \qquad (6.6.8)$$

where

$$H'_{rs}(X, Y) := \frac{1}{n+s} \sum_{\substack{m \geq 0 \\ p_i+q_i>0 \\ p_1+\cdots+p_m=n-1 \\ q_1+\cdots+q_m=s}} \frac{(-1)^m}{m+1} \frac{\text{ad}(X)^{p_1} \cdots \text{ad}(X)^{p_m} \text{ad}(Y)^{q_m} X}{p_1! \cdots p_m! q_m!}$$

$$\qquad (6.6.9')$$

and

$$H''_{rs}(X, Y) := \frac{1}{r+s} \sum_{\substack{m \geq 0 \\ p_i+q_i>0 \\ p_{m+1} \geq 0 \\ p_1+\cdots+p_{m+1}=r \\ q_1+\cdots+q_m=s-1}} \frac{(-1)^m}{m+1} \frac{\text{ad}(X)^{p_1} \cdots \text{ad}(Y)^{q_m} \text{ad}(X)^{p_{m+1}} Y}{p_1! \cdots q_m! p_{m+1}!}.$$

$$\qquad (6.6.9'')$$

These terms can be obtained by grouping together the terms of degree (r, s) in (X, Y) in both of the double series on the right-hand side of (6.6.4). Indeed, for example $\text{ad}(X)^{p_1} \cdots \text{ad}(Y)^{q_m} X$ is clearly a homogeneous polynomial of degree $p_1 + \cdots + p_m + 1$ in X and of degree $q_1 + \cdots + q_m$ in Y. Formulas (6.6.8), (6.6.9') and (6.6.9'') are known as the *Campbell–Hausdorff formula*. Since the summation conditions in (6.6.9') and (6.6.9'') imply $p_1 + q_1 + \cdots + p_m + q_m = r + s - 1$ and hence $m \leq r + s - 1$, these expressions are in fact finite sums, and (6.6.8) gives a *series expansion of $H(X, Y)$ whose terms are homogeneous polynomials in (X, Y)*, the total degree of the general term in (X, Y) of the series (6.6.8) being $r + s > 0$.

Exercise. Set

$$\boxed{[X_1 \cdots X_n] = \frac{1}{n} \text{ad}(X_1) \cdots \text{ad}(X_{n-1}) X_n}$$

$$\qquad (6.6.10')$$

$$= \frac{1}{n} [X1, [X2, \ldots, [X_{n-1}, Xn] \cdots]]$$

and

$$[X_1^{r_1} \cdots X_n^{r_n}] = [X_1 \cdots X_1 \cdots X_n \cdots X_n]$$

$$= \frac{1}{r_1 + \cdots + r_n} \text{ad}(X_1)^{r_1} \cdots \text{ad}(X_{n-1})^{r_n-1} X_n$$

$$\qquad (6.6.10'')$$

with r_1 factors $X_1, \ldots, r_n$ factors X_n (so that this expression is in fact trivial if $r_n > 1$). Show that in this notation, the Campbell–Hausdorff formula becomes

$$H(X, Y) = \sum_{m=1} \frac{(-1)^m}{m+1} \sum_{p_i+q_i>0} \frac{[X^{p_1} Y^{q_1} \cdots X^{p_m} Y^{q_m}]}{p_1! q_1! \cdots p_m! q_m!}. \tag{6.6.11}$$

Although some authors make a big deal of it, the Campbell–Hausdorff formula is not all that useful for the Advanced theory of Lie groups; it can be experimentally checked that thousands of pages may be read or written about Lie groups, for instance on their (finite or infinite-dimensional) representations, almost without any mention of it. The usefulness of the result is mostly of a pedagogical nature, and comes from the fact that it provides simple proofs for theorems that will be found later in this chapter. Besides, authors who address the subject refrain from manipulating the formula; some might go as far as giving the first few terms:

$$H(X, Y) = X + Y + \frac{1}{2}[X, Y] + \frac{1}{12}[X, [X, Y]]$$
$$+ \frac{1}{12}[Y, [Y, X]] - \frac{1}{24}[X, [Y, [X, Y]]] + \cdots, \tag{6.6.12}$$

but no one presents the computations leading to (6.6.12). It seems that, at the beginning of the century, some charitable character might have done the entire computation, and his anonymous contribution is remembered as rather convenient "remarks" or "exercises" for others.

Their discretion on this point will be easily understood when attempting to go from (6.6.8) to (6.6.12). The aim is to compute the terms of total degree ≤ 4 in

$$H(X, Y) = \left\{ X - \frac{1}{2} \sum_{p+q>0} \frac{u^p v^q X}{(p+q+1)p! q!} \right.$$
$$+ \frac{1}{3} \sum_{\substack{p_1+q_1>0 \\ p_2+q_2>0}} \frac{u^{p_1} v^{q_1} u^{p_2} v^{q_2} X}{(p_1 + \cdots + q_2 + 1) p_1! \cdots q_2!}$$
$$\left. - \frac{1}{4} \sum_{p_i+q_i>0} \frac{u^{q_1} v^{q_2} v^{p_2} v^{q_2} u^{p_3} v^{q_3} X}{(p_1 + \cdots + q_3 + 1) p_1! \cdots q_3!} + \cdots \right\}$$
$$+ \left\{ \sum_{0}^{\infty} \frac{u^r Y}{(r+1)!} - \frac{1}{2} \sum_{p+q>0} \frac{u^p v^q u^r Y}{(p+q+r+1) p! q! r!} \right.$$
$$+ \frac{1}{3} \sum_{p_i+q_i>0} \frac{u^{p_1} v^{q_1} u^{p_2} v^{q_2} u^r Y}{(p_1 + \cdots + r + 1) p_1! \cdots r!}$$
$$\left. - \frac{1}{4} \sum_{p_i+q_i>0} \frac{u^{p_1} v^{q_1} u^{p_2} v^{q_2} u^{p_3} v^{q_3} u^r Y}{(p_1 + \cdots + r + 1) p_1! \cdots r!} + \cdots \right\}, \tag{6.6.13}$$

where $u := \operatorname{ad}(X)$ and $v := \operatorname{ad}(Y)$. Attempting to systematically write down all the terms of total degree ≤ 4, this is what the author obtains:

$$
\begin{aligned}
\Big\{ X &- \frac{1}{2}\Big(\cancel{\frac{u^3X}{4.3!}} + \frac{u^2vX}{4.2!} + \cancel{\frac{u^2X}{3}} + \frac{uv^2X}{4.2!} + \frac{uvX}{3} + \cancel{\frac{uX}{2}} + \frac{v^3X}{4.3!} \\
&+ \frac{v^2X}{3.2!} + \frac{vX}{2}\Big) + \frac{1}{3}\Big(\cancel{\frac{u^2uX}{4.2!}} + \frac{u^2vX}{4.2!} + \cancel{\frac{uvuX}{4}} + \frac{uvvX}{4} \\
&+ \cancel{\frac{u.u^2X}{4.2!}} + \frac{u.uvX}{4} + \frac{u.v^2X}{4.2!} + \cancel{\frac{u.uX}{3}} + \frac{u.vX}{3} + \cancel{\frac{v^2uX}{4.2!}} \\
&+ \frac{v^2vX}{4.2!} + \cancel{\frac{v.u^2X}{4.2!}} + \frac{vuvX}{4} + \frac{v.v^2X}{4.2!} + \cancel{\frac{vuX}{3}} + \frac{vvX}{3}\Big) \\
&- \frac{1}{4}\Big(\cancel{\frac{uuuX}{4}} + \frac{uuvX}{4} + \cancel{\frac{uvuX}{4}} + \frac{uvvX}{4} + \cancel{\frac{vuuX}{4}} \\
&+ \frac{vuvX}{4} + \cancel{\frac{vvuX}{4}} + \frac{vvvX}{4}\Big)\Big\} + \Big\{ Y + \frac{uY}{2!} + \frac{u^2Y}{3!} + \frac{u^3Y}{4!} \\
&- \frac{1}{2}\Big(\frac{u^3Y}{4.3!} + \cancel{\frac{u^2vY}{4.2!}} + \frac{u^2Y}{3.2!} + \frac{u^2uY}{4.2!} + \cancel{\frac{uv^2Y}{4.2!}} + \frac{uvuY}{4} \\
&+ \cancel{\frac{uvY}{3}} + \frac{uu^2Y}{4.2!} + \frac{u.uY}{3} + \frac{uY}{2} + \cancel{\frac{v^3Y}{4.3!}} + \frac{v^2uY}{4.2!} + \cancel{\frac{v^2Y}{3.2!}} \\
&+ \frac{vu^2Y}{4.2!} + \frac{vuY}{3} + \cancel{\frac{vY}{2}}\Big) + \frac{1}{3}\Big(\frac{u^2uY}{4.2!} + \cancel{\frac{u^2vY}{4.2!}} + \frac{uvuY}{4} \\
&+ \cancel{\frac{uvvY}{4}} + \frac{uu^2Y}{4.2!} + \cancel{\frac{u.uvY}{4}} + \cancel{\frac{uv^2Y}{4.2!}} + \frac{uuuY}{4} + \frac{uuY}{3} \\
&+ \frac{uvuY}{4} + \cancel{\frac{uvY}{3}} + \frac{v^2uY}{4.2!} + \cancel{\frac{v^2vY}{4.2!}} + \frac{vu^2Y}{4.2!} + \cancel{\frac{vuvY}{4}} \\
&+ \cancel{\frac{vv^2Y}{4}} + \frac{vuuY}{4} + \frac{vuY}{3} + \frac{v.vuY}{4} + \cancel{\frac{v.vY}{3}}\Big) \\
&- \frac{1}{4}\Big(\frac{uuuY}{4} + \cancel{\frac{uuvY}{4}} + \frac{uvuY}{4} + \cancel{\frac{uvvY}{4}} \\
&+ \frac{vuuY}{4} + \cancel{\frac{vuvY}{4}} + \frac{vvuY}{4} + \cancel{\frac{vvvY}{4}}\Big)\Big\}.
\end{aligned}
$$

Since obviously $uX = vY = 0$, some of the terms are clearly trivial, and they have been crossed out. Then grouping together similar terms gives

$$X + Y - \frac{vX}{4} + uY\left(\frac{1}{2} - \frac{1}{4}\right) + v^2X\left(-\frac{1}{12} + \frac{1}{9}\right)$$

$$+ uvX\left(-\frac{1}{6} + \frac{1}{9}\right) + vuY\left(-\frac{1}{6} + \frac{1}{9}\right)$$

$$+ u^2Y\left(\frac{1}{6} - \frac{1}{12} - \frac{1}{6} + \frac{1}{9}\right) + v^3X\left(-\frac{1}{48} + \frac{1}{24} + \frac{1}{24} - \frac{1}{16}\right)$$

$$+ vuvX\left(\frac{1}{12} - \frac{1}{16}\right) + uv^2X\left(-\frac{1}{16} + \frac{1}{12} + \frac{1}{24} - \frac{1}{16}\right)$$

$$+ u^2vX\left(-\frac{1}{16} + \frac{1}{24} + \frac{1}{12} - \frac{1}{16}\right) + v^2uY\left(-\frac{1}{16} + \frac{1}{24} + \frac{1}{12} - \frac{1}{16}\right)$$

$$+ vu^2Y\left(-\frac{1}{16} + \frac{1}{24} + \frac{1}{12} - \frac{1}{16}\right) + uvuY\left(-\frac{1}{8} + \frac{1}{12} + \frac{1}{12} - \frac{1}{16}\right)$$

$$+ u^3Y\left(\frac{1}{24} - \frac{1}{48} - \frac{1}{16} - \frac{1}{16} + \frac{1}{24} + \frac{1}{24} + \frac{1}{12} - \frac{1}{16}\right).$$

Taking into account the equality $uY + vX = 0$, which states that $[X, Y] + [Y, X] = 0$, we first get

$$X + Y + \frac{1}{2}[X, Y] + \frac{1}{12}[X, [X, Y]] + \frac{1}{12}[Y, [Y, X]] + \frac{vuvX}{48} - \frac{uvuY}{48};$$

but equality $uY + vX = 0$ shows that $vuvX + uvuY = (vu - uv)vX = 0$ given that $vX = [Y, X]$ and

$$vu - uv = \text{ad}(Y)\text{ad}(X) - \text{ad}(X)\text{ad}(Y) = \text{ad}([Y, X]) \qquad (6.6.14)$$

because of the Jacobi identity. Formula (6.6.12) follows immediately.

Exercise. Imagine the text of an *Exercise* that would find its natural place at this point in the book, and provide the solution.

It might prove to be a solace to observe that it would have been possible to economize fifty percent of the work required for the previous calculation if we had started from another version of the Hausdorff formula, a version which the unnamed authors do not mention and which we have already come across in Chap. 3:

Exercise. Instead of the function $H(t)$ defined in Sect. 6.5, consider the function

$$F(t) = \log(\exp(X)\exp(tY)). \qquad (6.6.15)$$

Using the arguments of Sect. 6.5 show that $\phi(\text{ad}(F(t))F'(t) = Y$, and deduce that

$$H(X, Y) = X + \int_0^1 \psi(e^{\text{ad}(X)}e^{\text{ad}(tY)})Y \cdot dt. \qquad (6.6.16)$$

Deduce from this that

$$H(X, Y) = X + \sum_{\substack{m \geq 0 \\ p_i+q_i>0 \\ p_{m+1} \geq 0}} \frac{(-1)^m}{m+1}$$

$$\frac{\operatorname{ad}(X)^{p_1} \operatorname{ad}(Y)^{q_1} \cdots \operatorname{ad}(X)^{p_m} \operatorname{ad}(Y)^{q_m} \operatorname{ad}(X)^{p_{m+1}} Y}{(q_1 + \cdots + q_m + 1) p_1! q_1! \cdots p_{m+1}!}$$

(6.6.17)

and use it to check (6.6.12).

6.7 The Campbell–Hausdorff Formula. (Analyticity of the Function H)

Although the Campbell–Hausdorff formula was obtained from the Lie algebra *of a Lie group*, it can clearly be written down, at least formally, for *every* Lie algebra $\mathfrak{g}$, regardless of whether its dimension over a field k of characteristic 0 is finite or not. The expression $H(X, Y)$ that is then obtained must be regarded as a formal series over $\mathfrak{g} \times \mathfrak{g}$. The study of the algebraic properties of the expression thus obtained would carry us too far (see N. Bourbaki, *Lie Groups and Lie Algebras*, Chap. 2) and here we will confine ourselves to the case of a finite-dimensional Lie algebra over $\mathbb{R}$, the purpose of this section only being to show that, for X and Y sufficiently near zero, the series $H(X, Y)$ converges and represents an analytic function of the pair (X, Y).

Before stating the precise result we have in mind, a remark is in order: considering $\mathfrak{g}$ as a finite-dimensional vector space it is always possible to choose a norm $\|X\|$ on it satisfying the following property:

$$\|[X, Y]\| \leq \|X\| \cdot \|Y\|$$

(6.7.1)

for all X and Y. Indeed, in finite dimensions, every bilinear map is continuous, and so every norm on $\mathfrak{g}$ satisfies an inequality of the form

$$\|[X, Y]\| \leq M \|X\| \cdot \|Y\|$$

(6.7.2)

with a constant $M > 0$; it is then sufficient to multiply the chosen norm by M to get a norm satisfying condition (6.7.1).

Theorem 5 *Let $\mathfrak{g}$ be a finite-dimensional Lie algebra over $\mathbb{R}$ and endowed with a norm satisfying condition (6.7.1). Then the Hausdorff series*

$$H(X, Y) = \sum \frac{(-1)^m}{m+1} \sum_{p_i+q_i>0} \frac{[X^{p_1} Y^{q_1} \cdots X^{p_m} Y^{q_m}]}{p_1! q_1! \cdots p_m! q_m!}$$

(6.7.3)

is absolutely convergent and represents an analytic function of the pair (X, Y) on the open subset

$$\|X\| + \|Y\| < \log 2. \tag{6.7.4}$$

Convergence is immediate. Indeed, inequality (6.7.1) shows that, for all X, $\|\text{ad}(X)\| \leq \|X\|$. This implies

$$\|[X_1 \cdots X_n]\| \leq \frac{1}{n}\|X_1\| \cdots \|X_n\| \tag{6.7.5}$$

and so

$$\|[X^{p_1} Y^{q_1} \cdots X^{p_m} Y^{q_m}]\| \leq \frac{\|X\|^{p_1 + \cdots p_m}\|Y\|^{q_1 + \cdots q_m}}{p_1 + q_1 + \cdots + p_m + q_m} \tag{6.7.6}$$
$$\leq \|X\|^{p_1 + \cdots p_m}\|Y\|^{q_1 + \cdots q_m}.$$

Series (6.7.3) is therefore dominated by the series

$$\sum \frac{1}{m+1} \sum_{p_i + q_i > 0} \frac{\|X\|^{p_1}\|Y\|^{q_1} \cdots \|X\|^{p_m}\|Y\|^{q_m}}{p_1! q_1! \cdots p_m! q_m!}$$
$$= \sum \frac{1}{m+1}(e^{\|X\|}e^{\|Y\|} - 1)^m = \sum \frac{1}{m+1}(e^{\|X\| + \|Y\|} - 1)^m, \tag{6.7.7}$$

implying absolute convergence in the open subset (6.7.4).

To prove that the function is analytic on this open subset, we show that fixing an arbitrary basis (E_i) of $\mathfrak{g}$ and setting

$$X = \sum x_i E_i, \quad Y = \sum y_i E_i, \tag{6.7.8}$$

the series $H(X, Y)$ is represented by a convergent power series in the $2n$ variables x_i and y_i in the open subset defined by condition

$$\sum \|E_i\|(|x_i| + |y_i|) < \log 2. \tag{6.7.9}$$

The left-hand side being greater than $\|X\| + \|Y\|$, the open subset (6.7.9) is contained in the open subset (6.7.4) and need not be equal to it. But the union of the open subsets (6.7.9) associated to the various bases of $\mathfrak{g}$ is just the whole of the open subset (6.7.4). For if X and Y satisfy (6.7.4) there is obviously a basis (E_i) such that X and Y are either proportional to E_i, or else proportional to E_1 and E_2 respectively, so that X and Y trivially satisfy condition (6.7.9) in both cases. This point having been settled, the function $H(X, Y)$, being analytic on open subsets whose union is the whole of (6.7.4), is analytic on (6.7.4).

To expand $H(X, Y)$ into a power series in x_i and y_i, each term of the series $H(X, Y)$ has to be expanded. But as $[X_1 \cdots X_n]$ is obviously a multilinear function in X_k, the expansion of the general term of $H(X, Y)$ is obtained by writing

$$[X^{p_1} Y^{q_1} \cdots X^{p_m} Y^{q_m}] = [X \cdots X Y \cdots Y \cdots X \cdots X Y \cdots Y]$$

$$= \sum [E_{i_1^1} \cdots E_{i_{p_1}^1} E_{j_1^1} \cdots E_{j_{q_1}^1} \cdots E_{i_1^m} \cdots E_{i_{p_m}^m} E_{j_1^m} \cdots E_{j_{q_m}^m}] x_{i_1^1} \cdots y_{j_{q_m}^m} . \tag{6.7.10}$$

Then the expression obtained has to be divided by $(-1)^m (m+1) p_1! \cdots q_m!$, summed with respect to $p_1, \ldots, q_m$ and m, similar monomials being grouped together so as to put the result obtained in the form of a power series $\sum a_{\alpha\beta} x^\alpha y^\beta$ in x_i and y_i, and the latter has to be checked to converge absolutely in the open set (6.7.9). But the series obtained is obviously dominated, in absolute values, by the series that would be obtained by replacing each term of the right-hand side of (6.7.10) by its norm and by performing the same operations as above on the new series. However, by (6.7.5),

$$(p_1 + \cdots + q_m) \sum \| E_{i_1^1} \cdots E_{j_{q_m}^m} \| \cdot |x_{i_1^1} \cdots y_{j_{q_m}^m}|$$

$$\le \sum \| E_{i_1^1} \| \cdots \| E_{j_{q_m}^m} \| \cdot |x_{i_1^1}| \cdots |y_{j_{q_m}^m}|, \tag{6.7.11}$$

and defining the new norm on $\mathfrak{g}$ as

$$|X| = \sum \| E_i \| \cdot |x_i| \quad \text{for} \quad X = \sum E_i x_i, \tag{6.7.12}$$

the result obtained in (6.7.11) is clearly just

$$|X| \cdots |X| \cdot |Y| \cdots |Y| \cdots |X| \cdots |X| \cdot |Y| \cdots |Y|$$

$$= |X|^{p_1} |Y|^{q_1} \cdots |X|^{p_m} |Y|^{q_m}. \tag{6.7.13}$$

The convergence of the power series $\sum a_{\alpha\beta} x^\alpha y^\beta$ is therefore implied by that of the series

$$\frac{1}{m+1} \sum_{p_i + q_i > 0} \frac{|X|^{p_1} |Y|^{q_1} \cdots |X|^{p_m} |Y|^{q_m}}{(p_1 + \cdots + q_m) p_1! q_1! \cdots p_m! q_m!}, \tag{6.7.14}$$

and hence even more so by the convergence of the series following from (6.7.14) by removing the factors $p_1 + \cdots + q_m$ from the denominators; computation (6.7.7) then immediately leads to condition $|X| + |Y| < \log 2$, i.e. to (6.7.9), completing the proof.

Note that, were $\mathfrak{g}$ assumed to be the Lie algebra of a Lie group G (which by the way is not assumed...), the function $H(X, Y)$ expressing the composition law of G in the neighbourhood of the identity element would have to satisfy the associativity relation

$$H(X, H(Y, Z)) = H(H(X, Y), Z) \tag{6.7.15}$$

when $X, Y, Z \in \mathfrak{g}$ are sufficiently near zero. In fact, this relation can be proved applying purely algebraic arguments. It then enables us to construct a partially defined associative composition law with identity element and inverses in the neighbourhood of 0 in $\mathfrak{g}$. The outcome is an "analytic Lie group kernel" structure in the neighbourhood of 0 which could then be used to show that every finite-dimensional real Lie algebra is the algebra of a Lie group. This would however take us too far, and moreover is of limited interest since there is a much simpler method to establish the existence of a Lie group with a given Lie algebra by applying Ado's theorem which confirms that every (finite-dimensional real) Lie algebra can be realized as the subalgebra of some algebra $M_n(\mathbb{R})$. Here starting from the Campbell–Hausdorff formula for Lie algebras and groups, we will content ourselves to deducing properties of the groups in question.

6.8 Analyticity of Lie Groups

Its first consequence is the *possibility of endowing every C^∞ Lie group with an analytic Lie group structure compatible with its C^∞ structure* (i.e. such that locally, the C^∞ functions are those which, in an analytic chart, are expressed by C^∞ formulas in terms of the coordinates). For this, we first fix an open neighbourhood U of 0 in $\mathfrak{g}$ such that the exponential map induces a C^∞ isomorphism from U to an open neighbourhood U of e in G, and then we fix an open neighbourhood V $= -$V of 0 in $\mathfrak{g}$ such that

   (i)   the function $H(X, Y)$ is defined and analytic on V $\times$ V;
   (ii) $H(X, Y) \in$ U for all $X, Y \in$ V;
   (iii) $\exp(X)\exp(Y) = \exp(H(X, Y))$ for all $X, Y \in$ V.

To satisfy these conditions, it suffices to take U and V sufficiently small. Set $V = \exp($V$)$, so that V is an open neighbourhood of e in G satisfying $V = V^{-1}, V^2 \subset U$ by (iii), and endow U with the analytic structure that can be deduced from that of U by applying the exponential map. The following conditions then hold:

   (a) V is open in U and $V = V^{-1}, V^2 \subset U$;
   (b) the map $(x, y) \mapsto xy^{-1}$ from $V \times V$ to U is analytic;
   (c) for all $a \in G$, there is a neighbourhood $V' \subset V$ of e such that $aV'a^{-1} \subset U$ and such that the map $x \mapsto axa^{-1}$ from V' to U is analytic.

Property (a) is obvious since V is open in U, property (b) follows from conditions (i), (ii) and (iii) above; to prove (c), choose an open neighbourhood V$' \subset$ V of 0 in $\mathfrak{g}$ such that ad (a)V$' \subset$ U and observe that

$$a \cdot \exp(X) \cdot a^{-1} = \exp(\mathrm{ad}\,(a)X), \qquad (6.8.1)$$

so that by structure transfer, (c) holds if and only if the map ad (a) from V$'$ to U is analytic, which clearly is the case.

So the following general result remains to be established:

Lemma *Let U and V be two subsets of a group G containing e; assume U is endowed with an analytic structure satisfying the above conditions (a), (b), (c). Then*

there is a unique analytic group structure on G with respect to which V is open and
which induces on V the analytic structure induced by that given on U.

Once this lemma has been established, it will become clear that the analytic structure that ensues on the group G at hand is compatible with its C^∞ structure because
the compatibility will be obvious in V, and hence, applying arbitrary translations to
V, also in the neighbourhood of any point of G.

To prove the previous lemma, which we have taken from N. Bourbaki, *Lie Groups*
and Lie Algebras, chap. III, (the most unreadable presentation of the theory of Lie
groups ever published since Sophus Lie, but fortunately the chapters on semisimple
Lie groups and algebras make up for this), fix a neighbourhood W of e in U such
that $W = W^{-1}$, $W^3 \subset V$ and for all $a \in G$, using the translation $x \mapsto ax$ transfer
the analytic structure of W to the sets aW. We show that the analytic structures thus
obtained on the sets aW are mutually compatible, and define an analytic structure
on G answering the question.

1—We begin by showing that $aW \cap bW$ *is open in* aW *and* bW *for all* $a, b \in G$.
Indeed, $aW \cap bW = a(W \cap cW)$, where $c = a^{-1}b$, and so it is a matter of showing
that $W \cap cW$ is open in W for all c. This is obvious if $W \cap cW = \phi$. Otherwise,
$c \in W \cdot W^{-1} \subset V$ and, by (b), $x \mapsto c^{-1}x$ is an analytic, hence continuous, map
from V to U, and therefore also from W to U. As $W \cap cW$ is the inverse image of
the open subset W of U under this continuous map from W to U, the statement to be
proved is established. The same argument implies that cW *is open in* V *if* $cW \# W$.

2—We next show that *the analytic structures of* aW *and* bW *coincide in* $aW \cap$
bW. This amounts to showing that, as above, the analytic structures of W and cW
coincide in $W \cap cW$, and for this it is sufficient to prove that transferring the analytic
structure of W to cW by applying the map $x \mapsto cx$ provides cW with the analytic
structure induced by that of U. In other words, it amounts to showing that, if the open
subsets W and cW of U are endowed with the analytic structures induced by that of
U, then the map $x \mapsto cx$ is an isomorphism from the manifold W onto the manifold
cW. As c and W are contained in V, assumption (b) shows that $x \mapsto cx$ is an analytic
map from the open subset V (endowed with the induced analytic structure) to U.
Since it maps the open subset W to the open subset cW, it induces an analytic map
from the former to the latter. The inverse map from cW to W is $x \mapsto c^{-1}x$; but as
$c^{-1} \in V^{-1} = V$, the same argument shows that $x \mapsto c^{-1}x$ is analytic as a map from
V to U, and so perforce as a map from the open subset cW of V to the open subset
W of U.

We are now in a position to endow G with the analytic structure obtained by
gluing those of the subsets aW; each aW is then an open submanifold of G.

3—Let us show that V *is open in* G. It suffices to show that $V \cap aW$ is open
in aW for all $a \in V$, since V is the union of the sets $V \cap aW$ in question. It also
suffices to show that $a^{-1}V \cap W$ is open in W (with respect to the topology on U) for
all $a \in V$. However, since $a \in V$, the map $x \mapsto ax$ from V to U is analytic, hence
continuous, and as $a^{-1}V \cap W$ is the inverse image of the open subset W under this
map, the desired result follows.

4—We next show that *the analytic structures of U and G coincide on V*. It suffices to show that they coincide on the common open subset $V \cap aW$ for all $a \in V$. This amounts to showing that if $x \in W$ and $y \in V$ are connected by $y = ax$, then y is an analytic function of x and conversely (with respect to the given manifold structure on U), which is obvious since the maps $x \mapsto ax$ and $y \mapsto a^{-1}y$ from V to U are analytic when $a \in V = V^{-1}$.

5—The map $(x, y) \mapsto xy^{-1}$ from $V \times V$ to U being analytic, this must also be the case for the map $(x, y) \mapsto xy^{-1}$ from $W \times W$ to V. This and point 4 imply that *the map $(x, y) \mapsto xy^{-1}$ from $G \times G$ to G is analytic in the neighbourhood of the origin*.

6—Besides it is obvious by construction that *the left translations $x \mapsto ax$ are analytic maps from G to G*.

7—Let us now show that *the interior automorphisms $x \mapsto axa^{-1}$ are analytic maps from G to G*. As the analytic structure of G is the same on V and on U, assumption (c) gives the analyticity of $x \mapsto axa^{-1}$ in the neighbourhood of e. To prove it in the neighbourhood of an arbitrary point $b \in G$, write

$$axa^{-1} = ab \cdot b^{-1}x \cdot a^{-1} = aba^{-1} \cdot aya^{-1} \quad \text{where } y = b^{-1}x. \qquad (6.8.2)$$

So it is necessary to check that the map $x \mapsto b^{-1}x$ is analytic in the neighbourhood of b, which we already know to be the case, that $y \mapsto aya^{-1}$ is analytic in the neighbourhood of e, which we previous checked, and finally that the map $z \mapsto cz$ is analytic for all c, which point 6 of the proof consisted in.

8—To complete the proof, it remains to check that *the map*

$$(x, y) \mapsto xy^{-1} \qquad (6.8.3)$$

from $G \times G$ to G is analytic everywhere. It suffices to do so in the neighbourhood of a point (a, b). Setting $x' = a^{-1}x$ and $y' = b^{-1}y$, so that $xy^{-1} = ax'y'^{-1}b$, the question reduces to showing that the map $(u, v) \mapsto auv^{-1}b$ is analytic at the origin. It is, however, the composite of the map $(u, v) \mapsto uv^{-1}$, which is analytic at the origin as seen in point 5 of the proof, the left translation $z \mapsto az$, analytic everywhere according to point 6, and the right translation $g \mapsto gb$; hence analyticity of the latter remains to be checked. This follows from observing that $g \mapsto gb$ is the composite of $g \mapsto b^{-1}gb$, analytic by point 7, and $g \mapsto bg$, analytic by point 6.

As a consequence of the above results, the distinction between C^{∞} Lie groups and analytic Lie groups is pointless (all the more so because, as will shortly be seen, a locally compact group admits at most one analytic structure compatible with its topological group structure). In the remainder of this book, the expression "Lie group" will therefore denote an *analytic* group.

6.9 Limits of Products and Commutators

Because of the Campbell–Hausdorff formula, formulas (3.5.4) and (3.5.5) which, in Chap. 3, enabled us to endow the set of one-parameter subgroups of a locally linear group with a Lie algebra structure, can be extended to Lie groups:

Theorem 6 *Let G be a Lie group and X and Y be two elements of its Lie algebra. Then,*

$$\gamma_{X+Y}(t) = \lim(\gamma_X(t/n)\gamma_Y(t/n))^n, \tag{6.9.1}$$

$$\gamma_{[X,Y]}(st) = \lim(\gamma_X(s/n)\gamma_Y(t/n)\gamma_X(s/n)^{-1}\gamma_Y(t/n)^{-1})^{n^2} \tag{6.9.2}$$

for all $s, t \in \mathbb{R}$.

To prove this, write

$$\gamma_X(t)\gamma_Y(t) = \exp(tX)\exp(tY) = \exp(H(tX, tY)) \tag{6.9.3}$$

for sufficiently small $|t|$, with

$$H(tX, tY) = tX + \sum \frac{(-1)^m}{m+1} \frac{\mathrm{ad}\,(X)^{p_1} \cdots \mathrm{ad}\,(Y)^{q_m}\mathrm{ad}\,(X)^{p_{m+1}}Y}{(q_1 + \cdots + q_m + 1)p_1! \cdots q_m! p_{m+1}!} t^{p_1 + \cdots + p_{m+1}}$$

$$= \sum Z_r t^r, \tag{6.9.4}$$

where Z_r denotes the sum of the terms of total degree r in the series $H(X, Y)$. We thus get a power series in t, with coefficients in $\mathfrak{g}$ converging for sufficiently small $|t|$. Hence,

$$H(tX, tY) = Z_1 t + Z_2 t^2 + O(t^3) \quad \text{as } t \text{ tends to } 0 \tag{6.9.5}$$

with

$$Z_1 = X + Y, \quad Z_2 = \frac{1}{2}[X, Y], \tag{6.9.6}$$

as clearly follows from (6.6.12). By the way, the computation of the terms of degree 3 and 4 in (6.6.12) is unnecessary for our purpose here. We first deduce that

$$\gamma_X(t)\gamma_Y(t) = \exp(t(X + Y) + O(t^2)), \tag{6.9.7}$$

and so, for fixed t and indefinitely increasing n,

$$\left(\gamma_X\left(\frac{t}{n}\right)\gamma_Y\left(\frac{t}{n}\right)\right)^n = \left[\exp\left(\frac{t}{n}(X + Y) + O(1/n^2)\right)\right]^n$$

$$= \exp(t(X + Y) + O(1/n)) \tag{6.9.8}$$

and (6.9.1) is immediate. For (6.9.2), start from

$$\gamma_X(t)\gamma_Y(t) = \exp(U(t)) \quad \text{with} \quad U(t) = Z_1 t + Z_2 t^2 + O(t^3), \tag{6.9.9}$$

$$\gamma_X(t)^{-1}\gamma_Y(t)^{-1} = \exp(V(t)) \quad \text{with} \quad V(t) = -Z_1 t + Z_2 t^2 + O(t^3); \tag{6.9.10}$$

the latter follows from the former by changing X, Y into $-X, -Y$. As $U(t)$ and $V(t)$ are power series without constant term in t, for sufficiently small $|t|$,

$$\gamma_X(t)\gamma_Y(t)\gamma_X(t)^{-1}\gamma_Y(t)^{-1} = \exp(H(U(t), V(t))) \tag{6.9.11}$$

with a series expansion of the form

$$H(U(t), V(t)) = W_1 t + W_2 t^2 + \cdots = W_1 t + W_2 t^2 + O(t^3) \tag{6.9.12}$$

since $H(X, Y)$ is a power series without constant term in the coordinates of X and Y with respect to a basis of $\mathfrak{g}$ (substitution of power series without constant term into the variables of a convergent power series). The remainders $O(t^3)$ appearing in expressions (6.9.9) and (6.9.10) for $U(t)$ and $V(t)$ clearly have no influence on the computation of the first two terms W_1 and W_2. Neither do the terms in t^2 in $U(t)$ and $V(t)$ play any role in the computation of W_1. In other words, the term $W_1 t$ is equal to the first term of the expansion of

$$H(tZ_1, -tZ_1) = tZ_1 - tZ_1 + O(t^2) = O(t^2), \tag{6.9.13}$$

and so $W_1 = 0$. W_2 remains to be computed, and from what has been said, it is given by

$$H(tZ_1 + t^2 Z^2, -tZ_1 + t^2 Z_2) = W_2 t^2 + O(t^3). \tag{6.9.14}$$

As $H(X, Y)$ is a power series without constant term in the coordinates of X and Y, to compute the left-hand side of (6.9.14) mod $O(t^3)$ the power series $H(X, Y)$ can clearly be replaced by its terms of degree 1 and 2, i.e. by $X + Y + \frac{1}{2}[X, Y]$. Hence we have to calculate the coefficient of t^2 in

$$tZ_1 + t^2 Z_2 - tZ_1 + t^2 Z_2 + \frac{1}{2}[tZ_1 + t^2 Z_2, -tZ_1 + t^2 Z_2] = 2t^2 Z_2 + O(t^3). \tag{6.9.15}$$

Thus

$$W_2 = 2Z_2 = [X, Y] \tag{6.9.16}$$

and so (6.9.11) can now be written as

$$\gamma_X(t)\gamma_Y(t)\gamma_X(t)^{-1}\gamma_Y(t)^{-1} = \exp(t^2[X, Y] + O(t^3)). \tag{6.9.17}$$

Replacing t by $1/n$ and taking the n^2th power, the right-hand side becomes $\exp([X, Y] + O(1/n))$, and so, taking the limit, (6.9.2) follows for $s = t = 1$. The general case can be deduced from this particular one in Y by replacing X and Y by sX and tY, proving (6.9.2).

Exercise. Show that von Neumann's inequality ((3.1.26) of Chap. 3)

$$H(X, Y) = X + Y + O(\|X\| \cdot \|Y\|) \quad \text{as } X \text{ and } Y \text{ tend to } 0$$

continues to hold for all Lie groups. Could it have been used to compute expansion (6.9.12)?

6.10 Analyticity of Continuous Homomorphisms

We next establish the following result generalizing the similar statement of Chap. 3:

Theorem 7 *Let G and H be Lie groups. Every continuous homomorphism f from G to H is analytic.*

First suppose that the theorem has been proved for $G = \mathbb{R}$. Let $\mathfrak{g}$ and $\mathfrak{h}$ be the Lie algebras of G and H, and associate to each $X \in \mathfrak{g}$ the map

$$t \mapsto f(\gamma_X(t)) = f(\exp_G(tX)) \tag{6.10.1}$$

from $\mathbb{R}$ to H. It is obviously a continuous homomorphism, and so is analytic—from $\mathbb{R}$ to H. Hence there is a well-defined $Y = f'(X)$ in $\mathfrak{h}$ such that

$$f(\exp_G(tX)) = \exp_H(tY) \tag{6.10.2}$$

for all t. The map f' from the Lie algebra $\mathfrak{g}$ to the Lie algebra $\mathfrak{h}$ is a *homomorphism*. This follows, as in Chap. 3, Sect. 3.6, by applying f to (6.9.1) and (6.9.2) of the previous section, as well as to the trivial equality $\gamma_{sX}(t) = \gamma_X(st)$. Compatibility of f' with the algebraic operations on the elements of $\mathfrak{g}$ and $\mathfrak{h}$ then follows immediately from the continuity of the homomorphism f.

Then (6.10.2) can be written as (put $t = 1$)

$$f \circ \exp_G = \exp_H \circ f'. \tag{6.10.3}$$

This shows that fixing *canonical* charts of G and H in the neighbourhood of the identity, the coordinates of the point $y = f(x)$ of H are *linear* functions, hence perforce analytic, of x. The map f is therefore analytic in the neighbourhood of e, and as it is a homomorphism, f is trivially analytic everywhere.

To finish the proof it remains to show that every continuous one-parameter subgroup of a Lie group G is analytic. For this we need the following lemma:

Lemma *Let G be a Lie group and $\mathfrak{g}$ its Lie algebra. There is a neighbourhood U of 0 in $\mathfrak{g}$ having the following property: for $X, Y \in U$, the elements $\exp(X)$ and $\exp(Y)$ of G commute if and only if $[X, Y] = 0$.*

Let us fix a symmetric, convex, open neighbourhood V of 0 in the Lie algebra $\mathfrak{g}$ of G such that the exponential map induces an isomorphism from V onto an open neighbourhood V of e in G. As the map $(X, Y) \mapsto \mathrm{ad}\,(\exp(X))Y$ from $\mathfrak{g} \times \mathfrak{g}$ to $\mathfrak{g}$ is continuous, there is a symmetric, *convex*, open neighbourhood U of 0 in V such that, setting $U = \exp(U)$,

$$\mathrm{ad}\,(x)Y \in V \quad \text{for all } x \in U \text{ and } y \in U. \tag{6.10.4}$$

Denoting by $\log$ the inverse from V onto V of the exponential map, it then amounts to showing that

$$xy = yx \Rightarrow [\log(x), \log(y)] = 0 \quad \text{if} \quad x, y \in U. \tag{6.10.5}$$

Let us set $X = \log(x)$ and $Y = \log(y)$. Then $xyx^{-1} = \exp(\mathrm{ad}\,(x)Y) = y = \exp(Y)$, and as Y and $\mathrm{ad}\,(x)Y$ are in V, where the exponential map is injective, we first get $\mathrm{ad}\,(x)Y = Y$. But then $x \cdot \gamma_Y(t) = \gamma_Y(t)x$ for all t, where $\gamma(t) = \exp(tY)$ is a one-parameter subgroup. As U is convex and symmetric, $\gamma_Y(t) \in U$ if $|t| \le 1$, and the same argument (with $\gamma_Y(t)$ playing the role of x and x that of y) shows that similarly $\gamma_Y(t)\gamma_X(u) = \gamma_X(u)\gamma_Y(t)$ for all u and all $|t| \le 1$; in particular, $\gamma_X(t)$ and $\gamma_Y(1)$ commute for $|t| \le 1$. As above this implies that $\mathrm{ad}\,(\gamma_X(t))Y = Y$, since, by (6.2.5),

$$\mathrm{ad}\,(\gamma_X(t))Y = e^{\mathrm{ad}\,(tX)}Y = Y + t[X, Y] + \cdots \tag{6.10.6}$$

for all t; $[X, Y] = 0$ follows immediately. Hence the lemma.

We are now in a position to prove that every continuous homomorphism $f : \mathbb{R} \to G$ is analytic. Choose U and V as in the proof of the lemma and let J be an open interval centered at 0 in $\mathbb{R}$ such that $f(t) \in U$ for $t \in J$. Set $F(t) = \log(f(t))$ for $t \in J$. This gives a continuous map from J to U. Thanks to (6.10.5), the elements of U of type $F(t)$ are mutually commuting. Since, as seen in Sect. 6.2, $\exp(X)\exp(Y) = \exp(X + Y)$ if $[X, Y] = 0$,

$$f(s + t) = f(s)f(t) = \exp(F(s)) \cdot \exp(F(t)) = \exp(F(s) + F(t)) \tag{6.10.7}$$

for $s, t \in J$. If U is assumed to be sufficiently small so that

$$U + U \subset V, \tag{6.10.8}$$

then

$$F(s + t) = F(s) + F(t) \quad \text{if} \quad s, t, s + t \in J, \tag{6.10.9}$$

because $F(s) + F(t)$ and $F(s + t)$ then have the same image under the exponential map, while belonging to a subset of $\mathfrak{g}$, namely V, on which the exponential map is injective. As $\mathbb{R}$ is simply connected, F is the restriction to J of a homomorphism which like F is necessarily continuous from $\mathbb{R}$ to the additive group of $\mathfrak{h}$. Then $F(t) = tX$ for some $X \in \mathfrak{h}$. Hence $f(t) = \exp(tX)$ for $t \in J$, and so for all t, and f is analytic as claimed.

Theorem 7 applies notably to *linear representations* of G since these are the homomorphisms from G to the various groups $GL_n(\mathbb{R})$. On the other hand, it also shows that *if G is a locally compact group, there is at most one analytic group structure on G compatible with the topology on G.* Indeed, consider two such structures, and write G_1 and G_2 for the corresponding analytic groups. The identity map from G_1 to G_2 is continuous since the topologies on G_1 and G_2 are the same, and it is obviously a homomorphism. It is therefore analytic. The inverse map from G_2 to G_1 is also analytic for the same reasons, and so the analytic structures considered are identical, as claimed. This result makes it possible to give a *topological definition of Lie groups*: a topological group is defined to be a locally compact group admitting an analytic group structure compatible with its topology. Hence, it is then possible to speak of the analytic structure of a Lie group. On the other hand, Lie groups can be *characterized* by a property of a *topological* nature: they are *locally Euclidean* topological groups (i.e. admitting open subsets homeomorphic to open subsets of a space $\mathbb{R}^n$). The proof of this result (Hilbert's fifth problem, solved around 1950 by A.M. Gleason) is significantly more difficult than the constructions of the present chapter, and it can be found in the book by D. Montgomery and L. Zippin, *Topological Transformation Groups* (Interscience, 1955). It can also be inferred from exercises scattered in *Élements d'analyse* by J. Dieudonné, the problem consisting in localizing the relevant exercises, which in itself would be an excellent exercise for the reader (IQ + HARD WORK = MERIT).

6.11 Simply Connected Group Homomorphisms

We have once again convinced ourselves that for every Lie group homomorphism $f : G \to H$, there is a Lie algebra homomorphism $f' : \mathfrak{g} \to \mathfrak{h}$ corresponding to it such that

$$f \circ \exp_G = \exp_H \circ f'. \tag{6.11.1}$$

If G is connected, then this equality shows that f is fully determined by f'. But if instead we are given a Lie algebra homomorphism $F : \mathfrak{g} \to \mathfrak{h}$, is there a homomorphism $f : G \to H$ such that $F = f'$?

Theorem 8 *Let G and H be two Lie groups, $\mathfrak{g}$ and $\mathfrak{h}$ their Lie algebras and F a homomorphism from $\mathfrak{g}$ to $\mathfrak{h}$. If G is connected and simply connected then there is a uniquely defined homomorphism f from G to H such that $f' = F$.*

Remember that a Lie algebra homomorphism must be linear and satisfy

$$F([X, Y]) = [F(X), F(Y)] \quad \text{for all } X, Y \in \mathfrak{g}. \tag{6.11.2}$$

To show the existence of f, choose an open neighbourhood U of 0 in $\mathfrak{g}$ having the following properties:

(a) the Campbell–Hausdorff series converges in $\mathsf{U} \times \mathsf{U}$ and[3]

$$\exp_G(X) \exp_G(Y) = \exp_G(H_\mathfrak{g}(X, Y)) \quad \text{for all } X, Y \in \mathsf{U}. \tag{6.11.3}$$

(b) The exponential map is injective on the image of $\mathsf{U} \times \mathsf{U}$ under $H_\mathfrak{g}$ and induces an isomorphism from U onto an open neighbourhood U of e in G. Every sufficiently small neighbourhood of 0 in $\mathfrak{g}$ clearly satisfies these conditions. Since F, being linear, is continuous, it is even possible to choose an open neighbourhood V of 0 in $\mathfrak{h}$ in such a way that the following additional condition holds:

(c) $F(\mathsf{U}) \subset \mathsf{V}$ and V satisfies condition (a) with respect to the group H.

Next, let us consider $\log_G : U \to \mathsf{U}$, the inverse of the exponential, and define a map f from U to H by setting

$$f(x) = \exp_H \circ \log_G(x) \quad \text{for } x \in U. \tag{6.11.4}$$

It is obtained by taking the composite of three analytic maps and so is analytic. Then,

$$f(xy) = f(x)f(y) \quad \text{if} \quad x, y, xy \in U. \tag{6.11.5}$$

Indeed, let us next consider the elements $X = \log_G(x)$ and $Y = \log_G(y)$ of $\mathfrak{g}$. On the one hand, $xy = \exp_G(H_\mathfrak{g}(X, Y))$ and on the other $xy = \exp_G(Z)$ for some unique $Z \in \mathsf{U}$ since $xy \in U$. Now, according to condition (b), Z and $H_\mathfrak{g}(X, Y)$ belong to a subset of $\mathfrak{g}$ on which the exponential map is injective. So $Z = H_\mathfrak{g}(X, Y)$ and (6.11.4) then shows that

$$f(x) = \exp_H(F(X)), \quad f(y) = \exp_H(F(Y)), \quad f(xy) = \exp_H(F(H_\mathfrak{g}(X, Y))). \tag{6.11.6}$$

Hence setting $X' = F(X), Y' = F(Y)$ and $Z' = f(H_\mathfrak{g}(X, Y))$, it amounts to proving that $\exp_H(X') \exp_H(Y') = \exp_H(Z')$. But X' and Y' are in V, and V has been chosen so that $\exp_H(X') \exp_H(Y') = \exp_H(H_\mathfrak{h}(X', Y'))$ for $X', Y' \in \mathsf{V}$. So it is a matter of showing that $Z' = H_\mathfrak{h}(X', Y')$, i.e. that

$$F(H_\mathfrak{g}(X, Y)) = H_\mathfrak{h}(F(X), F(Y)). \tag{6.11.7}$$

---

[3] The Campbell–Hausdorff series of $\mathfrak{g}$ and $\mathfrak{h}$ are written $H_\mathfrak{g}$ and $H_\mathfrak{h}$.

But let us write the Campbell–Hausdorff series in its Dynkin form

$$H(X, Y) = \sum \frac{(-1)^m}{m+1} \sum_{p_i + q_i > 0} \frac{[X^{p_1} Y^{q_1} \cdots X^{p_m} Y^{q_m}]}{p_1! q_1! \cdots p_m! q_m!}. \tag{6.11.8}$$

Since F is a continuous linear map, $F(H_{\mathfrak{g}}(X, Y))$ can be computed in the domain where (6.11.8) is absolutely convergent, by applying F to each term of the series. It is therefore clearly a matter of showing that

$$F([X^{p_1} Y^{q_1} \cdots X^{p_m} Y^{q_m}]) = [X'^{p_1} Y'^{q_1} \cdots X'^{p_m} Y'^{q_m}] \tag{6.11.9}$$

where, as above, $X' = F(X)$, $Y' = F(Y)$. To do so it suffices to show that more generally

$$F([X_1 \cdots X_n]) = [X'_1 \cdots X'_n] \quad \text{where} \quad X'_k = F(X_k) \tag{6.11.10}$$

i.e. that

$$F(\mathrm{ad}\,(X_1) \cdots \mathrm{ad}\,(X_{n-1}) X_n) = \mathrm{ad}\,(X'_1) \cdots \mathrm{ad}\,(X'_{n-1}) X'_n \tag{6.11.11}$$

for all $X_k \in \mathfrak{g}$. But definition (6.11.2) of a Lie algebra homomorphism can also be formulated as

$$F(\mathrm{ad}\,(X)Y) = \mathrm{ad}\,(X')Y' \quad \text{where} \quad X' = F(X),\ Y' = F(Y). \tag{6.11.12}$$

Hence the proof of (6.11.11) follows readily by induction on n. This implies (6.11.5).

The assumption that G is connected and simply connected then enables us to apply Theorem 1 of Chap. 2 and to extend f to a homomorphism from G to H which obviously is the answer to the question, completing the proof.

Corollary 1 *Two connected and simply connected Lie groups are isomorphic if and only if their Lie algebras are isomorphic.*

Indeed, if G and H are simply connected and we consider mutually inverse isomorphisms $\mathfrak{g} \to \mathfrak{h}$ and $\mathfrak{h} \to \mathfrak{g}$, these isomorphisms can be "integrated" into mutually inverse analytic homomorphisms $G \to H$ and $H \to G$.

Corollary 1 shows that the classification of simply connected Lie groups reduces to that of Lie algebras (provided it is known that there is always a Lie algebra corresponding to a Lie group—see later), a problem that is far from being resolved although it is purely algebraic (how to classify semisimple Lie algebras has been known for a long time, but, probably it is ironical that the classification of Lie algebras called "solvable" poses intractable difficulties). If we wanted to deduce a classification of all connected Lie groups, we would also need to determine the centre of each simply connected Lie group and to classify all its discrete subgroups.

Corollary 2 *Let G be a connected and simply connected Lie group and $\mathfrak{g}$ its Lie algebra. The map which associates the linear representation $M' : \mathfrak{g} \to \mathscr{L}(E)$ of $\mathfrak{g}$ to each linear representation $M : G \to GL(E)$ of G is a bijection from the set of all linear representations of G onto the set of all linear representations of $\mathfrak{g}$.*

This is obvious. The point of this corollary is to show that the determination of linear representations (understood to be continuous, i.e. analytic, in finite-dimensional real vector spaces) of a simply connected group is a purely algebraic problem. When G is not simply connected it is also necessary to find among the representations of its universal covering $\tilde{G}$ those that are trivial on the kernel of the projection $\tilde{G} \to G$.

Corollary 3 *Every connected Lie group is locally linear.*

This corollary—which justifies the introduction of locally linear groups in Chap. 3 and which, seen from a different angle, could equally be used to discredit it—is immediate, provided the following statement is admitted to hold:

Ado's Theorem[4] *Every finite-dimensional Lie algebra $\mathfrak{g}$ over a field of characteristic zero is isomorphic to a matrix algebra, i.e. admits a "faithful linear representation".*

The proof of this result is purely algebraic and can be found almost everywhere. Here we admit it and show how it implies the corollary. Let G be a Lie group and $\mathfrak{g}$ its Lie algebra. By Ado, there is a faithful linear representation F of $\mathfrak{g}$ in $M_n(\mathbb{R})$ say. Apply the arguments of Theorem 8 by taking H to be the group $GL_n(\mathbb{R})$. This gives an open neighbourhood U of e in G and an analytic map f from U to $GL_n(\mathbb{R})$ satisfying (6.11.5) and which, by construction, is tangent at the origin to the homomorphism F from $\mathfrak{g}$ to $M_n(\mathbb{R})$. As F is injective, f is an immersion and thus a local embedding of G into $GL_n(\mathbb{R})$ in the sense of Chap. 3, Sect. 3.5; hence the corollary.

6.12 The Cartan–von Neumann Theorem

Following in von Neumann's footsteps, we showed in Chap. 3 that every locally compact group admitting a local embedding into a group $GL_n(\mathbb{R})$ is a Lie group (i.e. by Theorem 7, it can be uniquely endowed with an analytic structure compatible with its topological group structure). We are now in a position to show that von Neumann's result continues to hold for all Lie groups, not just $GL_n(\mathbb{R})$.[5]

---

[4] Bull. of the Physico-Mathematics Society of Kazan, 1935. Ado's complicated original proof has since then been greatly simplified.

[5] If we were willing to use Ado's theorem, which we are not, then we would observe that by taking the composite of j and a local embedding of G into a linear group gives a local embedding of H into a linear group; Theorem 9 would then follow from the analogous statement in Chap. 3.

Theorem 9 *Let G be a Lie group, H a locally compact group and j a local embedding of H into G, i.e. a continuous and injective map from a neighbourhood W of e in H to G such that $j(xy) = j(x)j(y)$ whenever x, y and xy are in W. Then H is a Lie group and j is analytic.*

The fact that j is analytic will become obvious because of Theorem 7 once H will have been shown to be a Lie group.[6] Hence we will content ourselves with justifying this latter point by arguing as in Chap. 3, Sect. 3.5. We keep the same notation whenever possible.

a—Choose a *compact* neighbourhood $V = V^{-1}$ of e in H such that $V^2 \subset W$. As j is injective, it induces a homeomorphism from V onto its image $j(V)$. Taking V sufficiently small, we may assume that $j(V) \subset U$ or $U = \exp(\mathsf{U})$ for some convex open neighbourhood U of 0 in the Lie algebra $\mathfrak{g}$ of G isomorphically mapped onto an open neighbourhood of e in G under the map $\exp = \exp_G$.

b—We show that if V is sufficiently small then U may be assumed *not to contain any non-trivial subgroup*. Indeed, taking V sufficiently small, in the above, U may be chosen to be a neighbourhood of 0 satisfying the lemma of Sect. 6.10: for $X, Y \in \mathsf{U}$, $\exp(X)\exp(Y) = \exp(Y)\exp(X)$ only if $[X, Y] = 0$. It can even be assumed that in addition the exponential map is injective on the set of sums $X + Y$, $X, Y \in \mathsf{U}$. These assumptions will enable us to show that, if U is bounded, then

$$x^n \in U \text{ for all } n \in \mathbb{N} \quad \text{implies} \quad x = e. \tag{6.12.1}$$

Indeed, set $x^n = \exp(x_n)$ with $x_n \in \mathsf{U}$. As $x^p x^q = x^q x^p$, $[X_p, X_q] = 0$, and so

$$\exp(X_{p+q}) = x^{p+q} = x^p x^q = \exp(X_p + X_q). \tag{6.12.2}$$

Since $\exp$ is injective on $\mathsf{U} + \mathsf{U}$, it follows that $X_{p+q} = X_p + X_q$ for all $p, q \in \mathbb{N}$, and thus $X_n = nX_1$. But assuming that U is bounded, nX_1 is in U for all n only if $X_1 = 0$, proving (6.12.1) and our claim.

As a result, likewise V *does not contain any non-trivial subgroup* of H. Indeed, if $y \in V$ satisfies $y^n \in V$ for all $n \in \mathbb{N}$, then $j(y^n) = j(y)^n$ would follow for all $n \geq 0$ (induction on n) and (6.12.1) would then force $j(y) = e$, in other words $y = e$ since j is injective.

We make a further remark. The argument used to prove (6.12.2) leads to a more general result which will be of use later: set $\log$ to be the inverse map to the exponential from U onto U; then

$$\log(xy) = \log(x) + \log(y) \quad \text{if} \quad x, y, xy \in U \text{ and if } xy = yx. \tag{6.12.3}$$

Indeed, the equality $xy = yx$ shows that $\log(x)$ and $\log(y)$ commute, hence that both sides of (6.12.3) have the same image under $\exp$, and as they are in $\mathsf{U} + \mathsf{U}$, the desired equality follows. This is the lemma of Sect. 6.10.

---

[6]Exercise: extend Theorem 7 to the homomorphisms of Chap. 2, Sect. 2.2.

c—To define the tangent vectors to G at e, we adopt the point of view of "deriving sequences" from Chap. 5, Sect. 5.1. We define a Lie subalgebra $\mathfrak{h}$ of $\mathfrak{g}$ by setting that $X \in \mathfrak{g}$ belongs to $\mathfrak{h}$ if and only if there exists a sequence (h_p) of elements of H tending to e and a sequence (ε_p) of numbers > 0 tending to 0 such that the sequence $(j(h_p), \varepsilon_p)$ is a deriving sequence in G defining X:

$$X(f) = \lim \frac{1}{\varepsilon_p}[f(j(h_p)) - f(e)] \qquad (6.12.4)$$

for any analytic function f in the neighbourhood of e in G. Before showing that $\mathfrak{h}$ is a Lie subalgebra of $\mathfrak{g}$, as in Chap. 3 (proof of Lemma 1) we first show that *for all* $X \in \mathfrak{h}$, $\exp(tX) \in j(V)$ *for sufficiently small* $|t|$.

Since by part (b) of the proof, V does not contain any non-trivial subgroup of H, for all sufficiently large p, there exists an integer $q_p > 0$ such that

$$h_p^q \in V \quad \text{for} \quad 0 \le q \le q_p, \quad h_p^{q_p+1} \notin V. \qquad (6.12.5)$$

As V is compact,

$$\lim h_p^{q_p} = h \qquad (6.12.6)$$

may be assumed to exist. As in Chap. 3, h belongs to the boundary of V, and so $h \ne e$ if H is not discrete, which will evidently be assumed, and so $\bar{h} = j(h) \ne e$. Hence, $\log(\bar{h}) \ne 0$. Moreover,

$$\log(\bar{h}) = \lim \log(\bar{h}_p^{q_p}) = \lim q_p \cdot \log(\bar{h}_p), \qquad (6.12.7)$$

because the assumptions on U are such that

$$g \in U, \quad g^2 \in U, \dots, g^q \in U \Rightarrow \log(g^q) = q \cdot \log(g) \qquad (6.12.8)$$

as can be seen from (6.12.3) by induction on q.

Comparing (6.12.4) and (6.12.7) shows that X *and* $\log(\bar{h})$ *are proportional.* Indeed, the log map being a chart of U, the arguments of Chap. 5, Sect. 5.1 show that if (g_p, ε_p) is a deriving sequence at e on G, the tangent vector it defines is just $\lim \frac{1}{\varepsilon_p} \log(g_p)$; hence putting $g_p = j(h_p) = \bar{h}_p$,

$$X = \lim \frac{1}{\varepsilon_p} \log(\bar{h}_p), \qquad (6.12.9)$$

and so X and $\log(\bar{h}) = \lim q_p \cdot \log(\bar{h}_p)$ are proportional.

To show that $\exp(tX) \in j(V)$ for small $|t|$, it therefore suffices to show that

$$\exp_G(t \cdot \log(\bar{h})) \in j(V) \qquad (6.12.10)$$

for small $|t|$, for example (since $V = V^{-1}$) for $0 \le t \le 1$. We apply the same arguments as in Chap. 3. According to (6.12.7), $\exp(t \cdot \log(\bar{h})) = \lim \exp(tq_p \cdot \log(\bar{h}_p)) = \lim \exp([tq_p] \cdot \log(\bar{h}_p))$ because of the continuity of the exponential map. Then

$$\exp(t \cdot \log(\bar{h})) = \lim \exp(\log(\bar{h}_p))^{[tq_p]} = \lim \bar{h}_p^{[tq_p]} \qquad (6.12.11)$$

and as $0 \le t \le 1$ implies $0 \le [tq_p] \le q_p$, the result obtained is in $j(V)$ by (6.12.5), hence so is its limit, $\exp(t \cdot \log(\bar{h}))$ for $0 \le t \le 1$.

d—As in Chap. 3, it follows that for all $X \in \mathfrak{h}$, *there exists a unique one-parameter subgroup* $t \mapsto \gamma_X^H(t)$ *in H such that* $j \circ \gamma_X^H(t) = \exp(tX)$ *for sufficiently small $|t|$*. By part (c) of the proof there indeed exists a $c > 0$ such that $\exp(tX) \in j(V)$ for $|t| < c$. Therefore, as j is injective on V, there is a unique continuous map γ_X^H from the interval $]-c, +c[$ to V such that $j \circ \gamma_X^H(t) = \exp(tX)$ for $|t| < c$. If s, t and $s + t$ belong to the interval $J =]-c, +c[$, then evidently

$$j \circ \gamma_X^H(s + t) = \exp((s + t)X) = j \circ \gamma_X^H(s) \cdot j \circ \gamma_X^H(t) = j(\gamma_X^H(s)\gamma_X^H(t)) \qquad (6.12.12)$$

since $j(xy) = j(x)j(y)$ whenever x, y and xy belong to the neighbourhood W containing V^2. Moreover, j is injective on W. So in conclusion,

$$\gamma_X^H(s + t) = \gamma_X^H(s)\gamma_X^H(t) \quad \text{if} \quad s, t, s + t \in J. \qquad (6.12.13)$$

This enables us to extend γ_X^H to a continuous homomorphism from $\mathbb{R}$ to H which answers the question, and as j is injective, it is clearly the only solution to the problem.

e—*Conversely, every one-parameter subgroup γ of H is a γ_X^H for some $X \in \mathfrak{h}$.* Indeed, the map $t \mapsto j \circ \gamma(t)$ is a local homomorphism from $\mathbb{R}$ to G, and so can be extended to a continuous and hence analytic one-parameter subgroup of G. Hence there exists a well-defined $X \in \mathfrak{g}$ such that $j \circ \gamma(t) = \exp(tX)$ for sufficiently small $|t|$. Setting $\gamma(1/p) = h_p$ and $\varepsilon_p = 1/p$, (6.12.9) clearly holds, and so $X \in \mathfrak{h}$. Then,

$$j \circ \gamma(t) = j \circ \gamma_X^H(t) \qquad (6.12.14)$$

for small $|t|$, hence $\gamma(t) = \gamma_X^H(t)$ for small $|t|$, and therefore for all t.

f—We are now in a position to show that $\mathfrak{h}$ *is a Lie subalgebra of $\mathfrak{g}$*. Since

$$\gamma_{sX}^H(t) = \gamma_X^H(st), \qquad (6.12.15)$$

$\mathfrak{h}$ is homothety invariant. If X and Y are in $\mathfrak{h}$, consider the elements $x_p = \gamma_X^H(1/p)$ and $y_p = \gamma_Y^H(1/p)$ of H, and set $\varepsilon_p = 1/p$. Then,

$$X = \lim \log(\bar{x}_p)/\varepsilon_p, \quad Y = \lim \log(\bar{y}_p)/\varepsilon_p \qquad (6.12.16)$$

and thus $X + Y$ and $[X, Y]$ are elements of $\mathfrak{g}$ defined by the deriving sequences $(\bar{x}_p \bar{y}_p, \varepsilon_p)$ and $(\bar{x}_p \bar{y}_p \bar{x}_p^{-1} \bar{y}_p^{-1}, \varepsilon_p^2)$ in G—to see this it suffices to use Theorem 6. But for large p,

$$\bar{x}_p \bar{y}_p = j(x_p y_p) \quad \text{and} \quad \bar{x}_p \bar{y}_p \bar{x}_p^{-1} \bar{y}_p^{-1} = j(x_p y_p x_p^{-1} y_p^{-1}) \qquad (6.12.17)$$

since j is a local homomorphism, so that the claimed result reduces to the definition of elements of $\mathfrak{h}$.

  g—Setting

$$\exp_H(X) = \gamma_X^H(1) \quad \text{for all} \quad X \in \mathfrak{h} \qquad (6.12.18)$$

defines a map $\exp_H : \mathfrak{h} \to H$. We show it is *continuous at the origin*. Otherwise, there would be a compact neighbourhood $V' \subset V$ of e in H and a sequence of vectors $X_p \in \mathfrak{h}$ tending to 0 in $\mathfrak{g}$ and such that $\gamma_{X_p}^H(1) \notin V'$ for all p. As the map $t \mapsto \gamma_{X_p}^H(t)$ is continuous, there is a larger interval $[0, t_p]$ originating at 0 in which $\gamma_{X_p}^H(t) \in V'$. Obviously, $t_p < 1$ and

$$\gamma_{X_p}^H(t_p) \in \partial V', \qquad (6.12.19)$$

the boundary of V'. This enables us, if need be by extracting a subsequence, to assume that the left-hand side of (6.12.19) tends to a limit $h \in \partial V'$. Now, $h \neq e$, and so $\log(\bar{h}) \neq 0$ since the log map is injective on U and thus necessarily also on $j(V')$. On the other hand,

$$j \circ \gamma_{X_p}^H(t) = \exp(t X_p) \quad \text{for} \quad 0 \le t \le t_p. \qquad (6.12.20)$$

This indeed holds for sufficiently small $|t|$, hence for t/n if n is sufficiently large. But setting $x = \gamma_{X_p}^H(t/n)$, we get $x^q \in V' \subset V$ for $1 \le q \le n$, and so $j(x^n) = j(x)^n$, and (6.12.20) follows since $\gamma_{X_p}^H(t) = x^n$ and $\exp_G(t X_p) = j(x)^n$. Then, applying log to both sides of (6.12.20) for $t = t_p$ gives

$$\log(\bar{h}) = \log(\lim j \circ \gamma_{X_p}^H(t_p)) = \lim \log(\exp_G(t_p X_p)) = \lim t_p X_p, \qquad (6.12.21)$$

because as X_p tends to 0 and t_p remains between 0 and 1, for large p, $t_p X_p$ clearly belongs to the neighbourhood U of 0 in $\mathfrak{g}$ in which $\log(\exp(X)) = X$. But (6.12.21) implies $\log(\bar{h}) = 0$. This contradiction proves the continuity of the map $\exp_H$ at the origin.

  h—We next show that $\exp_H$ *induces a homeomorphism from a neighbourhood of 0 in $\mathfrak{h}$ onto a neighbourhood of e in H*. For this, choose a compact convex neighbourhood S of 0 in $\mathfrak{h}$ such that $\exp_H(S) \subset V$; the existence of S follows from part (g) of the proof. For all $X \in S$, $tX \in S$ for $0 \le t \le 1$, and so

$$\gamma_X^H(t) = \gamma_{tX}^H(1) = \exp_H(tX) \in V. \qquad (6.12.22)$$

But the arguments that led to (6.12.20) show more generally that

$$j \circ \gamma_X^H(t) = \exp_G(tX) \text{ if } \gamma_X^H(t') \in V \quad \text{for} \quad 0 \le t' \le t. \tag{6.12.23}$$

As (6.12.22) holds for $0 \le t \le 1$, the first conclusion that follows is: for $t = 1$,

$$j \circ \exp_H(X) = \exp_G(X) \quad \text{for all} \quad X \in S. \tag{6.12.24}$$

Since j is a homeomorphism from V onto $j(V)$, and since the restriction of $\exp_G$ to S is continuous and injective, hence bicontinuous, if S is sufficiently small, then it becomes a matter of checking that $\exp_G(S)$ *contains* a neighbourhood of e in $j(V)$.

Instead of using the arguments of the proof of Theorem 5 of Chap. 3, let us choose a complement vector subspace $\mathfrak{k}$ to $\mathfrak{h}$ in $\mathfrak{g}$. If S and T are sufficiently small neighbourhoods of 0 in $\mathfrak{g}$ and $\mathfrak{k}$, the map $(X, Y) \to \exp(X)\exp(Y)$ from $S \times T$ to G is a homeomorphism from $S \times T$ onto a neighbourhood of e in G (Corollary 4 of Theorem 3). The property of interest to us will therefore be established once we will have shown that $j(V) \cap \exp(S)\exp(T) = \exp(S)$ provided S and T are sufficiently small. Besides, it suffices to show that, for fixed S and T,

$$j(V) \cap \exp\left(\frac{S}{n}\right) \exp\left(\frac{T}{n}\right) = \exp\left(\frac{S}{n}\right) \tag{6.12.25}$$

for large n.

Suppose this to be the case. Then there are vectors $X_n \in S$ and $Y_n \in T$ with $\exp(X_n/n)\exp(Y_n/n) \in j(V)$ and $Y_n \ne 0$ for all n. Choosing a neighbourhood V' of e in H such that $V' = V'^{-1}$ and $V'^2 \subset V$,

$$\exp(X_n/n)\exp(Y_n/n) \in j(V') \tag{6.12.26}$$

for large n since the left-hand side tends to e in $j(V)$. For the same reason, $\exp(X_n/n) \in j(V')$ for large n. It follows that $\exp(Y_n/n) \in j(V')^{-1}j(V') \subset j(V)$ for large n. Hence there is a sequence of points $h_n \in V$ such that $\exp(Y_n/n) = j(h_n)$ and so $Y_n/n = \log \circ j(h_n)$ for large n. As none of the Y_n are zero, it is possible to extract a subsequence of vectors $Y_n/\|Y_n\|$ (choose an arbitrary norm on $\mathfrak{g}$) converging to a limit $Y \ne 0$ in $\mathfrak{k}$. But as Y_n is proportional to $\log \circ j(h_n)$, which tends to 0, there is a sequence of numbers $\varepsilon_n > 0$ such that Y is the limit of a subsequence of the sequence having $\log(\bar{h}_n)/\varepsilon_n$ as general term, where as usual $\bar{h}_n$ denotes $j(h_n)$. This means that $Y \in \mathfrak{h}$ and as $\mathfrak{h} \cap \mathfrak{k} = 0$, this leads to a contradiction, proving (6.12.25) for large n.

i—We are now in a position to endow H with an analytic structure, first in the neighbourhood of the origin: it will be the one obtained by choosing an open neighbourhood of 0 in $\mathfrak{h}$ homeomorphically mapped under $\exp_H$ onto a neighbourhood of e in H, and by transferring to the latter the analytic structure of the neighbourhood under consideration in $\mathfrak{h}$ using $\exp_H$. Or, what comes to the same, it may be observed that by point (h) of the proof, for every sufficiently small neighbourhood

V' of e in H, the canonical chart $\log_G : U \to \mathsf{U}$ induces a homeomorphism from $j(V')$ onto a neighbourhood of 0 in $\mathfrak{h}$. As $\mathfrak{h}$ is a vector subspace of $\mathfrak{g}$, (by definition of submanifolds) this means that j induces a homeomorphism from V' onto a locally closed *submanifold* of G. The map $(x, y) \mapsto xy^{-1}$ from $H \times H$ to H, restricted to sufficiently small neighbourhoods of e in H, can therefore be identified with the restriction of the analytic map $(x, y) \mapsto xy^{-1}$ from $G \times G$ to G to neighbourhoods of e in the submanifold in question. It follows immediately that, in the neighbourhood of e, H can be endowed with an analytic structure such that the map $(x, y) \mapsto xy^{-1}$ from $H \times H$ to H is analytic in the neighbourhood of (e, e), and with respect to which j is obviously an immersion. (The Campbell–Hausdorff formula in $\mathfrak{h}$ could also be used to check compatibility between the analytic structure defined in the neighbourhood of e in H and the composition law in H).

The proof of the theorem will be complete once we know that the above construction can be carried out is such a way that it satisfies conditions (a), (b) and (c) of Sect. 6.8, *Lemma*. Conditions (a) and (b) are obvious. Condition (c), expressing the analyticity of $x \mapsto hxh^{-1}$ in the neighbourhood of e for each $h \in H$, is the one that requires checking. It would be obvious if $j(hxh^{-1}) = j(h)j(x)j(h)^{-1}$ since it would then only require checking in G. Unfortunately, this argument assumes h to be sufficiently near e. In the general case, H may be assumed to be connected (exercise: for a locally compact group to be a Lie group, it suffices that the identity component be open and a Lie group), in which case, $h = h_1 \cdots h_p$ with h_i such that $x \mapsto h_i x h_i^{-1}$ is analytic in the neighbourhood of e for all i. As composites of analytic maps in the neighbourhood of e leaving e fixed are analytic in the neighbourhood of e, condition (c) of Sect. 6.8 holds, completing the proof.

Corollary (E. Cartan) *Every closed subgroup of a Lie group is a Lie group.*

A closed subgroup endowed with the induced topology obvious admits a continuous injective homomorphism to a Lie group (the one it is contained in...), whence the theorem. It in fact means that a closed subgroup H of a Lie group G is always a *submanifold* of G. The proof of Theorem 9 indeed shows that if H is endowed with its analytic structure, the canonical injection j from H to G is an *immersion*. But it is also a homeomorphism from H onto its image. We still need to apply the following result:

Exercise. Let U and V be two manifolds and j an immersion from U to V. Assume that j is a homeomorphism from U onto its image. Show that the latter is a submanifold of V, isomorphic under j to U.

Note that the above corollary would enable us to quickly recover Theorem 8. Indeed, given a continuous homomorphism f from a Lie group G to a Lie group H, the graph of f is a closed subgroup R of $G \times H$, hence a Lie subgroup of $G \times H$. We let the reader deduce that f is analytic using elementary arguments.

6.13   Subgroups and Lie Subalgebras

Let G be a Lie group and $\mathfrak{g}$ its Lie algebra. A closed subgroup H in G is also a manifold, hence a Lie subgroup of G, and the Lie algebra $\mathfrak{h}$ of H can be canonically identified (up to the tangent linear map to the canonical injection from H to G at e) with a Lie subalgebra of $\mathfrak{g}$. Given the functorial character of the exponential map, the map $\exp_G : \mathfrak{g} \to G$ induces a map $\exp_H : \mathfrak{h} \to H$ on $\mathfrak{h}$, and so the image of $\mathfrak{h}$ under the exponential map from $\mathfrak{g}$ to G is contained in H, and contains a neighbourhood of e in H. *If H is connected*, then it follows that H *is the (abstract) subgroup of G generated by* $\exp(\mathfrak{h})$, which in theory makes it possible to reconstruct H from $\mathfrak{h}$.

Conversely, starting from a subalgebra $\mathfrak{h}$ of $\mathfrak{g}$ and denoting by H the subgroup generated (in a purely algebraic sense) by the elements $\exp(X)$, $X \in \mathfrak{h}$, the question that arises is whether H is a closed subgroup whose Lie algebra is $\mathfrak{h}$. The answer is no, as readily follows from the example of one-parameter subgroups that are not closed (geodesics of the torus). Indeed, let $t \mapsto \gamma(t) = \exp(tX)$ be an analytic one-parameter subgroup of G, and take $\mathfrak{h}$ to be the 1-dimensional subspace generated by X; it is a Lie subalgebra for trivial reasons: $[X, X] = 0 \in \mathfrak{h}$, and in this case the subgroup H generated by (and in fact identical to) the image of $\mathfrak{h}$ under the exponential map is just the image $\gamma(\mathbb{R})$ of $\mathbb{R}$ by the homomorphism γ; it may well not be closed, and might even be everywhere dense in G as shown by the example of $\mathbb{T}^n$ in Chap. 1.

Nevertheless, note that, in this case, the map γ from $\mathbb{R}$ onto H is bijective (otherwise $\gamma(\mathbb{R})$ would be compact and hence closed) and that if $H = \gamma(\mathbb{R})$ is endowed with the analytic structure of $\mathbb{R}$, then this gives a Lie group structure on H such that the canonical injection j from H to G is an *immersion*. We show that this situation can be generalized:

Theorem 10 *Let G be a Lie group, $\mathfrak{g}$ its Lie algebra, $\mathfrak{h}$ a subalgebra of $\mathfrak{g}$ and H the subgroup of G generated by $\exp_G(\mathfrak{h})$. Then there is a unique connected Lie algebra structure on H such that the injection j from H to G is analytic. Then j is an immersion and induces an isomorphism from the Lie algebra of H onto $\mathfrak{h}$.*

We begin with a remark. If there is a Lie group structure on H such that j is analytic, then j, being a homomorphism, is a subimmersion, and being injective, is necessarily an immersion given the standard form of subimmersions. As a consequence, $j'(e)$ is an isomorphism from the Lie algebra of H onto a subalgebra of $\mathfrak{g}$.

Consider two connected Lie algebra structures on H such that j is analytic with respect to them. Let H' and H'' be the Lie groups obtained by endowing H with these structures. We first show that the topologies of H' and H'' coincide. It suffices to show that the identity maps $H' \to H''$ and $H'' \to H'$ are continuous. We do so for the former. As it is a homomorphism, it even suffices to show that the identity map from H' to H'' is continuous somewhere. But let V' and V'' be compact neighbourhoods of e in H' and H''. As H' and H'' are connected,

$$H = H = H'' = \bigcup V'''^n = \bigcup V'''^n. \qquad (6.13.1)$$

Since $j : H \to G$ is injective and analytic with respect to H' and H'', j clearly induces homeomorphisms from the V'^n and the V''^n onto their images—in other words, on each V'^n the topology of H' coincides with that of G. Similarly the topology of H'' coincides on each V'''^n with that of G. As the V'^p and V''^q are moreover compact in G (being compact in H' and H''), so are the intersections $V'^p \cap V''^q$; these are closed with respect to the topologies of H' and H'' since these once again coincide with that of G on V'^p and V''^q. As H' is the union of the closed subsets $V'^p \cap V''^q$, Baire's lemma (Chap. 1, Sect. 1.2) shows that at least one of the sets $V'^p \cap V''^q$ has an interior point a in H'. Hence there is a neighbourhood of a in H' with the topology of H'' (namely $V'^p \cap V''^q$), implying the continuity of the identity map from H' to H'' at a.

These arguments readily show that if H is endowed with two analytic structures answering the question, they then coincide. Indeed, at the end of Sect. 6.10, it was shown that a topological group admits at most one Lie group structure (not necessarily connected) compatible with its topological group structure. As H' and H'' have been shown to have the same topology, they are identical as Lie groups.

Hence it remains to prove the existence of an analytic structure on H such that j is analytic and maps $\mathrm{Lie}(H)$ onto $\mathfrak{h}$.

For this, choose open neighbourhoods U and V of 0 in $\mathfrak{g}$ satisfying the conditions stated at the start of Sect. 6.8, in particular conditions (i), (ii) and (iii), and then set $U = \exp(\mathsf{U} \cap \mathfrak{h})$, $V = \exp(\mathsf{V} \cap \mathfrak{h})$. The exponential map induces a bijection from $\mathsf{U} \cap \mathfrak{h}$ onto U. This bijection enables us to endow U with the analytic structure that follows from that of the open subset $\mathsf{U} \cap \mathfrak{h}$ of $\mathfrak{h}$. As $\mathsf{V} = -\mathsf{V}$, $V = V^{-1}$. Moreover, V is open in U since so is $\mathsf{V} \cap \mathfrak{h}$ in $\mathsf{U} \cap \mathfrak{h}$, and $V^2 \subset U$ thanks to condition (iii) and to the fact that, $\mathfrak{h}$ being a Lie subalgebra of $\mathfrak{g}$, due to the very structure of the Hausdorff formula, $H(X, Y) \in \mathsf{U} \cap \mathfrak{h}$ whenever $X, Y \in \mathsf{V} \cap \mathfrak{h}$. Conditions (a), (b) and (c) from Sect. 6.8 are once again satisfied here: we have just seen this for (a); condition (b) simply expresses the fact that the map $(X, Y) \mapsto H(X, Y)$ from the manifold $(\mathsf{V} \cap \mathfrak{h}) \times (\mathsf{V} \cap \mathfrak{h})$ to $\mathsf{U} \cap \mathfrak{h}$ is analytic; so condition (c) remains to be checked for U, V and the group H. For this, as in Sect. 6.8, choose an open neighbourhood V' of 0 in $\mathfrak{g}$ such that $\mathrm{ad}\,(a)\mathsf{V}' \subset \mathsf{U}$ and $\mathsf{V}' \subset \mathsf{V}$, and set $V' = \exp(\mathsf{V}' \cap \mathfrak{h})$. We show that V' is indeed a neighbourhood of e in U, that $V' \subset V$, that $aV'a^{-1} \subset U$, and finally that the map $x \mapsto axa^{-1}$ from V' to U is analytic. The first two points are obvious. Since

$$aV'a^{-1} = \exp[\mathrm{ad}\,(a)(\mathsf{V}' \cap \mathfrak{h})] = \exp[\mathrm{ad}\,(a)\mathsf{V}' \cap \mathrm{ad}\,(a)\mathfrak{h}] \subset \exp[\mathsf{U} \cap \mathrm{ad}\,(a)\mathfrak{h}],$$
$$(6.13.2)$$

$aV'a^{-1} \subset U$ will become obvious once we have shown that $\mathrm{ad}\,(a)\mathfrak{h} \subset \mathfrak{h}$ for all $a \in H$. As H is generated by the elements $\exp(X)$ with $X \in \mathfrak{h}$, checking it when $a = \exp(X)$ for some $X \in \mathfrak{h}$ is sufficient; but this is then obvious since $\mathrm{ad}\,(a) = e^{\mathrm{ad}\,(X)}$ and $\mathrm{ad}\,(X)\mathfrak{h} \subset \mathfrak{h}$ since $\mathfrak{h}$ is a Lie subalgebra of $\mathfrak{g}$. The map $x \mapsto axa^{-1}$ from V' to

U requires investigating. If log denotes the inverse isomorphism to the exponential map from U onto $\mathsf{U} \cap \mathfrak{h}$, then

$$axa^{-1} = a \cdot \exp(\log(X)) \cdot a^{-1} = \exp(\text{ad}\,(a)(\log x)), \qquad (6.13.3)$$

and like the map log from V' to $V' \cap \mathfrak{h}$, the map ad (a) from $V' \cap \mathfrak{h}$ to $\mathsf{U} \cap \mathfrak{h}$ and the map exp from $\mathsf{U} \cap \mathfrak{h}$ to U are analytic; the desired result immediately follows.

Now that all this has been checked, the lemma of Sect. 6.8 can be applied. It shows that there is a unique Lie group structure on H with respect to which V is open and which induces on V the analytic structure induced by U.

This analytic structure answers the question. Indeed, let $\log_G$ from $\exp_G(\mathsf{V})$ onto V be the inverse of the exponential. It is a chart of the open subset $\exp_G(\mathsf{V})$ of G, and it transforms V, together with its analytic structure, into $\mathsf{V} \cap \mathfrak{h}$, i.e. into a submanifold of V. As a consequence, the canonical map j from H to G induces an isomorphism from the open submanifold V of H onto a submanifold of G; hence j is analytic. In addition, the image of $\text{Lie}(H)$ under $j'(e)$ is a tangent subspace in $\mathfrak{g}$ to the submanifold $V = \exp(\mathsf{V} \cap \mathfrak{h})$ of G at e; hence it is $\mathfrak{h}$, completing the proof.

It may be useful to warn the reader against the fact that in the so-called "international mathematical community", there is no consensus on what terminology to adopt for the subgroups H obtained by the method of Theorem 10. Some authors (for example S. Helgason, *Differential Geometry and Symmetric Spaces*, Academic Press, 1962) talk of "Lie subgroups", while for us Lie subgroups are closed; others (par example, C. Chevalley, *Theory of Lie Groups*, Princeton, 1946 or G. Hochschild, *The Structure of Lie Groups*, Holden Day, 1965) talk of "analytic subgroups", N. Bourbaki (*Lie Groups and Lie Algebras*, Chap. III, Springer, 1975) of "integral" subgroups, whereas J. Dieudonné talks of "immersed subgroups". Bourbaki's seemingly strange terminology is justified by the fact that the subgroups of Theorem 10 appear as integral manifolds of differential systems, a point of view that we have discarded in these Notes.[7] The terminologies used by Helgason and Chevalley seem to

---

[7]We give some brief indications about this point of view, leaving aside all superfluous subtleties. Let G be a C^r manifold with $r = \infty$ or $r = \omega$, and suppose that at each point x of G, there is vector subspace $\mathfrak{h}(x)$ of some fixed dimension p of the tangent vector space to G at x. Suppose that the map $x \mapsto \mathfrak{h}(x)$ is of class C^r in the following sense: in every sufficiently small open subset U of G, there exist p vector fields of class C^r on U whose values, at each point x of U, form a basis of the corresponding subspace $\mathfrak{h}(x)$.

An *integral manifold* of the field $\mathfrak{h}$ of subspaces is defined to be any submanifold H of G such that

$$T_x(H) = \mathfrak{h}(x) \quad \text{for all} \quad x \in H,$$

and $\mathfrak{h}$ is said to be *integrable* if there is an integral submanifold passing through every point of G. *Frobenius' Theorem* affirms that this is the case if and only if, for all open subsets $U \subset G$ and all vector fields L and M on U,

$$L(x), M(x) \in \mathfrak{h}(x) \quad \text{for all} \quad x \in U \quad \text{implies}$$

$$[L, M](x) \in \mathfrak{h}(x) \quad \text{for all} \quad x \in U. \qquad (*)$$

us dangerously ambiguous, so we will adopt N. Bourbaki's terminology and call any subgroup H for which there exists a Lie subalgebra $\mathfrak{h}$ of $\mathfrak{g}$ such that H is generated by $\exp_G(\mathfrak{h})$ an *integral subgroup* of G; $\mathfrak{h}$ will be said to be the *Lie algebra* of H.

To justify this terminology, it is necessary to show that the lie subalgebra $\mathfrak{h}$ of $\mathfrak{g}$ is fully determined by the subgroup H of G. Strictly speaking, this is a consequence of Theorem 10 since according to it, there is *at most* one *connected* Lie group structure on H such that the injection from H to G is analytic. But it is much more interesting to give a direct characterization of the elements of $\mathfrak{h}$ as a result of which $\mathfrak{h}$ will be definable in terms of the subgroup H without involving the latter's natural analytic structure, namely

$$X \in \mathfrak{h} \Leftrightarrow \exp_G(tX) \in H \quad \text{for all } t \in \mathbb{R}. \tag{6.13.4}$$

The implication $\Rightarrow$ is obvious since, by assumption, H is generated by $\exp_G(\mathfrak{h})$. Conversely, suppose that for some $X \in \mathfrak{g}$, $\exp_G(tX) \in H$ for all t and endow H with the connected Lie group structure given by Theorem 10. It is then a matter of showing that the map $t \mapsto \exp_G(tX)$ from $\mathbb{R}$ to the Lie group H is continuous. Indeed, if it is the case, then this map is a (continuous and hence analytic) one-parameter subgroup of H, and so is of the form $\exp_H(tY)$ for some $Y \in \mathrm{Lie}(H)$, and as the canonical injection $j : H \to G$ induces an isomorphism from $\mathrm{Lie}(H)$ onto $\mathfrak{h}$, $X = j'(Y) \in \mathfrak{h}$ and thus (6.13.3) follows.

The continuity of $t \mapsto \exp_G(tX)$ as a map from $\mathbb{R}$ to H endowed with its natural Lie group structure remains to be established. Since H is connected, it is the union of a sequence of compact subsets K_n. As $j : H \to G$ is continuous and injective, each K_n has in effect the topology of G, and is notably compact in G. Therefore

---

(Footnote 7 continued)
Besides it suffices to check that when there are vector fields L_i ($1 \le i \le p$) on an open subset U, which, at each $x \in U$, form a basis of $\mathfrak{h}(x)$, there exist functions c_{ij}^k on U such that

$$[L_i, L_j] = \sum c_{ij}^k L_k \tag{*}$$

for all i and j.

When $\mathfrak{h}$ is integrable, G can be endowed with a new manifold structure ($G(\mathfrak{h})$ will denote the manifold obtained by endowing the set G with this new structure of class C^r as in the case of G and $\mathfrak{h}$) having the following properties: the identity map from $G(\mathfrak{h})$ to G is an immersion; its tangent map at each point $x \in G$ maps the tangent space to $G(\mathfrak{h})$ at x isomorphically onto the given subspace $\mathfrak{h}(x)$ of $T_x(G)$. The connected components of the manifold $G(\mathfrak{h})$ are called the *maximal integral submanifolds of* $\mathfrak{h}$; they are of dimension p and each of them is canonically endowed with an injective immersion into G, but in general they cannot be identified with any veritable submanifold of G. To apply these results to the theory of Lie groups, take G to be a Lie group and $\mathfrak{h}$ the field of subspaces $x \mapsto x \cdot \mathfrak{h}$, where $\mathfrak{h}$ is a fixed Lie subalgebra of the Lie algebra $\mathfrak{g}$ of G. The integrability condition trivially holds and the maximal integral submanifold through the point e is then just the "Lie", "analytic", "integral" or "immersed" subgroup of G defined by $\mathfrak{h}$, endowed with the Lie group structure of Theorem 10.

Most presentations of differential geometry prove Frobenius' Theorem. For example, see M. Berger and B. Gostiaux, *Differential Geometry*, (Springer, 1988) or Cl. Godbillon, *Géométrie différentielle et mécanique analytique* (Herman, 1969).

the reals t such that $\exp_G(tX) \in K_n$ form a *closed subset* I_n of $\mathbb{R}$ and $\mathbb{R}$ being the union of these I_n, one of these I_n has an interior point (Baire). Since the restriction to I_n of the map $t \mapsto \exp_G(tX)$ is continuous as a map from I_n to K_n, and hence also as a map from I_n to H, there exists an $a \in \mathbb{R}$ in the neighbourhood of which the map $t \mapsto \exp_G(tX)$ from $\mathbb{R}$ to the Lie group H is continuous. As it is a group homomorphism, continuity everywhere follows trivially, qed.

Although Dieudonné always assumes his "immersed" subgroups to be connected, it would seem that it is reasonable to define an immersed subgroup, or more appropriately, an *immersed Lie group in* G, as a pair (H, j) consisting of a not necessarily connected Lie group H, and an injective analytic homomorphism $j : H \to G$. As already remarked, j is then necessarily an immersion and enables us to identify the Lie algebra of H with a subalgebra $\mathfrak{h}$ of G; the connected component H° of H being generated by the image of the exponential map, $j(H^\circ)$ is then the (integral!) subgroup of G whose Lie algebra is $\mathfrak{h}$.

However, one should not believe that in such a situation, the analytic structure of H is completely determined by the subgroups $j(H)$ of G. It is certainly the case when H is connected, as was seen while proving Theorem 10, but counterexamples can be readily constructed whenever non-connected topologies are admitted. Indeed, endowing an arbitrary subgroup H of G with a discrete topology turns it into a Lie group (in general not connected...) of dimension 0, and the canonical injection from H to G is then trivially analytic. If the given subgroup H happens to be an integral subgroup of a non-trivial Lie algebra, there are then two Lie group structures on H, differing from each other as much as desired, thereby answering the question! These trivial considerations might lead the reader to a pessimistic conclusion, namely that there is no "natural" method to endow a non-closed subgroup H of a Lie group G with a Lie group structure.

Following on N. Bourbaki's footsteps, we are going to see that, for any subgroup H of G, it is in fact possible to endow G with a Lie group structure "better than others" and which, in the case of an integral subgroup, will be precisely the one given in Theorem 10. The integral subgroups will then be those which, endowed with this "natural" analytic structure, are *connected*.

We first state a fundamental consequence of Theorem 10:

Corollary *Every finite-dimensional Lie algebra over $\mathbb{R}$ is the Lie algebra of a Lie group.*

Indeed, by Ado's theorem, it suffices to prove the corollary when the given Lie algebra $\mathfrak{g}$ is a subalgebra of the Lie algebra of some group $GL_n(\mathbb{R})$. The integral subalgebra of $GL_n(\mathbb{R})$ corresponding to $\mathfrak{g}$ by Theorem 10 then answers the question. (A simply connected group with Lie algebra $\mathfrak{g}$ can be obtained by replacing the integral subgroup considered with its universal covering, which justifies the remark made above that the classification of simply connected Lie groups is equivalent to that of Lie algebras).

6.14   The Natural Analytic Structure of a Subgroup

In accordance with the program stated above, we now establish the following result:

Theorem 11 *Let G be a Lie group, $\mathfrak{g}$ its Lie algebra and H an arbitrary subgroup of G. Let $\mathfrak{h}$ be the set of $X \in \mathfrak{g}$ satisfying the following property: there exist an open interval I centered at 0 in $\mathbb{R}$ and a C^1 map $\gamma : I \to G$ such that $\gamma(0) = e$, $\gamma(I) \subset H$ and $\gamma'(0) = X$. Then $\mathfrak{h}$ is a Lie subalgebra of $\mathfrak{g}$, and it is also the set of $X \in \mathfrak{g}$ such that*

$$\exp(tX) \in H \quad \text{for all } t. \tag{6.14.1}$$

There is a Lie group structure on H with the following properties:

(i)   *the canonical injection j from H to G is an immersion;*

(ii)   *the tangent map $j'(e)$ takes the Lie algebra of H onto $\mathfrak{h}$;*

(iii)   *the connected component of H is the subgroup H_o generated by $\exp(\mathfrak{h})$ endowed with the analytic structure resulting from Theorem 10;*

(iv)   *let M be a C^r $(r \geq 1)$ manifold and f a map from M to G such that $f(M) \subset H$; f is of class C^r as a map to the manifold G if and only if f is of class C^r as a map to the manifold H.*

The proof of this result, which can be skipped in a first reading, is rather long but involves interesting ideas. It can be divided into several parts.

a—We begin by showing that $\mathfrak{h}$ is a subalgebra of $\mathfrak{g}$. If $X_1, X_2 \in \mathfrak{h}$ are tangent to curves γ_1 and γ_2 drawn in H, $X_1 + X_2$ will clearly be tangent to the curve $t \mapsto \gamma_1(t)\gamma_2(t)$, which is also in H; it follows that $\mathfrak{h}$ is a vector subalgebra of $\mathfrak{g}$. As the internal automorphisms $x \mapsto hxh^{-1}$ preserve H, the same type of argument clearly shows that $\mathfrak{h}$ is $\operatorname{ad}(h)$-invariant for all $h \in H$. Hence if X and Y are in $\mathfrak{h}$ and $t \mapsto \gamma(t)$ is a fixed tangent curve to X at $t = 0$ in H, then $\operatorname{ad}(\gamma(t))Y \in \mathfrak{h}$ for all t; differentiating the left-hand side at $t = 0$ gives $[X, Y]$, and so $\mathfrak{h}$ is a Lie subalgebra of $\mathfrak{g}$.

b—Let H_o be the subgroup of G generated by $\exp_G(\mathfrak{h})$. We show that *if a C^1 curve in G passes through the origin and is contained in H, then it is contained in H_o.* Indeed, let $\gamma : I \to G$ be such a curve; hence

$$\gamma(0) = e, \quad \gamma(t) \in H \quad \text{for all} \quad t \in I. \tag{6.14.2}$$

Since H is a subgroup, more generally,

$$\gamma_s(t) = \gamma(s)^{-1}\gamma(s + t) \in H \tag{6.14.3}$$

whenever the right-hand side is well-defined. Differentiating for $t = 0$, by definition of $\mathfrak{h}$, we get $\gamma(s)^{-1}\gamma'(s) \in \mathfrak{h}$ for all $s \in I$. Hence it becomes a matter of showing that

$$\gamma(0) = e, \quad \gamma(t)^{-1}\gamma'(t) \in \mathfrak{h} \quad \text{for all} \quad t \in I \tag{6.14.4}$$

implies $\gamma(t) \in H_o$.

To do so choose a complement vector subspace $\mathfrak{k}$ of $\mathfrak{h}$ in $\mathfrak{g}$ and open neighbourhoods U and V of 0 in $\mathfrak{h}$ and $\mathfrak{k}$ such that the map

$$(Y, X) \mapsto \exp_G(Y) \exp_G(X) \tag{6.14.5}$$

induces an isomorphism from $U \times V$ onto an open neighbourhood W of e in G. Hence, for sufficiently small $|t|$, it is possible to set

$$\gamma(t) = \gamma_{\mathfrak{k}}(t)\gamma_{\mathfrak{h}}(t) \quad \text{with} \quad \gamma_{\mathfrak{k}}(t) \in \exp(V) \text{ and}$$
$$\gamma_{\mathfrak{h}}(t) \in \exp_G(U) \subset H_0; \tag{6.14.6}$$

$\gamma_{\mathfrak{k}}$ and $\gamma_{\mathfrak{h}}$ are clearly also of class C^1. Differentiating (6.14.6),

$$\gamma(t)^{-1}\gamma'(t) = \text{ad}\,(\gamma_{\mathfrak{h}})^{-1}[\gamma_{\mathfrak{k}}(t)^{-1}\gamma_{\mathfrak{k}}'(t)] + \gamma_{\mathfrak{h}}(t)^{-1}\gamma_{\mathfrak{h}}'(t) \tag{6.14.7}$$

for small $|t|$. By assumption, the left-hand side belongs to $\mathfrak{h}$; the second term on the right-hand side is also in $\mathfrak{h}$ since $\gamma_{\mathfrak{h}}$ takes its values in $\exp_G(U)$, an open subset of the Lie group H_o on which the analytic structure of H_o is that of the submanifold $\exp_G(U)$ of G (or, equivalently, can be derived from that of U via the exponential map). Since, moreover, $\mathfrak{h}$ is clearly ad (h)-invariant for all $h \in H_o$, (6.14.7) enables us to conclude that

$$\gamma_{\mathfrak{k}}(t)^{-1}\gamma_{\mathfrak{k}}'(t) \in \mathfrak{h} \tag{6.14.8}$$

for sufficiently small $|t|$. But let us consider the submanifold $U = \exp_G(U)$ and $V = \exp_G(V)$ of G. The map (6.14.5) being an isomorphism, so is the map $(x, y) \mapsto xy$ from $V \times U$ onto W. Hence, at a point yx of G, with $x \in U$ and $y \in V$,

$$T_{yx}(G) = T_y(V) \cdot x \oplus y \cdot T_x(U), \tag{6.14.9}$$

and so, multiplying on the left by y^{-1} and on the right by x^{-1}, we get

$$y^{-1} \cdot T_y(V) \cap T_x(U) \cdot x^{-1} = 0. \tag{6.14.10}$$

But U, endowed with its submanifold structure inherited from G, is also an open submanifold of H_o. So, for all $a \in U$, the translations $x \mapsto xa$ and $y \mapsto ya^{-1}$ map a neighbourhood of e in U onto a neighbourhood of a in U and conversely. Since these are analytic as maps from U to G, hence from U to U, in the neighbourhood of e and a respectively, for all $a \in U$ the differential of the map $x \mapsto xa$ at the origin induces an isomorphism onto $T_a(U)$ from the tangent space to U at e, namely $\mathfrak{h}$. Hence $T_x(U) \cdot x^{-1} = \mathfrak{h}$ in (6.14.10) and (6.14.8) then shows that $\gamma_{\mathfrak{k}}'(t) = 0$ for small $|t|$, so that $\gamma_{\mathfrak{k}}(t) = e$ in the neighbourhood of $t = 0$.

This argument (strongly influenced by the theory of differential systems…) shows that (6.14.4) implies $\gamma(t) \in H_o$ for small $|t|$. Applying this result to the curve (6.14.3) also satisfying (6.14.4), we see that more generally for all $s \in I$, $\gamma(t) \in \gamma(s) H_o$ for small $|t - s|$. Therefore, the set of $t \in I$ for which $\gamma(t)$ belongs to a given coset $g H_0$ is an open interval in I, and making g vary gives a partition of I into open sets. As an interval is connected, the claimed conclusion follows, namely that $\gamma(t) \in H_o$ for all $t \in I$.

c—We next show that H contains H_o. As H_o is generated by $\exp_G(\mathsf{U})$, where U is an arbitrary neighbourhood of 0 in $\mathfrak{h}$, it suffices to show that $\exp_G(\mathsf{U}) \subset H$ for such a neighbourhood. Let $(X_i)_{1 \leq i \leq p}$ be a basis of $\mathfrak{h}$. For each i, fix a curve γ_i in G with initial point e, of class at least C^1, tangent to the vector X_i at $t = 0$ and contained in H (hence in H_o by point (b) of the proof). The differential at the origin of the map

$$(t_1, \ldots, t_p) \mapsto \gamma_1(t_1) \cdots \gamma_p(t_p) \tag{6.14.11}$$

from I^p to G (where I is a sufficiently small interval centered at 0) is

$$(t_1, \ldots, t_p) \mapsto t_1 X_1 + \cdots + t_p X_p \tag{6.14.12}$$

and hence induces a bijection from $\mathbb{R}^p$ onto $\mathfrak{h}$.

But as seen above, if U is a sufficiently small neighbourhood of 0 in $\mathfrak{h}$, the map (6.14.11) takes its values in the submanifold $\exp_G(\mathsf{U})$ of G, at least when the $|t_i|$ are sufficiently small. Moreover, $\mathfrak{h}$ is precisely the tangent space to the latter at the origin. So for sufficiently small I, (6.14.11) is an isomorphism from I^p onto a neighbourhood of e in $\exp_G(\mathsf{U})$, hence in H_o. But (6.14.11) clearly takes its values in H. As a consequence, H contains a neighbourhood of e in H_o, and so $H_o \subset H$, as claimed.

d—We are now in a position to construct the analytic structure of H. First note that, since for all $h \in H$, ad (h) leaves $\mathfrak{h}$ invariant, the subgroup H_o is normal in H. Its analytic structure is also left invariant under the automorphisms $x \mapsto hxh^{-1}, h \in H$. It suffices to check this in the neighbourhood of $x = e$—but in the neighbourhood of e, the analytic structure of H_o is that of $\mathfrak{h}$ and the map $x \mapsto hxh^{-1}$ becomes the, necessarily analytic, linear map $X \mapsto$ ad $(h)X$ from $\mathfrak{h}$ to $\mathfrak{h}$.

As a result—we leave it to the reader to check the details—H can be endowed with a unique Lie group structure such that H_o (with its own analytic structure) becomes the connected component of e in H. Assertions (i), (ii) and (iii) of the statement are then obvious.

e—Part (iv) of the statement remains to be checked. The non-trivial part of the assertion is that if f is a C^r map from a manifold M to G, such that $f(M) \subset H$, then f remains C^r as a map from M to H. Since the injection j from H to G is an immersion, the property in question will follow from Chap. 2, Theorem 4, once we know that f is continuous as a map from M to the manifold H. Hence it is a matter of showing that f is continuous at every $a \in M$. Replacing f by $x \mapsto f(a)^{-1} f(x)$, it may be assumed that $f(a) = e$. Since the sets $\exp_G(\mathsf{U})$, where U is an arbitrary neighbourhood of 0 in $\mathfrak{h}$, form a fundamental system of neighbourhoods of e in H, it suffices to show that $f(x) \in \exp_G(\mathsf{U})$ for all $x \in M$ sufficiently near a.

For this, as in part (b) of the proof, choose a complement $\mathfrak{k}$ of $\mathfrak{h}$ in $\mathfrak{g}$. If U is a sufficiently small open neighbourhood of 0 in $\mathfrak{h}$ and if, likewise, V is a sufficiently small open neighbourhood of 0 in $\mathfrak{k}$, then the map (6.14.5) from $\mathsf{V} \times \mathsf{U}$ to G is an isomorphism onto an open neighbourhood W of e in G. Since f is continuous as a map from M to G, there is a connected open neighbourhood S of a in M such that $f(S) \subset W$. For all $x \in S$, there is an open neighbourhood $I \supset [0, 1]$ in $\mathbb{R}$ and a C^1 map $\lambda : I \to S$ such that $\lambda(0) = a$ and $\lambda(1) = x$. The curve $\gamma(g) = f \circ \lambda(t)$ is of class C^1 and is contained in $W \cap H$. Part (b) of the proof then shows that $\lambda(t) \in \exp_G(\mathsf{U})$ for sufficiently small $|t|$. The curve (6.14.3) implies that more generally the set of $t \in I$ for which $\gamma(t) \in \exp_G(\mathsf{U})$ is open in I. Indeed, every neighbourhood of e in the Lie group H has been seen to contain $\gamma(t)$ for sufficiently small $|t|$ (remember that these neighbourhoods of e in H are the sets $\exp_G(\mathsf{U}')$, where U' is a neighbourhood of 0 in $\mathfrak{h}$). However, $\exp_G(\mathsf{U})$ is itself open in H. Hence if $\gamma(s) \in \exp_G(\mathsf{U})$ for some given s, then it is possible to choose a neighbourhood U' of 0 in $\mathfrak{h}$ such that $\gamma(s) \exp_G(\mathsf{U}') \subset \exp_G(\mathsf{U})$, and as $\gamma(s)^{-1}\gamma(s + t) \in \exp_G(\mathsf{U}')$ for small $|t|$, as claimed, $\gamma(s + t)$ is in $\exp_G(\mathsf{U})$ for small $|t|$.

But γ is a continuous map from I to the open submanifold W of G and $\exp_G(\mathsf{U})$ is closed in W since U is closed in $\mathsf{V} \times \mathsf{U}$. Therefore the set of $t \in I$ for which $\gamma(t) \in \exp_G(\mathsf{U})$ is also closed. As I is connected, it follows that $\gamma(t) \in \exp_G(\mathsf{U})$ for all $t \in I$, and in particular, that $f(x) \in \exp_G(\mathsf{U})$ for all $x \in S$ when $t = 1$. Hence the continuity of f as a map from M to H. The fact that f is of class C^r as a map on the manifold M then follows trivially as above. This completes the proof of Theorem 11.

We let the reader check as an exercise that the analytic structure of H is fully determined by properties (i) and (ii).

Exercise. Let H be a subgroup of G. There exists a subalgebra $\mathfrak{h}$ of $\mathfrak{g}$ such that H is generated by $\exp_G(\mathfrak{h})$ (see Theorem 10) if and only if H is "arc-connected" in the following sense: for all $h \in H$, there is an open interval $I \supset [0, 1]$ and a C^1 map γ from I to G such that $\gamma(0) = e$, $\gamma(1) = h$ and $\gamma(I) \subset H$. [The result continues to hold if C^1 is replaced by C^0—in other words, *the subgroups of Theorem 10 are those which, endowed with the induced topology, are arc-connected*, but the proof is significantly more difficult; see M. Goto, *On an arcwise connected subgroup of a Lie group*, Proc. Amer. Math. Soc., XX(1969), pp. 157–162, or indications in N. Bourbaki, *Lie Groups*, Sect. 8, exer. 5). It is apparently not known whether integral subgroups of G can be characterized by the fact that they are connected (and not necessarily arc-connected) when they are endowed with the induced topology. It is difficult to believe this would not be the case...].

To conclude this section, we mention a curious consequence of Theorem 11:

Corollary *Let G be a Lie group and $\gamma : I \to G$ a C^1 curve in G such that $\gamma(0) = e$. Let $X = \gamma'(0)$ be the tangent vector to γ at the origin. Then, for all $t \in \mathbb{R}$, the elements $\exp_G(tX)$ of G belong to a subgroup H of G generated by $\gamma(I)$.*

Indeed, by construction, the Lie algebra $\mathfrak{h}$ of this subgroup clearly contains the vector X. It then suffices to apply (6.14.1).

6.15   Commutators in a Lie Group

In what follow, we set $(x, y) = xyx^{-1}y^{-1}$ when x and y are elements of a group G, and (M, N) will denote the subgroup of G generated by all commutators (m, n), where $m \in M$ and $n \in N$, when both M and N are subgroups of G. Similarly, if $\mathfrak{m}$ and $\mathfrak{n}$ are the subalgebras of a Lie algebra $\mathfrak{g}$, $[\mathfrak{m}, \mathfrak{n}]$ will denote the subspace (and not the subalgebra...) of $\mathfrak{g}$ generated by the expressions $[X, Y]$ for $X \in \mathfrak{m}$ and $Y \in \mathfrak{n}$.

Theorem 12 *Let G be a Lie group, $\mathfrak{g}$ its Lie algebra, $\mathfrak{m}$, $\mathfrak{n}$, and $\mathfrak{h}$ three subalgebras of $\mathfrak{g}$, M, N and H the subgroups of G generated by $\exp(\mathfrak{m})$, $\exp(\mathfrak{n})$ and $\exp(\mathfrak{h})$. Suppose that*

$$[\mathfrak{m}, \mathfrak{h}] \subset \mathfrak{h}, \quad [\mathfrak{n}, \mathfrak{h}] \subset \mathfrak{h}, \quad [\mathfrak{m}, \mathfrak{n}] \subset \mathfrak{h}. \tag{6.15.1}$$

Then H is invariant under M and N and $(M, N) \subset H$. If, moreover, $[\mathfrak{m}, \mathfrak{n}] = \mathfrak{h}$, then $(M, N) = H$.

The invariance of H under M and N follows immediately from the first two relations (6.15.1) and from

$$\exp(X)\exp(Y)\exp(X)^{-1} = \exp(e^{\text{ad}\,(X)}Y) = \exp(Y + \text{ad}\,(X)Y + \text{ad}\,(X)^2Y/2! + \cdots). \tag{6.15.2}$$

To show that $(M, N) \subset H$, suppose that $X \in \mathfrak{m}$ and $Y \in \mathfrak{n}$ in the latter formula; by (6.15.1), every term of the exponential series except Y is in $\mathfrak{h}$; hence

$$\exp(X)\exp(Y)\exp(X)^{-1} = \exp(Y + Z) \quad \text{with} \quad Z \in \mathfrak{h}. \tag{6.15.3}$$

To deduce that the commutator of the elements $m = \exp(X)$ and $n = \exp(Y)$ is in H, it therefore suffices to show that

$$\exp(Y + Z)\exp(-Y) \in H \quad \text{for} \quad Y \in \mathfrak{n}, Z \in \mathfrak{h}, \tag{6.15.4}$$

which we do for sufficiently small Y and Z. This will at least show that $(m, n) \in H$ if $m \in M$ and $n \in N$ are sufficiently near the identity in M and N endowed with their Lie group structures (Theorem 10).

But if X and Y, and hence Y and Z, are sufficiently small, it is possible to compute the left-hand side of (6.15.4) using the Campbell–Hausdorff formula. Then,

$$H(Y + Z, -Y) = Z + \frac{1}{2}[Y + Z, -Y] + \frac{1}{12}[Y + Z, [Y + Z, -Y]] + \cdots; \tag{6.15.5}$$

all the terms obtained are in $\mathfrak{h}$ because $[\mathfrak{n}, \mathfrak{h}] \subset \mathfrak{h}$. We leave it to the reader to check this for the general term of the formula. Hence (6.15.4) follows for sufficiently small Y and Z.

So indeed $(m, n) \in H$ for $m \in M$ and $n \in N$ sufficiently near e. As M and N are connected when they are endowed with their natural analytic structure, they are generated by any of their neighbourhoods of e. To deduce that $(m, n) \in H$ for all m and n, we then only need to use obvious formulas, for example

$$(m_1 m_2, n) = m_1 m_2 n m_2^{-1} m_1^{-1} n^{-1} = m_1 (m_2, n) n m_1^{-1} n^{-1}$$
$$= m_1 h n m_1^{-1} n^{-1} = h'(m_1, n), \tag{6.15.6}$$

and to take into account that H is invariant under M and N.

$(M, N) = H$ if $[\mathfrak{m}, \mathfrak{n}] = \mathfrak{h}$ remains to be shown. Let P be the subgroup (M, N); it amounts to showing that $H \subset P$, i.e. that $\exp(tZ) \in P$ for all t and $Z \in \mathfrak{h}$. To do so, apply Theorem 11 to P and let $\mathfrak{p}$ be the Lie subalgebra of $\mathfrak{g}$ associated to P, i.e. the set of tangent vectors to curves originating at e and contained in P. It is then a matter of showing that $\mathfrak{h} \subset \mathfrak{p}$, i.e. that $[X, Y] \in \mathfrak{p}$ for all $X \in \mathfrak{m}$ and $Y \in \mathfrak{n}$. Consider the one-parameter subgroups γ_X and γ_Y generated by X and Y. The subgroup P contains

$$\exp(sX)^{-1} \exp(tY) \exp(sX) \exp(tY)^{-1} = \exp(sX)^{-1} \exp(e^{t \cdot \mathrm{ad}\,(Y)} sX) \tag{6.15.7}$$

for all s and t. As a function of t with fixed s, (6.15.7) is an analytic curve contained in P, so that its tangent at the origin belongs to $\mathfrak{p}$. Setting

$$F(t) = e^{t \cdot \mathrm{ad}\,(Y)} sX = sX + st[Y, X] + \cdots, \tag{6.15.8}$$

so that $F'(0) = s[Y, X]$, the chain rule, i.e. formula (5.3.11) of Chap. 5, shows that the tangent vector to (6.15.7) at $t = 0$ equals

$$\exp(sX)^{-1} \cdot \exp'(sX) F'(0) = \sum \mathrm{ad}\,(-sX)^{n-1} F'(0)/n! \tag{6.15.9}$$

by Theorem 4. Hence for all s, $\mathfrak{p}$ contains the vector

$$F'(0) - s \cdot \mathrm{ad}\,(X) F'(0)/2! + \cdots = s[Y, X] - s^2[X, [Y, X]]/2! + \cdots. \tag{6.15.10}$$

The derivative of (6.15.10) at $s = 0$ being equal to $[Y, X]$, in conclusion, $\mathfrak{p}$ also contains $[Y, X]$ for all $X \in \mathfrak{m}$ and $Y \in \mathfrak{n}$, which completes the proof.

Exercise. The following proof of the fact that $(M, N) \subset H$ can be found in N. Bourbaki, *Groupes de Lie*, pp. 231–232; the exercise consists in formulating the arguments explicitly:

Suppose that $[\mathfrak{m}, \mathfrak{n}] \subset \mathfrak{h}$. The sum $\mathfrak{m} + \mathfrak{n} + \mathfrak{h}$ is a Lie subalgebra of $\mathfrak{g}$. Considering the integral subalgebra of G with Lie algebra $\mathfrak{m} + \mathfrak{n} + \mathfrak{h}$, the question reduces to the case where $\mathfrak{m} + \mathfrak{n} + \mathfrak{h} = \mathfrak{g}$ and G is connected. Then, $\mathfrak{h}$ is an ideal of $\mathfrak{g}$. First suppose that G is simply connected. Then H is a distinguished Lie subgroup of G. Let ϕ be the canonical morphism from G onto G/H. Then,

$$[\phi'(\mathfrak{m}), \phi'(\mathfrak{n})] = \{0\},$$

and so by the Hausdorff formula $\phi(M)$ and $\phi(N)$ commute; as a consequence, $(M, N) \subset H$. In the general case, let G' be the universal covering of G, and M', N', H' the integral subgroups of G' with Lie algebras $\mathfrak{m}$, $\mathfrak{n}$, $\mathfrak{h}$. Then, $(M', N') \subset H'$, and M, N, H are the canonical images of M', N', H' in G; hence $(M, N) \subset H$.

To understand this text (in which the notation has been changed to bring them in line with those of the present section), it is necessary to observe that for Bourbaki a "Lie" subgroup is *closed*, while, as mentioned above, an "*integral*" subgroup is obtained from Theorem 10. Furthermore an *ideal* of a Lie algebra g is a Lie subalgebra $\mathfrak{h}$ such that $[\mathfrak{g}, \mathfrak{h}] \subset \mathfrak{h}$. Hence the above text alludes to (in fact refers to) the following result:

Theorem 13 *Let H be a normal integral subgroup of a simply connected Lie group G. Then H is closed (hence a Lie subgroup of G) and the quotient Lie group G/H is also simply connected.*

Let $\mathfrak{g}$ be the Lie algebra of G and $\mathfrak{h} \subset \mathfrak{g}$ the subalgebra corresponding to H, so that H is generated by $\exp_G(\mathfrak{h})$. As H is normal, the subalgebra $\mathfrak{h}$ is an ideal of $\mathfrak{g}$; indeed, the elements $X \in \mathfrak{h}$ continue to be characterized by the fact that

$$\exp_G(tX) \in H \quad \text{for all } t, \tag{6.15.11}$$

so that if H is normal, then $\mathfrak{h}$ is invariant under the operators ad (g), $g \in G$ and hence also under ad (X), for all $X \in \mathfrak{g}$. This can be seen by taking $g = \exp(tX)$ and differentiating with respect to t at $t = 0$.

But if $\mathfrak{h}$ is an ideal of a Lie algebra $\mathfrak{g}$, the quotient space $\mathfrak{g}/\mathfrak{h}$ can be endowed with a Lie algebra structure by observing that, for all $X, Y \in \mathfrak{g}$, the class of $[X, Y]$ mod $\mathfrak{h}$ only depends on the classes of X and Y mod $\mathfrak{h}$ since $[X + \mathfrak{h}, Y + \mathfrak{h}] \subset [X, Y] + [X, \mathfrak{h}] + [\mathfrak{h}, Y] + [\mathfrak{h}, \mathfrak{h}] \subset [X, Y] + \mathfrak{h}$, with a notation which we hope is self-explanatory. The canonical map from $\mathfrak{g}$ to $\mathfrak{g}/\mathfrak{h}$ is then a Lie algebra homomorphism by construction.

However—assuming Ado's theorem...—, we know there exists a Lie group G' with Lie algebra $\mathfrak{g}/\mathfrak{h}$ (Corollary of Theorem 10), and G' may be assumed to be simply connected, if need be by replacing it by its universal covering. By Theorem 8, the canonical map $p : \mathfrak{g} \to \mathfrak{g}/\mathfrak{h}$ can be integrated into an analytic homomorphism f from the Lie group G to the Lie group G' such that $f'(e) = p$. The map f has constant rank, and as it is of maximal rank at the origin, it follows that it is a submersion. The image of G is therefore an open subgroup of G', and since G' is connected it follows that $f(G) = G'$. The kernel N of f is a Lie subgroup of G whose Lie algebra is ker $f'(e) = \ker p = \mathfrak{h}$. The connected component N° of N is also a Lie (i.e. closed) subgroup of G, and hence generated by $\exp_G(\mathfrak{h})$, which implies that $N^\circ = H$. This shows that H is a Lie subgroup of G.

To prove that G/H is simply connected, it suffices to show that $H = N$ since this implies that $G/H = G'$. But as $H \subset N$, the canonical map $G \to G/N$ is the composite of the canonical map $G \to G/H$ and a map $G/H \to G/N$ which is obviously a Lie group homomorphism.[8] This homomorphism is surjective since $f(G) = G'$, and its kernel is the image of N in G/H, i.e. a *discrete* subgroup of G/H since H is the connected component of N. As G/H is connected, G/H is a connected covering of the simply connected group $G' = G/N$. Hence $G/H = G/N$, and so $H = N$, completing the proof.

Note the following consequence of Theorems 12 and 13:

Theorem 14 *Let G be a connected Lie group. Then the derived series*

$$DG = (G, G), \quad D^2G = (DG, DG), \quad D^3G = (D^2G, D^2G), \ldots \quad (6.15.12)$$

consists of integral subgroups whose Lie algebras are the successive derived algebras

$$D\mathfrak{g} = [\mathfrak{g}, \mathfrak{g}], \quad D^2\mathfrak{g} = [D\mathfrak{g}, D\mathfrak{g}], \quad D^3\mathfrak{g} = [D^2\mathfrak{g}, D^2\mathfrak{g}], \ldots \quad (6.15.13)$$

of the Lie algebra $\mathfrak{g}$ of G. The same is true for the elements

$$CG = DG = (G, G), \quad C^2G = (G, CG), \quad C^3G = (G, C^2G), \ldots \quad (6.15.14)$$

of the lower central series of G, which correspond to the elements

$$C\mathfrak{g} = [\mathfrak{g}, \mathfrak{g}], \quad C^2\mathfrak{g} = [\mathfrak{g}C\mathfrak{g}], \quad C^3\mathfrak{g} = [\mathfrak{g}, C^2\mathfrak{g}], \ldots \quad (6.15.15)$$

of the lower central series of $\mathfrak{g}$. If, moreover, G is simply connected, the subgroups D^iG and C^iG are closed for all i.

The proofs can be left to the reader.

Exercise. Let G be a Lie group, A an integral subgroup of G, $\mathfrak{g}$ and $\mathfrak{a}$ the Lie algebras of G and A. (i) Let M be the set of $g \in G$ such that the operator $\mathrm{ad}\,(g) - 1$ maps $\mathfrak{a}$ to $D\mathfrak{a}$. Show that M is a closed subgroup whose Lie algebra $\mathfrak{m}$ consists of the elements $X \in \mathfrak{g}$ such that $\mathrm{ad}\,(X)$ maps $\mathfrak{a}$ to $D\mathfrak{a}$. (ii) For $g \in M$ and $X \in \mathfrak{a}$, set $\mathrm{ad}\,(g)X = X + Y$ with $Y \in D\mathfrak{a}$. Applying the Campbell–Hausdorff formula show that if X is sufficiently near 0, then the commutator $(g, \exp(X))$ belongs to the derived group DA of A (use the fact that it is an integral subgroup of G with Lie algebra $D\mathfrak{a}$). Using the fact that for $X \in \mathfrak{a}$, the elements $\exp(X)$ generate A, deduce that the subgroup (M, A) generated by the commutators (g, x) with $g \in M$ and $x \in A$ is contained in DA. In particular, $(\bar{A}, A) \subset DA$, where $\bar{A}$ is the closure of A. (iii) Let $\mathfrak{b}$ be the Lie algebra of $\bar{A}$ and N the set of $g \in G$ such that $\mathrm{ad}\,(g) - 1$ maps $\mathfrak{b}$ to $D\mathfrak{a}$. Show that it is a closed subgroup of G and find its Lie algebra $\mathfrak{n}$. Show that $\mathfrak{a} \subset \mathfrak{n}$ and that $\bar{A} \subset N$. (iv) Using the arguments of part (ii), show that for all $g \in N$, $(g, \exp(X)) \in DA$ for all $X \in \mathfrak{b}$ sufficiently near zero, then that $(N, \bar{A}) \subset DA$. Conclude that

$$(\bar{A}, \bar{A}) = (A, A) \quad (6.15.16)$$

---

[8]See Theorem 15 below.

for every integral subgroup A of G. In particular, A is normal in $\bar{A}$ and $\bar{A}/A$ is commutative. (For a slightly different proof, see N. Bourbaki, *Groupes de Lie*, p. 232.)

Exercise. (Alternative proof of (6.15.16), taken from chap. XVI of G. Hochschild, *The Structure of Lie Groups*, where significantly more difficult results can be found.) Replacing G by $\bar{A}$, A may be assumed to be dense in G, and it is then a matter of showing that $DG = DA$. (i) Show that the normalizer in G of any integral subgroup of G is closed (show that the normalizer of an integral subgroup H consists of the elements $g \in G$ such that ad (g) leaves the Lie algebra of H invariant). Any integral subgroup normalized by A is therefore normal in G; in particular, A and DA are normal in G and their Lie algebras are ideals of $\mathfrak{g}$. (iii) Let $\widetilde{G}$ be the simply connected covering of G, p the canonical map from $\widetilde{G}$ onto G, Z its kernel and B the integral subgroup of $\widetilde{G}$ with Lie algebra $\mathfrak{a}$. Show that DB is closed in $\widetilde{G}$, then that $D\widetilde{G} = DB$ (use the fact that ZB is dense in $\widetilde{G}$). As asked to do, deduce that $DG = DA$.

Exercise. A Lie algebra $\mathfrak{g}$ is said to be *nilpotent* if $C^i\mathfrak{g} = \{0\}$ for large i and a connected Lie group G is called nilpotent if so is its Lie algebra, i.e. if $C^iG = \{e\}$ for large i. Let G be a connected nilpotent Lie group and $\mathfrak{g}$ its Lie algebra. (i) Show that the Campbell–Hausdorff series $H(X, Y)$ reduces to a polynomial in X and Y (use the fact that ad (X) is nilpotent for all $X \in \mathfrak{g}$). (ii) Show that $\exp(X)\exp(Y) = \exp(H(X, Y))$ for all $X, Y \in \mathfrak{g}$ (replace X and Y by tX and tY and observe that both sides are analytic maps from $\mathbb{R}$ to G identical in the neighbourhood of 0). Deduce that the exponential map is surjective. (iii) Show that $\mathfrak{g}$, endowed with the composition law $(X, Y) \mapsto H(X, Y)$, is a simply connected Lie group with Lie algebra $\mathfrak{g}$. Deduce that if G is simply connected, then the map exp is a homeomorphism from $\mathfrak{g}$ onto G. (iv) Show that the tangent linear map to the exponential map is bijective at all points X of $\mathfrak{g}$. Deduce that, *if G is simply connected, then the exponential map is an isomorphism from the manifold $\mathfrak{g}$ onto the manifold G.* (v) Show that *if G is simply connected, then every integral subgroup A of G is closed and simply connected* (this result continues to hold if G is *solvable*, i.e. if $D^i\mathfrak{g} = 0$ for large i).

Exercise. Let V be a finite-dimensional vector space over a commutative field k. The purpose here is to show that for every subgroup G of $GL(V)$ whose elements are all *unipotent*, there is a basis of V with respect to which every $g \in G$ is represented by strictly triangular matrices. (i) First assume that k is algebraically closed and that V does not contain any non-trivial G-invariant subspace. Using Burnside's theorem (Chap. 5, Sect. 5.14), show that $\mathrm{Tr}[u(g - 1)] = 0$ for all $u \in \mathcal{L}(V)$ and $g \in G$. Deduce that $g = 1$ for all $g \in G$ and that $\dim(V) = 1$. (ii) Now returning to the general case, make G act on the vector space $\bar{V}$ derived from V by extending the base field to the algebraic closure of k. Applying (i) to a G-invariant subspace of $\bar{V}$ and of minimal dimension, show that G fixes all non-trivial vectors in $\bar{V}$. (iii) Show that there is a non-trivial vector in V fixed by G (note that every $x \in \bar{V}$ fixed by G is obtained by solving linear equations with coefficients in k). (iv) Complete the proof by an induction argument on the dimension of V. (v) A connected Lie group G is nilpotent if and only if ad (g) is unipotent for all g. (vi) Let G be an integral subgroup of $GL_n(\mathbb{R})$ consisting of unipotent matrices. Then G is closed and simply connected (use the fact that the group of strictly triangular matrices is nilpotent and simply connected, hence that all its integral subgroups are closed and simply connected by the previous exercise).

6.16 Belated Regrets About Quotient Groups

In the proof of Theorem 13, we used the existence of a Lie group structure on G/H when H is a normal Lie (hence closed) subgroup of a Lie group G. This "obvious" result ought to have been given much earlier in these Notes and, more precisely, at the end of Chap. 4 where the case of homogeneous spaces was treated. Putting things (and people) in their natural place in the hierarchy is one of the basic rules of correct behaviour in scientific matters: Science does not like disorder. Before permanently sealing this majestic pyramid, let us make use of the little time we still have left and run to fetch the fragile basket of reeds that we forgot to drop off at the canonical instant in the burial chamber where, freshly embalmed, the deceased pharaoh is impatiently awaiting before confronting the swirling dark waters of the Styx:

Theorem 15 *Let G be a Lie group and H a normal (closed) Lie subgroup in G. Then the quotient analytic structure on G/H turns G/H into a Lie group. If $\mathfrak{g}$ and $\mathfrak{h} \subset \mathfrak{g}$ are the Lie algebras of G and H, the tangent map to the canonical projection $p : G \to G/H$ at the origin induces an isomorphism from the Lie algebra $\mathfrak{g}/\mathfrak{h}$ onto the Lie algebra of G/H. A homomorphism from the group G to a Lie group G', trivial on H, is analytic if and only if it is the composite of p and an analytic homomorphism from G/H to G'. If K is a Lie subgroup of G containing H, then $p(K)$ is a Lie subgroup of G/H and p induces an isomorphism from the quotient Lie group K/H onto the Lie group $p(K)$. If K is normal, the canonical map from G/K onto $(G/H)/(K/H)$ is a Lie group isomorphism.*

To begin with, by Theorem 5 of Chap. 4, if there are regular equivalence relations R and S on two manifolds X and Y, then $R \times S$ is clearly an equivalence relation on $X \times Y$ and the natural map from $X \times Y/R \times S$ onto the product manifold $(X/R) \times (Y/S)$ is a manifold isomorphism. In the case of a Lie group G and of a closed subgroup H of G, the manifolds $G \times G/H \times H$ and $(G/H) \times (G/H)$ are clearly seen to be isomorphic. Denote by p the canonical map from G onto G/H, which is a submersion, and observe that, if H is normal, then the isomorphism in question transforms the map $(x, y) \mapsto xy^{-1}$ from $(G/H) \times (G/H)$ to G/H into the map q from $G \times G/H \times H$ to G/H derivable from the map

$$(x, y) \mapsto p(xy^{-1}) \tag{6.16.1}$$

from $G \times G$ to G/H. So, by passing to the quotient   mod $H \times H$, to show that G/H is a Lie group becomes a matter of showing that $q : G \times G/H \times H$ to G/H is analytic. By the universal property of quotient manifolds or of submersions (Chap. 4, Theorem 4), it suffices to show that (6.16.1) is an analytic map from $G \times G$ to G/H. This is obvious since it is the composite of the map $(x, y) \mapsto xy^{-1}$ from $G \times G$ to G and the canonical projection p. As p is a submersion, the tangent subspace of $T_e(G) = \mathfrak{g}$ to the inverse image of $p(e) \in G/H$, i.e. the tangent subspace to H, in other words the subalgebra $\mathfrak{h}$, is necessarily the kernel of the tangent map $p'(e)$. The projection p being a submersion, $p'(e)$ is surjective and since ker $p'(e) = \mathfrak{h}$ as already seen, we can conclude that $p'(e)$ induces an isomorphism from the vector space $\mathfrak{g}/\mathfrak{h}$ onto the Lie algebra of G/H; it is a Lie algebra isomorphism since p is a homomorphism. If a homomorphism f from G to a Lie group G' is analytic and maps H onto the identity element, then it is the composite of p and a homomorphism g from G/H to G'. Since p is a submersion and $g \circ p$ is analytic, g is analytic thanks to[9] Theorem 4 of Chap. 4. The converse is trivial.

Finally, let us consider a Lie subgroup K of G containing H. The map p being open takes the complement of K onto an open subset of G/H. As $K \supset H$, the complement of this open subset is just $p(K)$; it is therefore a closed subgroup of G/H. The obvious map from K/H to G/H is analytic since taking its composition with the canonical projection from K onto K/H gives the restriction to K of the map p from G to G/H, which is analytic. On the other hand, since the map under consideration from K/H to G/H commutes with the actions of K on these two manifolds and as K acts transitively on K/H, the map in question is of constant rank, and hence is a subimmersion and in fact an immersion since it is clearly injective. On the other hand, it is a homeomorphism from K/H onto its image $p(K)$ for topological reasons unrelated to the theory of Lie groups. We have already observed in an exercise, at the end of Sect. 6.12 of the present chapter, that when an immersion is a homeomorphism from the initial manifold onto its image, it is a submanifold of the final manifold and the given immersion an isomorphism from the initial manifold onto its image manifold. Applying this result here, we conclude that $p(K)$ is a submanifold, i.e. a Lie subgroup, of G/H and that p induces an isomorphism from K/H onto $p(K)$. (Of course, since $p(K)$ is a closed subgroup of G/H, it should be a Lie subgroup of G/H, but using this difficult theorem would destroy the elementary nature of our arguments and, for example, would not enable us to relocate Theorem 15 in its proper place in Chap. 4 in a revised and corrected posthumous edition of these Notes.)

Rather than showing that G/K is isomorphic to $(G/H)/(K/H)$ when K is normal, let us abstain from applying this assumption and instead show that by making K act *on the right*[10] on $G/H = X$ gives a regular equivalence relation on X with respect to which the quotient manifold is precisely the homogeneous space G/K. As K contains H, there is a canonical bijection j from X/K onto G/K (where X/K is the quotient set of X by the equivalence relation whose classes are the orbits of K).

---

[9]In the statement of Theorem 4 of Chap. 4, the continuity assumption on h is in fact superfluous. Indeed, if the map $h \circ \pi$ is a morphism, then it is perforce continuous: but a submersion is an open map; the continuity of $h \circ \pi$ on X then implies the continuity of h on the open subset $\pi(X)$ of Y.

[10]As H is a normal subgroup of K, $gHk = gkH$ for all $g \in G$ and $k \in K$. This is why K can be made to act on the right on G/H. The actions thus obtained on the group G/H are obviously right translations by elements of the subgroup $p(K)$.

The quotient manifold structure of X/K, if it exists, is fully characterized by the fact that the canonical map q from X onto X/K is a submersion (Chap. 4, Theorem 5). We thus need to show that by transporting the manifold structure of G/K to X/K using j gives a manifold structure on X/K such that q is a submersion. Proving directly that $j \circ q : X = G/H \to G/K$ is a submersion comes to the same. As the canonical map $p : G \to G/H$ is a surjective submersion, to do so, it is sufficient to check that the composite map $(j \circ q) \circ p : G \to G/K$ is itself a submersion— which is obvious since it is the canonical map from G onto the quotient manifold G/K. This completes the proof.

Note that the last part of the proof shows more generally that if G contains a Lie subgroup K and a Lie subgroup H normal in K (but not necessarily in G), then the equivalence relation on G/H obtained by making K act on the right on G/H is regular. It may be wondered if, likewise, given two arbitrary Lie subgroups K and H of G, the equivalence relation obtained on G/H by making K act *on the left* is always regular. This cannot be the case because were it so, the orbits of K in G/H would be closed submanifolds of G/H. In particular, the doubles cosets KgH in G, namely the inverse images of these orbits under the canonical map from G onto G/H, would be closed in G. Hence, to obtain a counterexample, it suffices to choose K and H in such a way that the class KH is not closed. For example, take G to be the group $SL_2(\mathbb{R})$, H the subgroup $\begin{pmatrix} 1 & * \\ 0 & 1 \end{pmatrix}$, so that G/H is the pointed plane $\mathbb{R}^2 - \{0\}$, and K the subgroup of matrices of type $\begin{pmatrix} * & 0 \\ * & * \end{pmatrix}$. The class KH, evidently far from being closed in G, consists of the matrices

$$\begin{pmatrix} a & b \\ c & d \end{pmatrix} \quad \text{with} \quad ad - bc = 1, \quad a \neq 0, \tag{6.16.2}$$

in other words it is open in G. When K is made to act on the pointed space, it is easy to see that we obtain two equivalence classes, namely the open subset consisting of points (x, y) such that $x \neq 0$ and the closed one consisting of points $(0, y)$ with $y \neq 0$.

Index

© Springer International Publishing AG 2017
R. Godement, *Introduction to the Theory of Lie Groups*,
Universitext, DOI 10.1007/978-3-319-54375-8